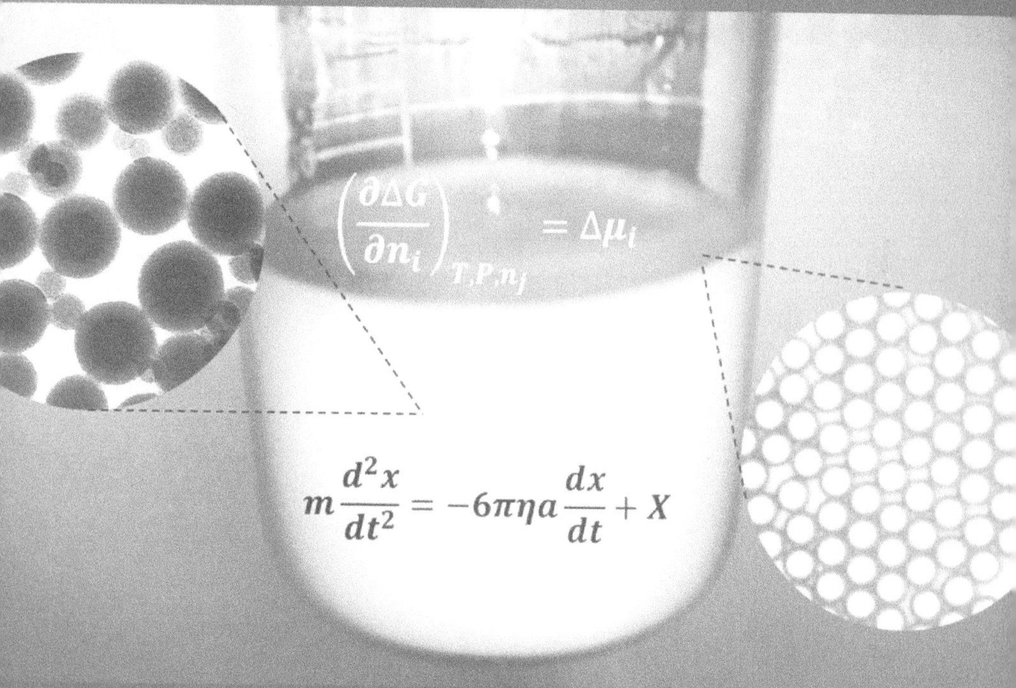

Heterophase Polymerization

Heterophase Polymerization
Basic Concepts and Principles

Hugo Hernandez
Klaus Tauer

JENNY STANFORD
PUBLISHING

Published by

Jenny Stanford Publishing Pte. Ltd.
Level 34, Centennial Tower
3 Temasek Avenue
Singapore 039190

Email: editorial@jennystanford.com
Web: www.jennystanford.com

British Library Cataloguing-in-Publication Data
A catalogue record for this book is available from the British Library.

Heterophase Polymerization: Basic Concepts and Principles

ISBN 978-981-4877-32-9 (Hardcover)
ISBN 978-1-003-11929-6 (eBook)

Contents

Preface

by Hugo Hernandez

By the time I graduated as a chemical engineer, I never imagined working with polymers, much less writing a book on this field. My first job was at a chemical company where different types of polymers, including polymer latexes, were manufactured. At first, polymers appeared to me as fascinating although mysterious materials, behaving completely different from the usual chemical substances I knew so far. I still remember the time when, by curiosity, I read the first scientific paper on emulsion polymerization. At that moment I was not able to understand much, but it was clear to me that I had found a fascinating but challenging topic. Then, I set a goal and began a journey toward understanding such complex systems.

Eventually, I had the chance of working in research and development, particularly on different research projects involving emulsion polymerization. Even though I had learned after reading a lot of books and papers, as well as from my own experience in the lab, pilot, and industrial plant, I still wanted to obtain a deeper knowledge. That is when I decided to pursue my doctoral studies working on emulsion polymerization.

In February 2006, I received an acceptance letter by Dr. Klaus Tauer, group leader of the "*Heterophase Polymerization*" group at the Colloid Chemistry Department of the Max Planck Institute of Colloids and Interfaces, for beginning my studies at the International Max Planck Research School on Biomimetic Systems. During the 4 years that I stayed at the Max Planck Institute, first as a doctoral student and then as a postdoc, I had the chance to dedicate myself to learning, in as much depth as possible, about polymers, colloids, physical chemistry, and mathematical modeling. It was also during that time when I realized that science is unfinished, and *nothing* from what I had learned so far *was written in stone*. Following the principles of good scientific practice established by the Max Planck Society, and with the guidance of Dr. Tauer, I officially began publishing my contributions to science in 2007.

In 2015, inspired by the ideas of freedom of research, transparency, and public sharing of knowledge promoted at the

Max Planck Institute, I decided to create ForsChem Research (www.forschem.org). The main goal of this not-for-profit initiative is searching for a better understanding of different natural phenomena from a molecular perspective (and involving a lot of math, of course). All results obtained in this initiative are open and available to anyone interested in reading them.

The idea of writing a book on emulsion polymerization was first mentioned to me by Dr. Tauer in 2010. Even though we did not have the chance to finish the project then, we decided to resume it again in 2019, but this time considering the broader field of heterophase polymerization. Thus, we tried our best to summarize our vision on heterophase polymerization, so that future generations might find it easier to understand and achieve even further progress.

I am very much grateful to Dr. Klaus Tauer, my doctoral supervisor and co-author of this book, who devoted his lifework and curiosity to the science and technology of heterophase polymerization. I also want to thank Silvia, my beloved wife, for continuously supporting and encouraging me toward working on this book. I also want to thank my mother and my late father for all their teachings and support. I am also grateful to Jaime Aguirre, my master's thesis advisor, for being my scientific peer and dear friend. I also want to acknowledge all the scientists that have contributed to this fascinating field (some of which I had the opportunity to meet), as well as all other colleagues that helped me in one way or another along this road. Finally, I would like to thank all the readers of this book for your patience (I hope you read it all) and, most of all, for your scientific curiosity.

November 2020

Preface

by Klaus Tauer

When a graduate starts career in the field of heterophase polymerization, she or he faces quite a challenge, because to be successful in this field, she or he has to deal intensively with two scientific areas, namely colloid science and polymer science. Typically, the beginner will find the basics of heterophase polymerization in standard textbooks that focus primarily on emulsion polymerization. One of its pillars is the Smith–Ewart model, which is more than 70 years old. That is exactly how I started my professional career in a research institute in 1977. I quickly noticed some problems with this model, particularly regarding its applicability to various variants of emulsion polymerization. Following Renè Descartes "*dubium sapientiae initium*" ("*to doubt is the start on the path to wisdom*"), I started new projects to investigate particle nucleation and swelling of latex particles.

All knowledge is preliminary, and hence with the availability of new experimental data, the modification of older models is necessary. Clearly, the same must apply to emulsion polymerization. We tried to do so in this book, at least partly.

However, the history of emulsion polymerization research shows that more and more specific mechanisms to overcome some of the problems have been developed. Exemplarily, some new mechanistic assumptions for particle nucleation and radical entry appear, at least to me, to be detached from the generally valid scientific ideas. Just when nucleation is considered, this is a very general natural phenomenon and there is no reason to consider for emulsion polymerization a principally different framework.

Visiting many scientific meetings over the last 40 years, I made the following remarkable observation concerning the relationship between heterophase polymerization and polymer science, on the one hand, and colloid science, on the other hand. For research polymer chemists, this topic is too boring and not hot enough mainly because it is a well-established industrial polymerization technique. For colloid chemists, heterophase polymerization is a "horrible" system because the parameters such as surface area and composition are constantly changing due to the chemical reactions. This is probably the main reason that an own scientific community was founded in 1972, the International Polymer Colloids Group (IPCG), which organizes biannual international meetings.

Retired since spring 2017 and looking back on my professional carrier, I come to the conclusion that heterophase polymerization gave me an extremely interesting working life, with only a few desperate hours but more often with joy due to new and surprising results in the border area between colloid and polymer chemistry. In this context, I acknowledge deeply the cooperation of former PhD students, technical staff, and colleagues at the Max Planck Institute of Colloids and Interfaces.

Hugo Hernandez, one of the most talented students in my group, surprised me one day with a Brownian dynamics simulation to improve modeling radical entry in emulsion polymerization. This was for me the ignition to think about a new text on heterophase polymerization. Hugo did most of the work in finishing this book, and as his former supervisor, I am feeling honored and particularly thankful to him for the great job he did.

Particular thanks is due to Prof. Dr. Markus Antonietti, Head of the Department of Colloid Chemistry, who gave me the freedom and the financial support to follow my research ideas.

Finally, I am extremely thankful to my beloved wife, Birgit, who spent 50 years on my side and patiently tolerated my second big love, chemistry.

The topics in the present book are our personal insight into this fascinating marriage of colloid and polymer science, and we offer our sincere apologies to all the colleagues whose excellent contribution to the field we had to omit.

November 2020

Introducing an Alternative Understanding of Heterophase Polymerization

A huge number of books already exist dealing with one or the other heterophase polymerization technique, specifically mainly with emulsion polymerization. An enormous amount of knowledge has been accumulated over the past 110 years of practical heterophase polymerization. So why is there a need for another book on heterophase polymerization? There is one main reason why this seems necessary to us. All books on emulsion polymerization deal in the kinetics sections almost exclusively with deterministic models. But in our opinion, there is the need and nowadays also the computational possibility to apply stochastic methods to elucidate heterophase polymerization. In addition, it is high time to make a step out of the box of the basic ideas established decades ago. We are discussing heterophase polymerization not only in the light of a different modeling strategy but also on the base of new concepts and experimental results, which allow drawing a more consistent general frame of heterophase polymerization. The general frame is set by complementary principles of classical thermodynamics and Brownian dynamics.

In the following subsections, we will highlight some alternative ideas on how to deal with heterophase polymerization instead of relying on the ideas that have been established for more than 70 years, particularly swollen micelles as nuclei for particles and instantaneous swelling of polymer particles, without being experimentally confirmed.

Before starting, we wanted to include some additional thoughts for approaching this or any other scientific text. The scientific knowledge at a given time is provisional and imperfect. Due to the presence of uncertainties, we will never be able to reach the absolute truth of things [1]. Thus, we can only continuously improve our set of working theories and models for describing the world. Knowing and accepting this, we are not claiming to write the absolute truth, but we are trying to discuss trustworthy results representing our current state-of-the-art knowledge on the field of *heterophase polymerization*.

On the other hand, all natural sciences and all scientific results must rely on a set of generally accepted methodical rules [2]:

1. A scientific fact must be confirmed by observations and experiments on real objects. No assertion that defies experience is scientific. (*uncontradicted*)
2. Observations and experiments must be conducted repeatedly and if necessary analyzed by statistical methods. (*confirmed*)
3. Observations and experiments must be reproducible and yield the same result (within statistical tolerance) under the same conditions. (*reproducible*)
4. The implementation must be impartial and without presupposition. (*independent of any person*)
5. All natural sciences are based upon the rule of uniformity of nature, stating that the same natural laws apply everywhere. (*universal*)
6. Everything has a cause that yields the same effect under the same circumstances. With only statistically ascertainable events, the singular event can seem noncausal; with increasing repetition of the measurement, causality becomes evident for the whole of all events. (*causal*)

Although these rules are obvious and logical, it is necessary to keep reminding them and to validate scientific results with their help. To the best of our knowledge, we selected the presented data accordingly.

I. Thermodynamics and Brownian Dynamics

With classical thermodynamics and Brownian dynamics (BD), we have two powerful tools available to explore chemical reactions and physical processes, which are important for heterophase polymerizations at different relevant scales, from the macroscopic to the colloidal and molecular scales.

Thermodynamics is a powerful tool typically describing systems consisting of very large numbers of entities interacting with each other in very different ways and complexity (typically single- or multi-component bulk systems, but also systems as big as our earth or even bigger can be described). For equilibrium condition, even such vast systems can be described with quite a small number of quantities or parameters such as the mass, pressure, volume, and internal energy of the system or other equivalent quantities without knowing all the details of the individual entities. A detailed knowledge of the behavior of the individual objects of the system is not required. Thermodynamics allows, from a given starting state, to decide whether it is stable or not under given conditions. Furthermore, assuming equilibrium, it is possible to predict which final state might be achieved. Note, for evaluations with thermodynamics, the variable "time" plays no role. In order to decide whether a reaction/transition is feasible, thermodynamics offers two possibilities [3]: the Gibbs function (ΔG, constant temperature and pressure) or Helmholtz function (ΔH, constant temperature and volume) and the entropy (S). The Gibbs function is preferred for chemical reactions, and its partial differential quotient with respect to n, the number of molecules of a species "i" involved,

$$\left(\frac{\partial \Delta G}{\partial n_i}\right)_{T,P,n_j} = \Delta\mu_i \tag{i}$$

corresponds to the change in the chemical potential ($\Delta\mu_i$) of that species at constant temperature, pressure, and number of other species. A change in the starting system can only happen when ΔG (or $\Delta\mu$, respectively) decreases. In a brilliant paper on the "elusive chemical potential," Baierlein discussed the meaning of the chemical potential [4]. One meaning is the tendency to diffuse. Consequently, the chemical potential as a function of position measures the tendency of particles to diffuse. Additionally, the chemical potential can be used to measure the rate of change and a characteristic energy.

The transition or reaction comes to halt when ΔG ($\Delta\mu$) reaches zero. This state is the most probable one and is characterized by an equalized chemical potential throughout the reaction space.

Assuming that a reaction takes place under perfect control of temperature and pressure, i.e., $\Delta T = \Delta P = 0$, and that no work is transferred, then the entropy change for this process is $\Delta S_{\mathrm{tu}} = \Delta G/T$. Note, ΔS_{tu} is the sum of entropy changes in the environment (ΔS_{e}) and the reaction system (ΔS_{s}), whereby one or the other may rule ΔS_{tu}.

ΔG and ΔS are related and, hence, both can be used for predictions, but "...the most general statement appropriate to any observed change in an assembly of materials is that the total entropy of the assembly must increase. By the total entropy change is meant the sum of all entropy changes in any way related to the observed change in state for an assembly large enough to encompass all aspects of the change. The crucial distinction at this point is then between total energy change and total entropy change." [3]

The main reason for the entropy increase during any chain-growth polymerization reaction is the enormous gain in configurational entropy (cf. Fig. i) because the double bond is stiff and the single bond much more flexible. For polycondensation reactions, the main entropy gain results from the release of small molecules. In both cases, ΔS_s rules the overall entropy change. Additionally, entropy can be released to the environment as in most polymerization reactions, heat is released to the environment (exothermic reactions).

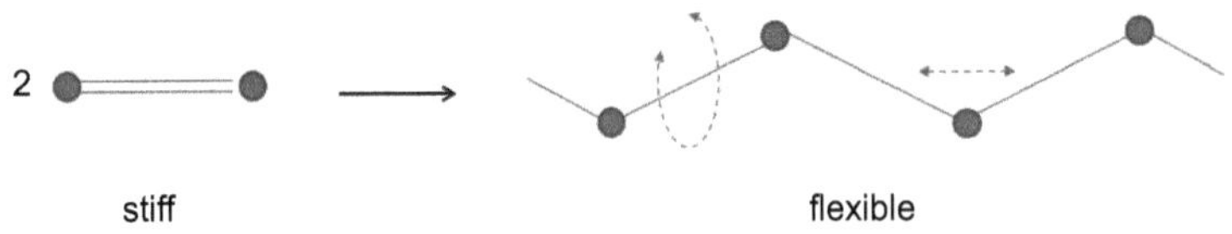

Figure i Illustration of configurational entropy gain during chain-growth polymerization.

Thermodynamics for heterophase polymerization turned out to be very powerful because it allowed deriving important equations explaining the behavior of colloidal systems, which are characterized by huge interfacial areas and energies. Probably the most meaningful relations for colloidal systems are the *Kelvin equation* in combination with the *Laplace equation*. However, both equations are also valid for one-component systems, which are, at given conditions, present in two phases and purely rely only on thermodynamic principles, meaning they do not need to consider the specific chemical properties of materials but the geometrical shape of the phases. For any colloidal system, including heterophase polymerizations, particles try to reach the spherical shape as long as the mobility of the atoms/molecules involved allows it. This is because a sphere has, at given volume and interfacial tension, the lower interfacial energy compared to any other geometrical forms.

Monomer droplets (exclusively) and polymer particles (mainly) are the most important spherical colloidal objects in heterophase polymerizations, and hence the following discussion will be restricted to these entities (i.e., bubbles are neglected).

In any two-phase system, the phases are in contact and an interfacial tension or surface tension exists between them. Discussing surface tension means that one considers not necessarily the geometrical dividing surface, but for nonplanar surfaces, it is the surface of tension, which should be considered. For deriving the Kelvin equation, one has to consider a vapor phase in equilibrium with its liquid, being the liquid of spherical shape, i.e., a drop (cf. Fig. ii). Each phase is governed by a Gibbs–Duhem equation (Eqs. ii and iii), where S, T, P, v, μ, and δ stand for entropy, temperature, pressure, molar volume, chemical potential, and equilibrium displacement, respectively; d denotes the dispersed and c the continuous phase. Equilibrium of the drops requires considering mechanical and chemical equilibrium. The mechanical equilibrium of the drop requires $\Delta P \cdot dV = \sigma \cdot dA$ and the chemical equilibrium $\mu^d = \mu^c = \mu^\sigma$, where $\Delta P = P^d - P^c$, V is the drop volume, A is the drop surface, μ's are the chemical potentials in the drop, in the continuous phase and on the surface, respectively. P^c is the pressure of the material forming the dispersed phase in the continuous phase, e.g. for a liquid drop, the vapor pressure outside the drop. Note that the tension is pulling, but the pressure is pushing the interface, and so both forces keep the droplet in shape.

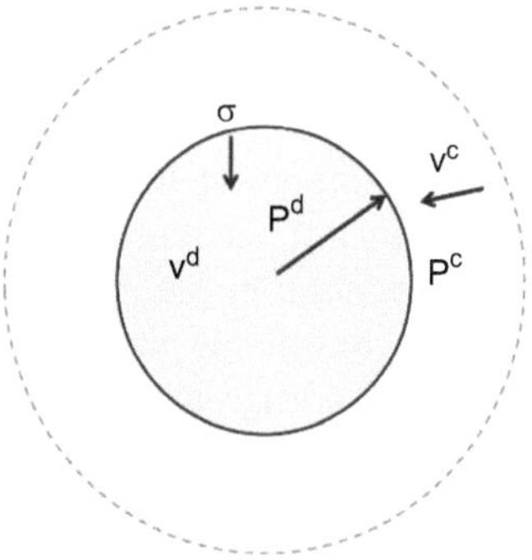

Figure ii Sketch of a colloidal particle dispersed in the continuous phase; P^d, v^d are pressure and molar volume inside the dispersed phase; P^c, v^c are pressure and molar volume in the continuous phase, and σ is the interfacial/surface tension between the dispersed and the continuous phase.

$$S^{d}dT - v^{d}\delta p^{d} + \delta\mu^{d} = 0 \tag{ii}$$

$$S^{c}dT - v^{c}\delta p^{c} + \delta\mu^{c} = 0 \tag{iii}$$

If the number of molecules and the temperature are unchanged, the Laplace equation (Eq. iv) follows from the mechanical equilibrium condition:

$$\Delta P = \frac{2\sigma}{r} \tag{iv}$$

In the equilibrium state, the droplet experiences some fluctuations (δ denotes the equilibrium displacements) in ΔP and the chemical potentials: $\delta P^{d} - \delta P^{c} = \delta(2\sigma/r)$, $\delta\mu^{d} = \delta\mu^{c}$. Applying the ideal gas assumption and integrating from $r = \infty$ to r, the Kelvin equation for a drop (Eq. v) is obtained, where P_0 is the pressure on a flat surface ($r = \infty$) and P^c the pressure on the drop.

$$\ln\frac{P^{c}}{P_0} = \frac{2\sigma}{r}\frac{v^{d}}{RT} \tag{v}$$

Equation v shows that the pressure over the particles of the dispersed phase increases with decreasing drop size. In fact, it is the surface/interfacial tension leading to an increase in the chemical potential of the molecules inside the droplet, which finds expression in increased P^c.

As the pressure is proportional to the number of atoms/molecules, the pressures in Eq. v can be replaced, namely, P_0 by the solubility of the bulk material (S_0) in the continuous phase and P^c by the solubility of a drop/particle of size r ($S(r)$), cf. Eq. vi.

$$\ln\frac{S(r)}{S_0} = \frac{2\sigma}{r}\cdot\frac{v^{d}}{RT} \tag{vi}$$

The consequences of Eqs. iv–vi for colloids and heterophase polymerizations are significant. Smaller drops and particles have a clear tendency to release material (i.e., a tendency to shrink), and since the solubility in the continuous phase remains unchanged, the released matter re-condenses on larger particles. This scenario within the molecular picture means that due to the higher curvature of the smaller particles, the contacts of the surface atoms/molecules are effectively reduced leading to a higher pressure of these outer layer objects to leave. Leaving of matter from smaller and re-condensing onto larger colloidal particles is important for proper understanding of the colloidal aspects of heterophase polymerizations.

For colloid chemistry of heterophase polymerizations, the interfacial tension is an important parameter. Remember, any interface is subjected to an interfacial tension. However, there are important differences in the details depending on the state of the matter of the phases forming the interface. All combinations of matter except gas–gas systems form interfaces, and the details depend, besides the chemical properties, on the mobility of each phase. Clearly, due to the absence of mobility at a solid surface, the solid–liquid or solid–gas interface shows a principally different behavior than the liquid–gas or liquid–liquid interface. This means for solid surfaces, such as polymer particles in heterophase polymerizations, it is impossible to measure directly its surface tension, whereas it is possible for the monomer–water interface.

The above discussion is based on two excellent textbooks [5, 6] on interfaces and thermodynamic properties of various colloidal objects (assembly of molecules such as bubbles, droplets, and particles), which we recommend for further reading.

In addition, the following points are important for heterophase polymerizations. First, matter flows spontaneously from a region of high chemical potential to a region of low chemical potential, which is important for swelling of polymer particles with monomer. Second, mass moves from a region of high gravitational potential to a region of low gravitational potential, which is important for the spatial uniformity during polymerization (monomer droplets and polymer particles) and during storage of polymer dispersions. Third, for colloidal entities, the chemical potential depends inversely on the size, meaning that a tiny monomer drop has a higher chemical potential than a larger one and the bulk monomer phase. The chemical potential can, therefore, be used to determine how a reaction system, also a heterophase polymerization system, has to react in order to achieve the thermodynamically favored state, i.e., equilibrium. A system is in the state of equilibrium when the chemical potential of each substance is the same in all phases present in the system. The important point is that a system always moves toward the equilibrium state, but it does not necessarily reach equilibrium.

The term *Brownian dynamics* was coined in the memory of the Scottish naturalist Robert Brown who observed in 1827 the trembling movement of pollen grains on top of a water pool, shortly after the performance of optical microscopes was considerably improved [7]. This motion is perfectly irregular, and it could only be explained first by Albert Einstein in 1905 who realized that it was caused by the tiny water molecules randomly hitting the much bigger pollen grains. The motion of all the water molecules hitting the suspended particle is quite complicated, and hence their effect on its movement is described probabilistically in terms of exceedingly frequent statistically independent impacts [8]. Consequently, a suspended particle motion is perfectly irregular in any direction, back and forth; it may stop a while and then continue with turns and twists. Each particle in the system follows its own specific path, but nevertheless, Brownian motion causes an overall net movement of the particles in the system depending on the particular conditions. In contrast to the deterministic physics (including thermodynamics), this is a completely different behavior because limited predictability arises due to stochastic fluctuations and it requires a new approach, namely, stochastic differential equations and averaging over time and space for successful description of the system. A stochastic variable (like the particle's travelling path from the initial point in given time) represents not a single value but a whole range of possible values all connected with a certain probability. Thus, a stochastic variable is quite abstract. However, describing the expected or average behavior of a large sample of particles is possible.

In contrast to thermodynamics, where time and the actual position of a particle is unimportant, the space-temporal behavior of a Brownian entity is crucial. The Langevin equation is a famous example of this kind of equations describing the irregular motion of a Brownian entity [9].

The difference between treating an issue with thermodynamics and BD can be illustrated as follows. If a particle suspended on the top of a pool has a higher density than the surrounding liquid, it will settle to the ground of the container due to the action of gravity. This is the final equilibrium state derived by thermodynamics. Within the frame of BD, a persistent average downward drift is calculated as well. However, the particle does not settle straightforwardly over any time interval, since the irregular hits by the water molecules cause deviations from the direct path within smaller time intervals, but over larger time intervals, the particle moves downward and finally settles on the bottom.

For the conditions during heterophase polymerizations, the following aspect of BD is of considerable importance. Due to the erratic/irregular/chaotic movements of all entities present in the continuous phase, a certain probability exists of encounters between all objects, i.e., between particles and particles, droplets and droplets, dissolved molecules and dissolved molecules, particles and droplets, dissolved molecules and particles, and dissolved molecules and droplets. Clearly, these encounters are important during heterophase polymerization with respect to colloidal stability of the dispersion but also with respect to the entry of molecular species into the particles/droplets (and also exit out of the particles/droplets when considering BD inside the dispersed objects).

For heterophase polymerizations, entry events (from the continuous phase into the colloidal objects) and exit events (from the colloidal objects into the continuous phase) of the various entities present must be triggered due to differences in the chemical potential in the different phases. For instance, thermodynamics tells us that according to the difference in the chemical potential of the monomer in the droplets and the polymer particles, a flux of monomer from the drops into the particles will happen. On the other hand, BD tells us that the path of the species entering or leaving the colloidal objects do not follow a straight line but is a random walk.

II. Swollen Micelles: Fact or Fancy

The idea that nucleation in classical emulsion polymerization (surfactant concentration above the critical micelle concentration, CMC) happens when radicals enter monomer-swollen micelles was historically considered a key step on the way to understand the process. Nowadays we have to conclude that it is just a nice idea that, however, cannot withstand serious scientific arguments.

Swollen micelles only exist at very low oil content, i.e., when a small amount of oil (monomer) is present in a micellar solution. Harkins, one of the great pioneers of emulsion polymerization research, whose results are not appreciated according to their meaning nowadays, proved such behavior. With regard to the micelles as the locus of polymer particle initiation, Harkins described very interesting experimental results by means of X-ray scattering [10–12]. He found that styrene at its saturation concentration in water increases the diameter of fatty acid micelles by about 1.2 nm (under starved conditions, i.e., styrene is present only dissolved in water and solubilized in micelles but not as bulk phase or droplets), but upon polymerization initiated with peroxodisulfate, the size of the micelles decreased to its initial value. This process could be repeated several times by consecutive swelling and polymerization until the surface of the polymer particles was grown so large that the soap concentration decreased due to adsorption on the particles below the critical micelle concentration. These results led to the important conclusion that a growing polystyrene chain is incompatible with the alkyl chains in the interior of micelles and, hence, polymer chains grow out of micelles, i.e., **nucleation happens outside micelles**. The incompatibility between polystyrene and alkyl chains was repeatedly proved more than 50 years later by attempts to polymerize the monomer inside double layers of dioctadecyldimethyl ammonium bromide vesicles. Transmission electron microscopy pictures revealed that phase separation takes place during the polymerization in a way that the polystyrene molecules gather at a particular place, whereas the monomer was uniformly distributed over the whole bilayer [13].

In a typical emulsion polymerization, the oil (monomer) volume fraction is about 50% and the coexistence of swollen micelles with monomer drops is, for thermodynamic reasons, unlikely. If a big enough interface is added to a micellar solution, micelles will dissolve and surfactants will redistribute according to the adsorption equilibrium among all interfaces (droplet–water, water–air, and container–water interface). Thus, during the emulsification process, micelles convert to oil drops.

Not only the surfactant, also will the oil (monomer) redistribute among the droplets in order to equilibrate their chemical potential. Note that the chemical potential of emulsion droplets as colloidal objects depends inversely on the drop size, according to the Lifshitz–Slezov–Wagner theory [14, 15] for Ostwald ripening and Kabalnov's extension [16]. Note that Kabalnov's extension holds also for swollen polymer particles.

Spontaneous emulsification (cf. next paragraph) counteracts Ostwald ripening as it generates much smaller drops than forced emulsification. Hence, it generates larger surface area per oil volume and triggers surfactant redistribution. The higher the surfactant concentration, the smaller the drop size due to emulsification, and moreover spontaneous emulsification is reinforced.

If the overall interface in the emulsion is larger than $a_S N_L V_W (S_{app}$-CMC), micelles will vanish; here a_S is the surface area covered per surfactant molecule in a saturated adsorption layer, N_L is Loschmidt's number, V_W is the volume of water, S_{app} is the applied surfactant concentration, and CMC is the critical micelle concentration. This is exactly the argumentation as used in the micellar nucleation theory of emulsion polymerization (Smith–Ewart theory) when the already nucleated particles grow and attract more and more surfactant molecules.

Coming back to the initial question: Are swollen micelles fact or fancy? Depending on the conditions, the answer is YES and NO. YES holds for the absence of a free monomer phase, i.e., starved conditions with respect to monomer, a situation which is untypical for heterophase polymerization. To the contrary, NO holds for the typical scenario of heterophase/emulsion polymerization under flooded conditions with respect to monomer.

According to the data of Harkins, even under starved conditions with respect to monomer, swollen micelles surely do not precede polymer particles because even if a radical enters, the polymer chains grow out of the micelles.

III. Spontaneous Emulsification: Principles and Meaning

In most cases, heterophase polymerization can start only after two not completely miscible liquids, typically monomer(s) and continuous phase, are brought in contact. Typically, after contacting monomer and continuous phase, both liquids will be vigorously mixed by the input of mechanical energy. Provided the mixing is intensive enough, any combination of immiscible and mutually nonreactive liquids can be broken up into an emulsion. However, except for microemulsions, the efficiency of the energy input for emulsification is very low since most of the energy is wasted to heat [17]. The question which liquid forms the dispersed phase and which will be the continuous phase is an interesting one. It is not only important for heterophase polymerizations but generally for making emulsions in cosmetics, shower gels, and cleaning supplies. Wilder Bancroft summarized the results of emulsification experiments from the late 19th to the early 20th century in a rule of thumb (known as Bancroft's rule), which meant that in order to have a kinetically stable emulsion, the emulsifying agent must be soluble in the continuous phase [18]. In principle, this rule finds its expression in the empirical system of the HLB values (hydrophilic–lipophilic balance), which was developed in the middle of the 20th century to allow the selection of the most effective emulsifier for a desired type of emulsion [19]. The HLB values were determined in tedious emulsification experiments. In addition, the result of the emulsification process is influenced by the volume ratio of the liquids, the concentration of the emulsifier, and, particularly for nonionic emulsifiers, the temperature.

Interestingly, a poly(ethylene glycol)-based surfactant may stabilize either an oil-in-water or the inverse emulsion in dependence on the temperature. This behavior is the consequence of a lower critical solution temperature or a cloud point at approximately 100°C in aqueous solutions [20]. The change in the properties with temperature is considered in the concept of the *phase inversion temperature* (PIT) [21]. This is very useful in predicting properties of nonionic surfactants based on poly(ethylene glycol), which is the most important hydrophilic group in nonionic emulsifiers also for heterophase polymerizations. It is interesting to note that at the PIT, both the interfacial tension and the droplet size are at the minimum, an effect which can be utilized during the emulsification step to create smaller emulsion drops.

Phase inversion is a process where for a given stabilizer the continuous phase becomes the dispersed one and vice versa. This is mainly observed in the case of polymeric surfactants with a stabilizing moiety possessing a critical solution temperature. Prominent examples are surfactants with poly(ethylene glycol) units (PEG) and poly(N-isopropylacryl amide). For PEG, increasing the temperature causes the HLB of the surfactant to decrease and it may subsequently, in accordance with Bancroft's rule, promote the stabilization of water-in-oil instead of oil-in-water emulsion. Furthermore, whether or not phase inversion occurs depends on the polarity of the oil phase, the kind of electrolyte and its concentration, other additives as for instance organic, water-soluble solvents increasing the oil solubility in water, and on the oil volume fraction.

However, another scenario is possible. What happens when both liquids are contacted and kept quiescently without the input of external mechanical energy?

The standard textbook answer is equilibration of the chemical potentials in either phase. This means that a certain number of molecules of both liquids dissolve in the counter phase according to the respective solubility at the given temperature. This scenario leads to molecular solutions of monomer in the continuous phase and vice versa. Particularly, the monomer concentration in the continuous phase is crucial for the polymerization reactions therein.

Surprisingly, attempts to follow the equilibration process by means of time-dependent multi-angle laser light scattering (MALLS) revealed a completely different scenario. For the combination water (continuous phase) and styrene (monomer), the MALLS data, acquired in the water phase, showed during the equilibration phase (before initiating the polymerization) an angle-dependent increase in the scattering intensity over about 48 h, which corresponded to a growing average size of the scattering objects with quite a broad size distribution (cf. Fig. iii) [26, 27]. After about 2 days, a kind of dynamic equilibrium was reached.

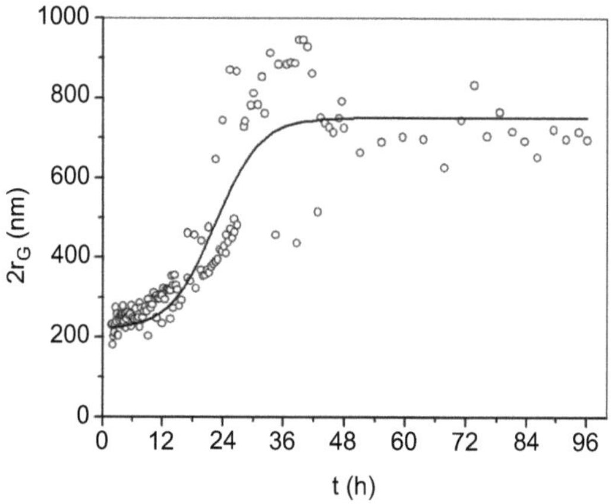

Figure iii Development of the particle size during the equilibration of a quiescent styrene–water system at 25°C; r_G, the radius of gyration, was determined from the model interpretation of MALLS scattering curves acquired in the water phase with monomodal spheres as described in Refs. [22–25]; data points represent a summary from three independent repeats; experimental setup: the scattering cells are standard quartz glass cuvettes with an inner diameter of 1.8 cm and a height of 8 cm filled with degassed water (about 10 ml) and styrene (2 ml) was carefully placed on top; the cuvettes were closed with a Teflon stopper.

That the scatters detected in the MALLS studies are really drops was confirmed by light microscopy. Figure iv shows styrene droplets as observed in the water phase (for details, see Ref. [26]).

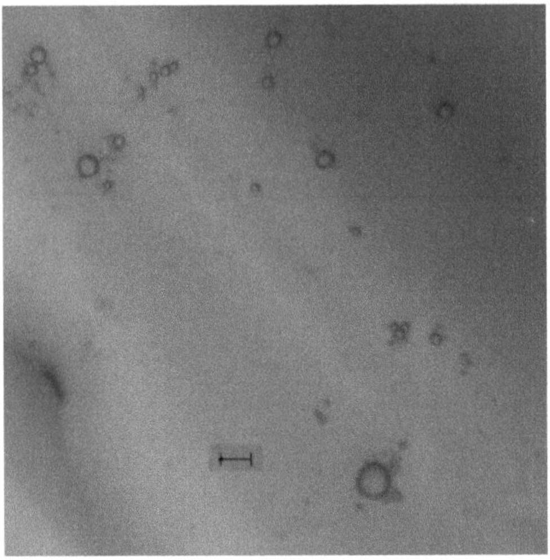

Figure iv Light microscopy images of the aqueous phase in the system styrene on water showing styrene drops; the bars indicate 10 μm; image was taken a few hours after sample preparation; experimental setup: styrene layer on top of pure water in a sealable optical cuvette of 2 mm thickness.

A behavior like this was observed for all combinations of (partly) immiscible liquids so far investigated (cf. Fig. v). Further examples with water as one component include tert-butyl styrene, acrylates and methacrylates (even laurylmethacrylate), dodecane and cyclodecane. Moreover, droplet formation takes place on either side of the interface with the water drops in the organic continuous phase larger than the oil drops in water [28].

These results suggest that the equilibrium state after establishing quiescent contact between two partly immiscible liquids is not two molecular solutions on either side of the interface but two emulsions with quite low volume fraction of the dispersed droplets, and hence both emulsions appear transparent to the naked eye. So the question is not only regarding the monomer concentration in the continuous phase, but it must be extended regarding the solution state.

Further, though indirect, support for the results of the MALLS and microscopy experiments comes from the comparison with the temporal development of the gross styrene content in water (C_{STY}) determined by gas chromatography (GC) (Fig. vi) [29]. For this investigation, the volume of the experiment is much larger than that for MALLS and microscopy studies (about a factor of 100 higher) and also the geometry of the experimental setup differs. All these might cause no issues for molecular solutions, but for emulsions, the situation is clearly different.

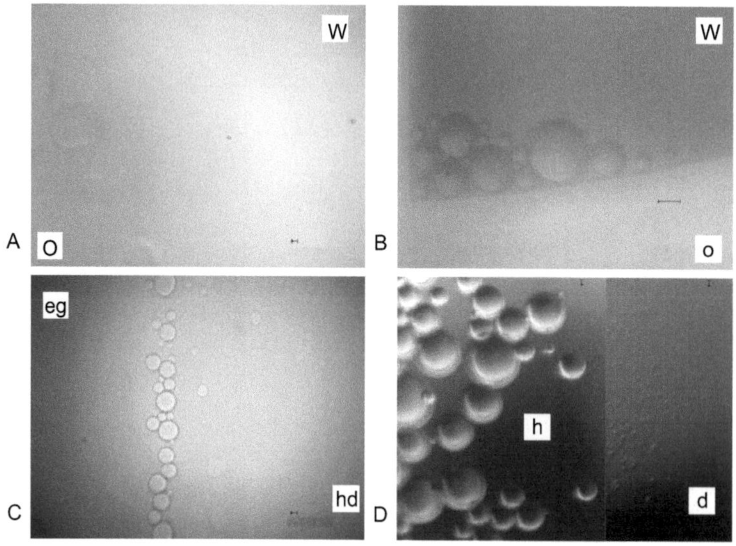

Figure v Light microscopy images of various combinations of immiscible fluids proving spontaneous droplet formation; the bars indicate 10 μm; A: hexane – water, B: polystyrene solution in ethyl benzene – water, C: ethylene glycol (eg) – hexadecane (hd), D: hexane – dimethylformamide; o — organic phase; w — water phase; image C shows ethylene glycol droplets and image D shows hexane drops in dimethylformamide (d) and dimethylformamide drops in hexane (h); experimental setup: optical glass cuvettes of 2 mm thickness.

The classical, for more than 70 years, most widely accepted solubility values of styrene in water at different temperatures were published by Lane already in 1946 based on the assumption of molecular solutions on either side, i.e., styrene in water as well as water in styrene [30]. Lane determined the styrene concentration in water by means of the formaldehyde–sulfuric acid reagent and by cloud point determinations. He assumed that equilibrium conditions were obtained by shaking styrene and water at a given temperature and then allowing it to stand for 24 h to completely separate. He determined the styrene concentration in water at 25°C and 70°C to be about 3 mM and 6 mM, respectively. Notably, the concentrations of water in styrene are much higher with about 37 mM and 67 mM at 25°C and 50°C, respectively, determined by Karl–Fischer titration and cloud point determination.

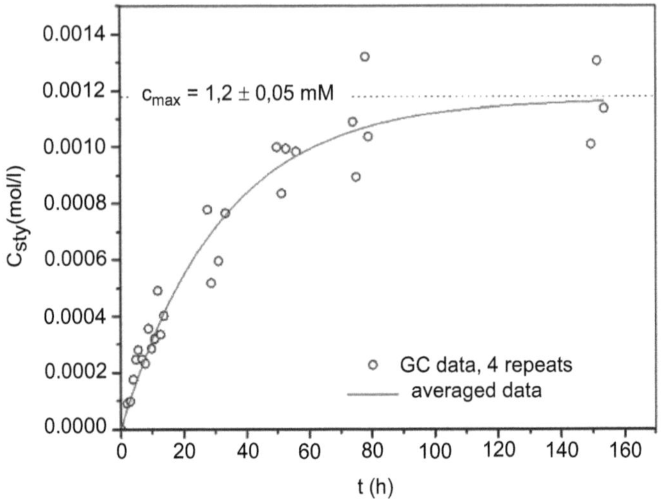

Figure vi Temporal development of the apparent styrene concentration in water during quiescent contact between styrene and water at 25°C in a Teflon reactor determined with gas chromatography; experimental details: 3.3 g of styrene (31.7 mmol) was placed in a reversed glass funnel with an inner diameter of 4.6 cm (area of 16.62 cm^2) 390 ml pure, degassed water.

The results presented in Figs. iii–vi are not in line with Lane's study and put into question his experimental procedure. First, he was unable to check whether or not styrene drops were present in the aqueous phase (and water drops in the styrene phase). Second, his assumption reaching equilibrium after 24 h is surely not correct. A general experimental difficulty results from the fact that it is necessary to avoid even slight temperature fluctuations because in the very vicinity of the equilibrium, the solution is prone to phase separation phenomena. Particularly, even a minimal decrease in temperature causes de-mixing recognizable by strong turbidity and destroys equilibrium [28]. In addition, an experimental study shows that the complete phase re-separation after emulsion formation (ethyl benzene in water) lasted, depending on the nature and concentration of the surfactant, several days (up to more than 200 days have been experimentally confirmed) [28]. Even in the absence of surfactant, the phase separation (only by visual inspection) was not completed within 24 h.

Furthermore, the formation of emulsions instead of molecular solutions may cause problems with respect to the homogeneity of the samples. In the absence of stirring, emulsions with broad droplet size distribution and low volume fraction of the dispersed phase exhibit a stronger spatial composition gradient than molecular solutions. Thus, the position of the sampling in the container may contribute to additional scattering of the results. The growing average size of the droplets (cf. Fig. iii) suggests that with increasing time (increasing drop size), the reproducibility should decrease as proven by the data in Fig. vi.

Still another point should be considered. The experimental requirements for determining the "equilibrium" concentration of the solute in the complementary phase are different for forced and spontaneous emulsification. Meaningful sampling after forced emulsification requires waiting long enough until de-mixing is completed. For spontaneous emulsification, the crucial step is allowing the system to equilibrate long enough.

Consequently, the value C_{max} = 1.2 mM (cf. data of Fig. vi) does not necessarily mean that the equilibrium is already reached in the particular experiment after 150 h. The solubility value for styrene in water obtained by Lane at 25°C is by a factor of 2.5 higher, but very likely due to the fact that after forced emulsification, the phase separation was not yet complete.

In summary, there is clear experimental evidence that **spontaneous emulsification is fact and not fancy**. It is particularly important for heterophase polymerization because the drops generated by spontaneous emulsification are smaller than those arising from forced nucleation. Smaller drops move faster and should possess a higher frequency of encounters with other objects (dissolved molecules, radicals, droplets, and particles). Consequently, particularly nucleation and swelling of polymer particles should be influenced. Encounter with a radical may cause a kind of droplet nucleation whereby the drops are not completely covered with surfactant, which is an important difference to Harkins experiments with swollen micelles. On the other hand, encounter with a polymer particle promotes its swelling.

IV. Swelling: Newer Experimental Results

Swelling of polymer particles is the last step for the monomer on its way to the main reaction loci, i.e., the polymer particles. It is a crucial step in heterophase polymerization because more than 90% of the monomer conversion in a typical emulsion polymerization takes place inside the monomer-swollen polymer particles. Since the rate of polymerization is extremely high, it requires also a high monomer concentration inside the particles. The most important question in this context is: How does the monomer get from the droplets into the latex particles?

It is commonplace to assume that the monomer supply is fast enough, even so fast that the used monomer is instantaneously replaced.

Again, we touch here the role of emulsification, i.e., the role of monomer drops in heterophase polymerization. Already in 1947 Harkins stated that the role of monomer drops is "*to act as a storehouse of monomer from which its molecules diffuse into the aqueous phase and from this into either soap micelles or polymer monomer latex particles*" [11]. Until today, this idea persists and is still state of the art for emulsion polymerization: "*In the presence of monomer droplets, the monomer-swollen particles grow and the monomer concentration within these particles is* ***kept constant*** *by monomer diffusing through the water phase from the monomer droplets*" [31]. This suggests that the particles are homogeneously swollen. However, since the 1960s, experimental evidence is available proving a monomer-rich shell—polymer-rich core structure of the growing polymer particles [32–38]. Conclusions were drawn based on various experimental techniques, including kinetic studies [32–35], sophisticated electron microscopy studies [35, 37], radioactively labeled monomer [37], and small angle neutron scattering investigations [38]. The experimental data have been obtained with monomers of various hydrophilicity such as styrene [32, 33, 35, 37], methacrylates [34, 38], and vinyl acetate [36]. The experimental data doubtlessly prove that the growth of particles does not happen homogeneously across the particles' volume but inside a surface layer. Additionally, there is thermodynamic reasoning that a monomer-rich shell guarantees higher conformational entropy of the polymer close to the interface compared with the higher polymer concentration for the assumption of homogeneously swollen particles [39]. This so-called "*wall-repulsion effect*" [40–43] is an additional reason for a polymer-depleted layer (or a monomer-rich shell) close to the interface in swollen polymer particles also at equilibrium. Thus, also an entropy-driven reorganization inside swollen particles supports the formation of the polymer-core–monomer-shell morphology. Clearly, this conclusion is extremely important for the mechanism of heterophase polymerization. However, it stayed forgotten for too long and was almost completely ignored by the scientific community.

The assumption that the monomer concentration inside the polymerizing particles is constant as long as monomer drops exist is still persisting even though experimental data of the monomer concentration inside the latex particles during the course of emulsion polymerization do not support it, but in fact show the opposite behavior [35, 44–46]. Remarkably, the corresponding results have been obtained with both water-soluble potassium peroxodisulfate [35, 44] and oil-soluble (2,2′-azo-bis(2-methylpropionitril)) initiators [45, 46] whereby, regardless of the initiator, typical emulsion polymerization kinetics has been observed. These results clearly show that the monomer concentration inside the particles is not constant but decreases with increasing conversion in the course of the reaction. Thus, there is neither monomer saturation nor an equilibration established inside the particles.

The state of the art assigns only a passive role to the monomer drops in emulsion polymerization. However, both, some basic considerations within the frame of classical thermodynamics and newer experimental results, prove that the role of monomer drops is everything but passive.

The instantaneous replenishment of the monomer inside the active particles containing a propagating radical requires that the monomer uptake frequency should correspond to at least the propagation frequency [47]. This requirement can be expressed by Eq. vii where $C_{M,P}$ is the monomer concentration inside the particles, k_p is the propagation rate constant, $\tilde{D}$ is the monomer diffusion coefficient, and x is the distance inside the particle ($x = 0$ is the center of the spherical particle with radius r_0 and $x = r_0$ is the distance from the center to the interface). A relation such as Eq. vii is also known as Thiele modulus (ϕ_{Th}) [48, 49], which is a characteristic number, typically describing the ratio between the reaction and the diffusion rate in catalytic reactions.

$$k_p C_{M,P} = \frac{\tilde{D}}{x^2} \tag{vii}$$

However, a detailed look at the scene reveals a serious problem with this apparently quite logical assumption of easy monomer diffusion through the aqueous phase. In general, when the specific moment interactions between components of the reaction mixture are neglected, diffusion is the net flow of matter from a more concentrated region to a less concentrated region with the aim to equilibrate the chemical potential, here that of the monomer inside the reaction system. Hereinafter, the reaction system comprises only droplets, particles, and water but neglects the gas phase. Figures vii and viii sketch the situation with respect to the monomer concentration across the emulsion polymerization space and illustrate the problem to be addressed.

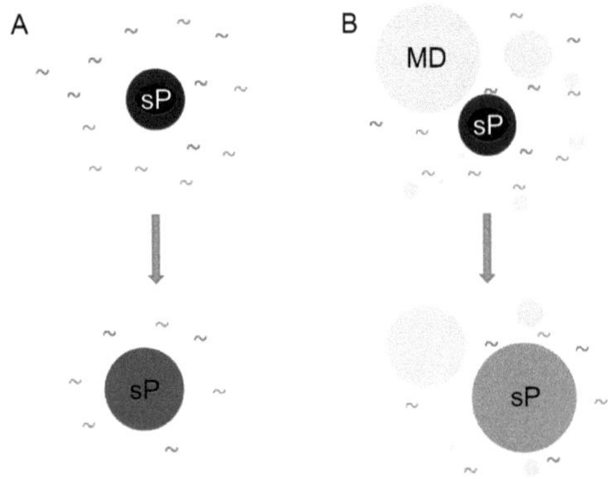

Figure vii Sketch of swelling of latex particles without (A) and with (B) emulsification; sP — swelling polymer particles, MD — monomer droplets, ~ — dissolved monomer molecules; not to scale.

Note that due to the colloidal nature of the drops, the chemical potential inside the monomer drops is larger than that of the bulk phase [45], which is the driving force for de-emulsification after stopping forced emulsification (Ostwald ripening or the growth of larger drops on the expense of smaller ones).

The assumption that during emulsion polymerization the monomer molecules simply diffuse through the aqueous phase into the latex particles is commonplace. However, simulation results based on Fick's diffusion laws show that the instantaneous replenishment of the consumed monomer during emulsion polymerization requires a close contact between the monomer and the polymer particles [47]. In summary, from the simulation results using Fick's second diffusion law, the following conclusions result. First, a high degree of swelling in the molar concentration range as observed for aqueous latex particles requires a high concentration of swelling agent (monomer during emulsion polymerization) immediately at the particle–water interface. Second, the concentration of swelling agent (monomer during emulsion polymerization) at the particle interface determines the influx into the particle interior. This means that for the situation during EP, there is a critical monomer concentration above which monomer diffusion is fast enough instantaneously to replenish the consumed monomer. Third, as a logical consequence of the simulations, all situations or measures that reduce the concentration of the swelling agent (monomer) in an immediate proximity of the particles surface are of detrimental influence on swelling.

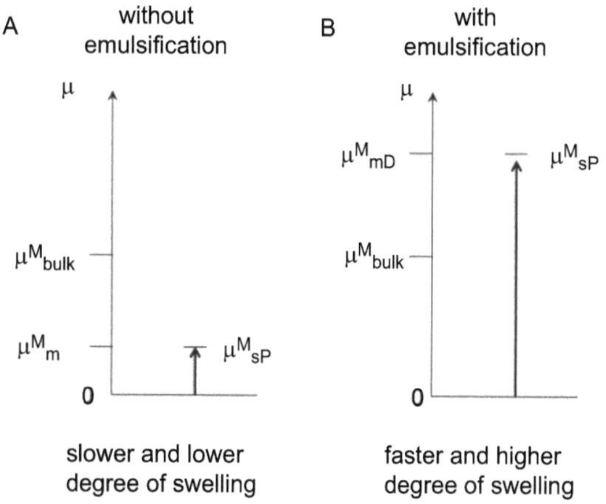

Figure viii Situation describing, with respect to the chemical potential (μ) of the monomer, the swelling scenarios without (A) and with (B) emulsification; μ^M_{bulk} — chemical potential of the bulk monomer, μ^M_m — chemical potential of the molecularly dissolved monomer, μ^M_{mD} — chemical potential of the monomer inside droplets, μ^M_{sP} — chemical potential of the monomer inside the swelling particles.

These basic considerations and simulation results triggered new experimental studies on swelling of polymers in aqueous systems (polystyrene as polymer and ethyl benzene as swelling agent) by means of rather unorthodox swelling experiments [50]. The results of these experiments disprove the classical idea that during emulsion polymerization, the swelling of latex particles with monomer to the extent required for typical emulsion polymerization kinetics can solely happen via diffusion of molecularly dissolved monomer molecules through the aqueous phase. Furthermore, and as a consequence of intimate droplet–particle contacts, it was experimentally proven that drops of the swelling agent containing a hydrophobic dye dissolved can act as cargo and tint the polymer particles. The experimental data from both swelling pressure studies and emulsion polymerization of styrene with varying stirring rate prove consistently a strong influence of droplet–particle collisions.

For heterophase polymerizations, it is generally assumed that the stirrer speed is only of minor influence and modern textbooks even completely ignore this topic. Only a few results have been published indicating an influence of the mixing intensity:

- If the inert gas for purging contains traces of oxygen or if a chain transfer agent has to diffuse out of the monomer droplets into the particles [51],
- If the monomer diffusion to the reaction loci is influenced [52],
- If the emulsifier distribution between the interfaces is changed considerably [53],
- For surfactant-free polymerizations, the hydrodynamic conditions are important for the reproducibility with respect to particle size and particle size distribution [54],
- If too high shear forces are applied, the latex may be prone to coagulation [55].

Most of the investigations have been carried out by increasing the stirrer speed starting from well-mixed conditions, cf. Ref. [53].

In a comprehensive study, quite a strong influence of the stirrer speed, varied between 50 and 500 revolutions per minute, on the rate of polymerization and latex and polymer properties has been observed for emulsion polymerization of styrene and methyl methacrylate, emulsifier concentration above the critical micelle concentration, and potassium peroxodisulfate as initiator [56]. The lowest stirrer speed is unable to completely disperse the monomer, i.e., a bulky monomer phase remains on top of the water phase. The higher the stirring speed, the faster the polymerization, the lower the particle size, and the higher the average molecular weight. Thus, increasing stirrer speed causes smaller monomer drops (larger droplet surface) and higher rate of swelling. All these effects can be explained considering that increasing the stirring rate, or the rate of mechanical mixing in general (e.g., recycling pump speed), increases the rate and energy of molecular and colloidal collisions promoting chemical reactions and mass transfer events, and reducing the spatial heterogeneity in the composition of the system.

All these findings **prove the importance of emulsification, both forced and spontaneous emulsification, for the kinetics of emulsion polymerization,** and particularly, swelling of polymer particles to that high degree necessary for fast polymerization during emulsion polymerization requires droplet–particle collision with subsequent coalescence.

References

1. Hernandez, H. (2020). Formulation and testing of scientific hypotheses in the presence of uncertainty. *ForsChem Research Reports*, **5**(1), pp. 1–16. doi: 10.13140/RG.2.2.36317.97767.
2. Theimer, W. (1978). *Handbuch naturwissenschaftlicher Grundbegriffe*. (Deutscher Taschenbuch Verlag GmbH & Co. KG, Germany). Stichwort: Naturwissenschaftliche Methode, pp. 334–340.
3. Strong, L. E. and Halliwell, H. F. (1970). An alternative to free energy for undergraduate instruction. *Journal of Chemical Education*, **47**(5), pp. 347–352.
4. Baierlein, R. (2001). The elusive chemical potential. *American Journal of Physics*, **69**(4), pp. 423–434.
5. Defay, R., Bellemans, A., and Prigogine, I. (1966). *Surface Tension and Adsorption*. (Longmans, Grun & Co Ltd, UK).
6. Hill, T. L. (1994). *Thermodynamics of Small Systems*. (W. A. Benjamin, Inc., Publishers, USA).
7. Perrin, J. (2013). *Brownian Movement and Molecular Reality*. Translated by F. Soddy. (Dover Publications, USA).
8. Einstein, A. (1905). Motion of suspended particles in stationary liquids required from the molecular kinetic theory of heat. *Annalen der Physik*, **17**, pp. 549–560.
9. Peters, E. A. J. F. (2000). *Polymers in Flow: Modelling and Simulation*. Doctoral Dissertation. (Technische Universiteit Delft, The Netherlands). Chapter 2.
10. Harkins, W. D. (1945). A general theory of the reaction loci in emulsion polymerization. *The Journal of Chemical Physics*, **13**(9), pp. 381–382.

11. Harkins, W. D. (1947). A general theory of the mechanism of emulsion polymerization. *Journal of the American Chemical Society*, **69**(6), pp. 1428–1444.
12. Harkins, W. D. (1950). General theory of mechanism of emulsion polymerization. II. *Journal of Polymer Science*, **5**(2), pp. 217–251.
13. Jung, M., van Casteren, I., Monteiro, M. J., van Herk, A. M., and German, A. L. (2000). Pulsed-laser polymerization in compartmentalized liquids. 1. Polymerization in vesicles. *Macromolecules*, **33**(10), pp. 3620–3629.
14. Lifshitz, I. M. and Slezov, V. V. (1958). About kinetics of supersaturated solid solution diffusion disintegration. *Jounal of Experimental and Theoretical Physics (JETP)*, **35**, pp. 479–492.
15. Wagner, C. (1961). Theorie der alterung von niederschlägen durch umlösen (Ostwald-reifung). *Zeitschrift für Elektrochemie, Berichte der Bunsengesellschaft für physikalische Chemie*, **65**(7–8), pp. 581–591.
16. Kabal'nov, A. S., Pertzov, A. V., and Shchukin, E. D. (1987). Ostwald ripening in two-component disperse phase systems: Application to emulsion stability. *Colloids and Surfaces*, **24**(1), pp. 19–32.
17. Ludwig, A., Flechtner, U., Prüss, J., and Warnecke, H. J. (1997). Formation of emulsions in a screw loop reactor. *Chemical Engineering & Technology*, **20**(3), pp. 149–161.
18. Bancroft, W. D. (1912). The theory of emulsification. I. *The Journal of Physical Chemistry*, **16**, pp. 177–233.
19. Griffin, W. C. (1949). Classification of surface-active agents by HLB. *Journal of the Society of Cosmetic Chemists*, **1**, pp. 311–326.
20. Harris, J. M. (1992). Introduction to biotechnical and biomedical applications of poly (ethylene glycol). In: *Poly (ethylene glycol) Chemistry: Biotechnical and Biomedical Applications* (Harris, J. M., Ed.) (Springer, USA). pp. 1–14.
21. Jönsson, B., Lindman, B., Holmberg, K., and Kronberg, B. (1998). *Surfactants and Polymers in Aqueous Solution* (Wiley, UK).
22. Dautzenberg, H. and Rother, G. (1988). Interpretation of light scattering from supermolecular structures in liquid systems by master curves. *Journal of Polymer Science Part B: Polymer Physics*, **26**(2), pp. 353–366.
23. Dautzenberg, H. and Rother, G. (1992). Supermolecular structures in polymer solutions interpretation of static light scattering data. *Macromolecular Symposia*, **61**(1), pp. 94–113.
24. Dautzenberg, H. and Rother, G. (2004). Response of polyelectrolyte complexes to subsequent addition of sodium chloride: Time-dependent static light scattering studies. *Macromolecular Chemistry and Physics*, **205**(1), pp. 114–121.

25. Rother, G. (2000). Model interpretation in static light scattering. *Macromolecular Symposia*, **162**(1), pp. 45–62.

26. Kozempel, S., Tauer, K., and Rother, G. (2005). Aqueous heterophase polymerization of styrene: A study by means of multi-angle laser light scattering. *Polymer*, **46**(4), pp. 1169–1179.

27. Tauer, K., Kozempel, S., and Rother, G. (2007). The interface engine: Experimental consequences. *Journal of Colloid and Interface Science*, **312**(2), pp. 432–438.

28. Tauer, K., Wei, C., Tripathi, A., and Kiryutina, O. (2017). Experiments and thoughts on mass transfer during emulsification. *Advances in Polymer Science*, **281**, pp. 23–43.

29. Kozempel, S. (2005). *Emulgatorfreie Emulsionspolymerisation: Monomerlösungszustand und Teilchenbildung.* Doctoral Dissertation. (MPI of Colloids and Interfaces/Universität Potsdam, Germany).

30. Lane, W. H. (1946). Determination of solubility of styrene in water and of water in styrene. *Industrial & Engineering Chemistry Analytical Edition*, **18**(5), pp. 295–296.

31. van Herk, A. and Gilbert, R. (2005). Emulsion polymerisation. In: *Chemistry and Technology of Emulsion Polymerisation* (van Herk, A., Ed.) (Blackwell Publishing Ltd, UK), pp. 46–78.

32. Bereshnoi, G. D., Chomikowski, P. M., and Medvedev, S. S. (1961). Izledovanie emulsionnoi (lateksnoi) polimerisatii stirola (Investigation of emulsion (latex) polymerization of styrene). *Vysokolmol. Soedin.* **111**(12), pp. 1839–1846.

33. Medveded, S. S. (1969). Some problems on emulsion polymerization. In: *IUPAC International Symposium on Macromolecular Chemistry*, Budapest 1969. (Akadémiai Kiadó, Hungary), pp. 39–63.

34. Brodnyan, J. G., Cala, J. A., Konen, T., and Kelley, E. L. (1963). The mechanism of emulsion polymerization. I. Studies of the polymerization of methyl methacrylate and n-butyl methacrylate. *Journal of Colloid Science*, **18**(1), pp. 73–90.

35. Grancio, M. R. and Williams, D. J. (1970). The morphology of the monomer–polymer particle in styrene emulsion polymerization. *Journal of Polymer Science Part A-1: Polymer Chemistry*, **8**(9), pp. 2617–2629.

36. Elgood, B. G., Gulbekian, E. V., and Kinsler, D. (1964). Emulsion polymerization of vinyl acetate. *Journal of Polymer Science Part B: Polymer Letters*, **2**(3), pp. 257–261.

37. Keusch, P. and Williams, D. J. (1973). Equilibrium encapsulation of polystyrene latex particles. *Journal of Polymer Science: Polymer Chemistry Edition*, **11**(1), pp. 143–162.

38. Bolze, J. and Ballauff, M. (1995). Study of spatial inhomogeneities in swollen latex particles by small-angle X-ray scattering: The wall-repulsion effect revisited. *Macromolecules*, **28**(22), pp. 7429–7433.

39. Keusch, P., Graff, R. A., and Williams, D. J. (1974). Polymer segment density distributions in saturated polystyrene latex systems. *Macromolecules*, **7**(3), pp. 304–310.

40. Casassa, E. F. (1967). Equilibrium distribution of flexible polymer chains between a macroscopic solution phase and small voids. *Journal of Polymer Science Part B: Polymer Letters*, **5**(9), pp. 773–778.

41. Casassa, E. F. and Tagami, Y. (1969). An equilibrium theory for exclusion chromatography of branched and linear polymer chains. *Macromolecules*, **2**(1), pp. 14–26.

42. De Gennes, P. D. (1981). Polymer solutions near an interface. Adsorption and depletion layers. *Macromolecules*, **14**(6), pp. 1637–1644.

43. De Gennes, P. G. (1987). Polymers at an interface: A simplified view. *Advances in Colloid and Interface Science*, **27**(3–4), pp. 189–209.

44. van der Hoff, B. M. E. (1962). Kinetics of emulsion polymerization. In: *Polymerization and Polycondensation Processes. Advances in Chemistry Series 34*. (Platzer, N. A. J., Ed.) (American Chemical Society, USA), pp. 6–31.

45. Ryabova, M. S., Sautin, S. N., and Smirnov, N. I. (1975). Number of particles in the course of the emulsion polymerization of styrene initiated by an oil-soluble initiator. *Journal of Applied Chemistry of the USSR (Zh Prikl Khim)*, **48**(7), pp. 1577–1582.

46. Stähler, K. (1994) *Einfluß von Monomer-Emulgatoren auf die AIBN-initierte Emulsionspolymerisation von Styren*. Doctoral Dissertation. (Universität Potsdam, Germany).

47. Tripathi, A., Wei, C., and Tauer, K. (2017). Swelling of latex particles: Towards a solution of the riddle. *Colloid and Polymer Science*, **295**(1), pp. 189–196.

48. Hill, C. G. and Root, T. W. (2014). *An Introduction to Chemical Engineering Kinetics and Reactor Design*, 2nd Ed. (John Wiley & Sons, USA).

49. Thiele, E. W. (1939). Relation between catalytic activity and size of particle. *Industrial & Engineering Chemistry*, **31**(7), pp. 916–920.

50. Wei, C. and Tauer, K. (2016). Features of emulsion polymerization: How does the monomer move from the droplets into the latex particles? *Macromolecular Symposia*, **370**(1), pp. 99–109.

51. Gerrens, H. (1959). Kinetik der emulsionspolymerisation. *Fortschritte der Hochpolymeren-Forschung*, **1**, pp. 234–328.

52. Arai, K., Arai, M., Iwasaki, S., and Saito, S. (1981). Agitation effect on the rate of soapless emulsion polymerization of methyl methacrylate in water. *Journal of Polymer Science: Polymer Chemistry Edition*, **19**(5), pp. 1203–1215.

53. Nomura, M., Harada, M., Eguchi, W., and Nagata, S. (1972). Effect of stirring on the emulsion polymerization of styrene. *Journal of Applied Polymer Science*, **16**(4), pp. 835–847.

54. Goodwin, J. W., Hearn, J., Ho, C. C., and Ottewill, R. H. (1974). Studies on the preparation and characterisation of monodisperse polystyrene latices. III. Preparation without added surface active agents. *Colloid and Polymer Science*, **252**(6), pp. 464–471.

55. Aslamazova, T. R. (1999). The role of polymer surface in the hydrodynamic stability of particles of emulsifier-free poly (methacrylate) latexes. *Colloid Journal*, **61**(2), pp. 158–163.

56. Tauer, K., Schellenberg, C., and Zimmermann, A. (2000). Nucleation in heterophase polymerizations. *Macromolecular Symposia*, **150**(1), pp. 1–12.

Chapter 1

Molecular Description of Heterophase Polymerization

1.1 Introduction to Heterophase Polymerization

Heterophase polymerization is a generic term, which describes polymerization reactions under nonhomogeneous conditions with respect to physical and chemical properties of the reaction mixture. Heterophase polymerizations are best defined as a generic process for making polymer dispersions, i.e., resulting in a state of matter where polymers are finely dispersed in a continuous phase [1]. An example of a polymer dispersion prepared by heterophase polymerization is presented in Fig. 1.1.

> **Heterophase polymerization**: Polymerization reaction under heterogeneous conditions, resulting in polymer dispersions.

Even though products of natural heterophase polymerization in the *Hevea brasiliensis* tree (i.e., natural rubber) have been used by humanity for at least 3 millennia, synthetic heterophase polymerization dates back to about 1909 with the birth of emulsion polymerization [2]. The term *heterophase polymerization* was already mentioned by Flory in his book *Principles of Polymer*

Heterophase Polymerization: Basic Concepts and Principles
Hugo Hernandez and Klaus Tauer

ISBN 978-981-4877-32-9 (Hardcover), 978-1-003-11929-6 (eBook)
www.jennystanford.com

Chemistry in 1953 [3]. On the other hand, the term *heterogeneous polymerization* has also been employed since the 1940s [4]. In the 1960s, Frenkel [5] proposed to differentiate heterogeneous polymerization (polymerization taking place in a heterogeneous system) from heterophase polymerization (polymerization reactions simultaneously taking place in different phases of the system). However, this difference is questionable, since emulsion polymerization of water-insoluble monomers would correspond to a heterogeneous, *homophase polymerization*. Thus, nowadays, heterophase polymerization is considered equivalent to heterogeneous polymerization [6]. Generally, distinguishing between homogeneous and heterogeneous reactions is useful by careful inspection of whether an interface is involved or not. Note that even if the main reaction is homogeneous (for instance solution polymerization), side reactions taking place at the borders of the reaction volume, i.e., at the reactor walls or at the interface to the air are heterogeneous in nature.

Figure 1.1 Polymer dispersion of poly(vinyl acetate-co-butyl acrylate) prepared by semi-batch seeded emulsion polymerization.

In 2011, the Polymer Division of the International Union for Pure and Applied Chemistry (IUPAC) published its recommendations on the terminology for polymers and polymerization processes in dispersed systems [7]. It defined different types of heterophase polymerization, such as emulsion polymerization, dispersion polymerization, and suspension polymerization; it might be considered that an overall picture of these heterophase polymerization variants was lacking. We will discuss the different heterophase polymerization variants in more detail in Section 1.6.

From its beginnings, heterophase polymerization in general, and emulsion polymerization in particular, has been considered by many of its practitioners "more an art than a science." The main reason behind this generalized idea is the extremely complex and nonlinear influence of process variables on the final properties of the product, which makes it very difficult to understand and predict. Two important characteristics of heterophase polymerization processes are its multiscale nature (where relevant phenomena take place at different time and length scales) and its inherent randomness (evident in the emergence of molecular weight, chain conformation, particle size, and particle composition distribution just to mention a few) [8]. Such complexity of heterophase polymerization is the result of the interaction of all different types of molecules present in the different phases of the system. And thus, the apparent chaotic behavior of heterophase polymerization can be understood and even predicted, considering the behavior and interaction of individual molecules in the system.

Therefore, the purpose of this chapter is presenting a fundamental molecular picture of heterophase polymerization. It will be shown that all different types of existing heterogeneous polymerizations can be explained by the use of essentially the same molecular interpretation. Furthermore, homogeneous polymerization can be regarded as a particular case of the more general heterophase polymerization when only one single phase is present in the system. It is, therefore, possible to describe a generalized mechanism of heterophase polymerization, valid not only for heterogeneous processes but also, under some simplifications, for homogeneous reactions.

Let us picture a general ***ab initio heterophase polymerization*** process. The system initially contains a mixture of monomer molecules (chemical building blocks of a polymer chain), active molecules (chemical compounds involved in polymerization reactions), and inert molecules (e.g., solvents, dispersing media, stabilizers, inert additives, etc.). However, there are no polymer chains in the system. In certain cases, monomer molecules may also act as active molecules (capable of starting the polymerization either thermally or photochemically), or they may also play a role as stabilizers and/or additives for the polymerization.

***Ab initio* polymerization**: Polymerization process where the system is initially absent of polymer chains.

Because of the different magnitude of the interaction forces between all types of molecular species in the system, certain molecules with less affinity with the majority of the molecules in the mixture will tend to segregate forming small dispersed phases in the continuous phase. Sometimes, these small dispersed phases are not evident to the naked eye. With time, and as a result of molecular motion, the small segregated phases eventually collide and aggregate forming even larger entities. Molecular diffusion may also result in the loss of material from the segregated phase, reducing its size. The leaving molecules may enter another segregated phase, or they can form a new separate entity. In the absence of shear forces, this whole process is known as *spontaneous emulsification*. As those segregated phases grow larger, external forces (e.g., gravitational) become relevant, promoting the accumulation and further aggregation of the dispersed phase eventually forming an evident additional phase in the system. Mechanical stirring (including ultrasound), when applied with enough intensity, may eventually break those large dispersed phases again into smaller ones. In any case, a balance between aggregation and breakage (or disintegration) of dispersed phases is expected to occur eventually. The size of the dispersed phases at dynamic equilibrium will be largely determined by the presence of stabilizer molecules in the system, which avoid or at least slow down re-aggregation of droplets and complete phase separation (e.g., the presence of a monomer "pool"). The stabilizer molecules have the property of presenting lower interaction energy when located at interfaces. Thus, by placing themselves at the interface between the dispersed phase and the continuous phase, they retard aggregation by means of intermolecular repulsive forces. Simultaneously, active molecules present in the system (or added some time later) initiate the polymerization. Those active molecules (e.g., initiators, starters, or catalysts, depending on the particular polymerization mechanism) can be present in the continuous phase or inside the dispersed phases. They can also eventually move from one phase to the other, depending on their kinetic energy and thermodynamic barriers for crossing the interface. They then react with monomer molecules present in any phase of the system, forming and growing polymer chains. The rate of polymerization is mainly determined by the local monomer concentration and the local temperature, but might also be influenced by the presence of active compounds (as they can promote side reactions). The final molecular mass distribution will then be determined by the reaction path followed by each chain in the system, and these paths are determined by the local environment and their particular trajectories (e.g., moving from one phase to another). As polymerization proceeds, polymer chains might eventually segregate forming an independent dispersed phase, or they tend to aggregate forming segregated polymer particles, which also grow to a certain size determined by the interaction among polymer chains, monomer molecules, and inert molecules. As it can be seen, all macroscopic properties of polymer dispersions (particle size, molecular weight, and rate of polymerization) are ultimately determined as the result of all the physical and chemical processes taking place at the molecular scale. Since each molecule will have a different history and a different behavior, the final macroscopic properties are the result of distributions of molecular properties.

In the following sections, the most basic phenomena observed at the molecular scale will be described. These phenomena comprise the conceptual background required for a better understanding of the molecular mechanism of heterophase polymerization, which will be presented in Chapter 2.

1.2 Molecular Forces

The most fundamental phenomenon taking place at a molecular scale is the interaction between different molecules. Those interactions can be either attractive or repulsive depending on the nature of the interacting molecules. According to their nature, intermolecular forces can be classified into three categories [9–12]:

1. *Electrostatic or Coulomb forces* between charged particles (ions) and between permanent dipoles, quadrupoles, and higher multipoles. They arise from the interaction between the static charge distributions of the two molecules. They are strictly pairwise additive and may be either attractive or repulsive.
2. *Induction forces* between a permanent dipole (or multipole) and an induced dipole, i.e., a dipole induced in a molecule with polarizable electrons. The dipole moments are induced in atoms and molecules by the electric fields of nearby charges and permanent dipoles. Induction effects arise from the distortion of a particular molecule by the electric field of all its neighbors, or by internal charge separation in a vibrating non-symmetrical molecule, which subsequently interacts with the external electric fields. Induction effects are always attractive and nonadditive. The polarizability represents how easily the electrons in the molecule can be displaced by an electric field.
3. *Quantum mechanical* forces, which give rise to *specific* or *chemical forces* (covalent or chemical bonding, hydrogen bonds and charge-transfer complexation), *dispersion* forces and to the *repulsive* steric or exchange interactions (due to *Pauli's exclusion principle*) that balance the attractive forces at very short distances. Dispersion forces are always attractive and arise because the charge distributions of the molecules fluctuate as electrons move.

The interaction potential (ψ) between two molecules (or particles in general), also known as *pair potential,* is related to the force between the two molecules (F) by:

$$\vec{F}(r_{ij}) = -\frac{d\psi(r_{ij})}{dr_{ij}}\hat{r}_{ij} \tag{1.1}$$

where r_{ij} is the intermolecular separation. The work that must be done to separate two molecules from the intermolecular distance r_{ij} to infinite separation is $-\psi(r_{ij})$. By convention, attraction forces (and potentials) are negative and repulsion forces (and potentials) are positive.

Purely electrostatic interaction potentials can be determined using **Coulomb's equation**:

$$\psi^{C}(r_{ij}) = \frac{q_i q_j}{4\pi\varepsilon_0 r_{ij}} \tag{1.2}$$

where q_i and q_j are the net electric charges of molecules i and j, respectively, and ε_0 is the dielectric permittivity of free space (8.854 $\times 10^{-12}$ $C^2/J \cdot m$).

For nonionic molecules, there are three main types of attractive interactions: *Keesom* (dipole–dipole) interactions (Eq. 1.3), *Debye* (dipole–nonpolar) interactions (Eq. 1.4), and *London dispersion* (nonpolar–nonpolar) interactions (Eq. 1.5). All these interactions are usually denoted as *van der Waals* interactions and have in common the same dependence with respect to the intermolecular separation (r_{ij}^{-6}):

$$\psi^{\mathrm{K}}(r_{ij}) = -\frac{u_i^2 u_j^2}{3(4\pi\varepsilon_0)^2 k_{\mathrm{B}} T}\frac{1}{r_{ij}^{\ 6}} \tag{1.3}$$

$$\psi^{\mathrm{D}}(r_{ij}) = -\frac{u_i^2 \alpha_j}{(4\pi\varepsilon_0)^2}\frac{1}{r_{ij}^{\ 6}} \tag{1.4}$$

$$\psi^{\mathrm{L}}(r_{ij}) = -\frac{3}{4}\frac{h\nu_{ij}\alpha_i\alpha_j}{(4\pi\varepsilon_0)^2}\frac{1}{r_{ij}^{\ 6}} \tag{1.5}$$

where u is the electric dipole moment, k_B is the Boltzmann constant (1.381×10^{-23} J/K), α is the electric polarizability, h is the Planck constant (6.626×10^{-34} J·s), and ν is the electronic ionization frequency. The **van der Waals interaction** is always attractive and can, in general, be expressed as:

$$\psi^{vdW}(r_{ij}) = -\frac{C}{r_{ij}^6} \tag{1.6}$$

where C is the corresponding van der Waals interaction parameter.

The **repulsive interaction** for nonionic molecules is given by an exponential potential of the form:

$$\psi^{\mathrm{R}}(r_{ij}) = Ae^{-Br_{ij}} \tag{1.7}$$

where A and B are interaction parameters.

Additional intermolecular potentials correspond to the interaction of ionic molecules with permanent dipoles (Eq. 1.8) or with nonpolar molecules (Eq. 1.9):

$$\psi^{\mathrm{ID}}(r_{ij}) = -\frac{q_i^2 u_j^2}{6(4\pi\varepsilon_0)^2 k_{\mathrm{B}} T}\frac{1}{r_{ij}^{\ 4}} \tag{1.8}$$

$$\psi^{\mathrm{IN}}(r_{ij}) = -\frac{q_i^2 \alpha_j}{2(4\pi\varepsilon_0)^2} \frac{1}{r_{ij}^4} \tag{1.9}$$

The general expression for the interaction potential is then:

$$\psi^{\mathrm{T}}(r_{ij}) = \psi^{\mathrm{C}}(r_{ij}) + \psi^{\mathrm{K}}(r_{ij}) + \psi^{\mathrm{D}}(r_{ij}) + \psi^{\mathrm{L}}(r_{ij}) + \psi^{\mathrm{R}}(r_{ij}) + \psi^{\mathrm{ID}}(r_{ij}) + \psi^{\mathrm{IN}}(r_{ij}) \tag{1.10}$$

The inclusion of additional potentials in Eq. 1.10 is also possible, such as external potentials (e.g., gravity, electromagnetic fields, etc.), or specific or chemical interaction potentials (hydrogen bonds, metallic bonds, etc.).

For nonionic molecules, a commonly used approximation is the **Lennard–Jones interaction potential** (Eq. 1.11), where the exponential repulsion interaction is approximated by the 12th power of the intermolecular distance:

$$\psi^{\mathrm{LJ}}(r_{ij}) = 4\varepsilon\left[\left(\frac{\sigma}{r_{ij}}\right)^{12} - \left(\frac{\sigma}{r_{ij}}\right)^{6}\right] \tag{1.11}$$

where σ is the equilibrium distance and ε is the energy well depth. The 6th power used for the attractive potential is very reasonable, based on the dependence of the van der Waals interactions. However, the use of a 12th power repulsive potential is not always a good approximation. Sometimes, a **Buckingham interaction potential**, using an exponential repulsive interaction together with the van der Waals attractive interactions, fits better:

$$\psi^{\mathrm{B}}(r_{ij}) = \varepsilon\left[\frac{6}{\alpha - 6} e^{\alpha\left(1 - \frac{r_{ij}}{\sigma}\right)} - \frac{\alpha}{\alpha - 6}\left(\frac{\sigma}{r_{ij}}\right)^{6}\right] \tag{1.12}$$

where α is an interaction parameter representing the steepness of the repulsive interaction.

Ultimately, intermolecular forces are electrostatic in nature, since all molecules present a certain degree of polarization, depending on the particular spatial arrangement of the electrons of the atoms. Table 1.1 summarizes the main types of intermolecular interaction potentials depending on the nature of the interacting molecules.

Table 1.1 Types of interacting molecules and their interaction potentials

Interacting molecules	Ion	Permanent dipole	Induced dipole
Ion	Electrostatic or Coulomb (Eq. 1.2)	Eq. (1.8)	Eq. (1.9)
Permanent dipole	Eq. (1.8)	Keesom (Eq. 1.3)	Debye (Eq. 1.4)
Induced dipole	Eq. (1.9)	Debye (Eq. 1.4)	London dispersion (Eq. 1.5) Short-range repulsion (Eq. 1.7)

Figure 1.2 shows the typical behavior of the interaction potential as a function of intermolecular distance for some representative forces: electrostatic interaction (Eq. 1.2) between charges of the same sign, van der Waals attraction (Eq. 1.6), and repulsion (Eq. 1.7). At larger separations, only the electrostatic forces are significant, whereas at short distances, van der Waals and repulsion forces become increasingly stronger.

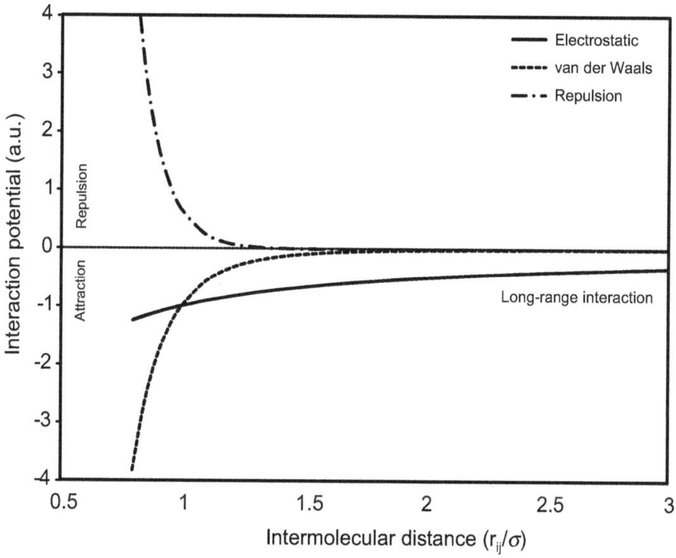

Figure 1.2 Interaction potential for the most representative intermolecular forces. Solid line: electrostatic force; dotted line: van der Waals force; dashed-dotted line: repulsive force.

In Fig. 1.3, the net interaction force between two nonionic molecules is presented. Three different models are considered: (1) the sum of van der Waals and repulsive interactions, (2) the Lennard–Jones interaction potential (Eq. 1.11), and (3) the Buckingham interaction potential (Eq. 1.12). All three models predict the formation of a potential well, i.e., a minimum in the interaction potential, where the net interaction force becomes zero (Eq. 1.1). The distance at which the potential well is observed corresponds to the equilibrium distance (σ) between the two interacting molecules. However, in the case of the Lennard–Jones approximation, the interaction potential at distance σ is zero, but it is not the minimum. The Buckingham potential satisfactorily corrects the drift in potential well of the Lennard–Jones approximation but predicts a deeper potential well.

1.3 Molecular Diffusion

1.3.1 Brownian Motion and the Laws of Diffusion

The most important consequence of intermolecular forces is the continuous, changing motion of the molecules, characterized by an apparently random and chaotic behavior, which in fact is the result of the permanent collisions between them. Every molecule at any particular moment is influenced by the different attraction and repulsion forces exerted by its neighbor molecules, whatever kind of molecules they are. Following the classical law of motion (Eq. 1.13), a molecule i with mass m_i will accelerate in space according to the force resulting from the summation of all attractive and repulsive forces. In addition, rotation, stretching, and some other internal types of motion are the result of the interaction forces between the different atoms in a molecule, as well as the neighboring atoms from different molecules.

$$\vec{a}_i(t) = \frac{\sum_{j \neq i} \vec{F}_{ij}(t)}{m_i} \tag{1.13}$$

The resulting chaotic motion of individual molecules is usually denoted as ***Brownian motion***, random motion, or simply as ***diffusion***.

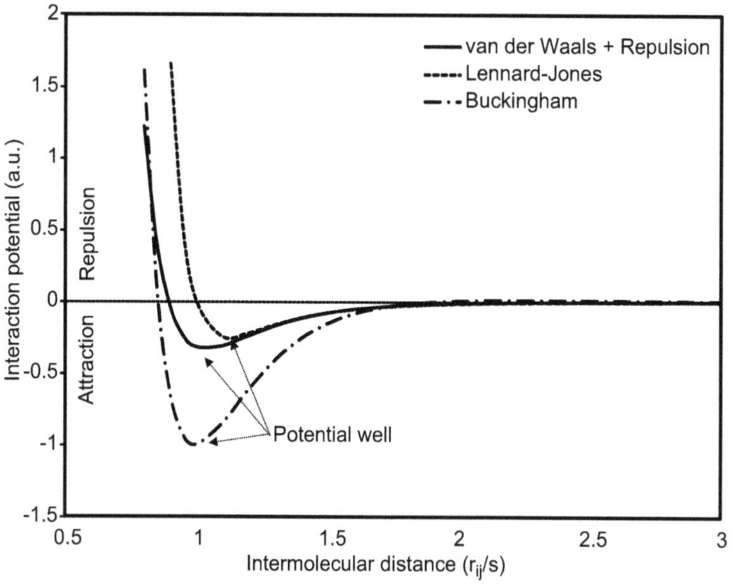

Figure 1.3 Net interaction potential according to different models. Solid line: van der Waals + repulsive forces; dotted line: Lennard–Jones interaction model; dashed-dotted line: Buckingham interaction model.

Brownian motion: Also denoted as random motion or random walk. It denotes the typical random behavior of individual molecules or particles in a given medium. It was first evidenced and investigated by the Scottish botanist Robert Brown in 1827. He concluded that this random motion was not related to living motion since it could be observed also in suspensions of inorganic materials.

Diffusion: Process by which matter is transported from one part of a system to another as a result of random molecular motions.

From a purely statistical point of view, the path followed by a body subject to random motion may be described by the following relationship:

$$\langle x^2 \rangle \propto t \tag{1.14}$$

In Eq. 1.14, $\langle x^2 \rangle$ is the squared displacement of the body averaged over an elapsed time t. The proportionality factor is closely related to the diffusivity or diffusion coefficient, which is a measure of the speed of the displacement (or diffusion) of the body.

An example of a trajectory described by a particle under Brownian motion in three dimensions is presented in Fig. 1.4.

In a molecular system, the random motion of the molecules is caused by the frequent collision with other molecules in the proximity. However, a purely Brownian motion can only be observed if the forces exerted by other molecules are randomly distributed in direction and magnitude in such a way that the average net force acting on the molecule over relatively long periods of time is zero. Otherwise, a drift will be observed with significant consequences on the molecular behavior of the system.

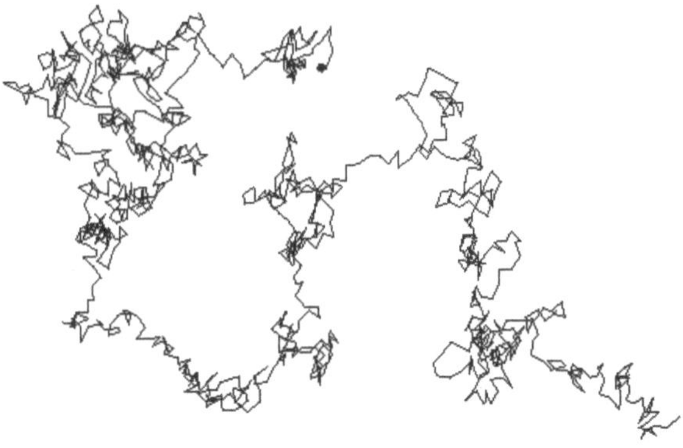

Figure 1.4 Example of a trajectory obtained by three-dimensional Brownian motion projected in two dimensions; the third dimension is perpendicular to the page.

The force exerted on a single molecule will vary in direction and magnitude throughout time. The force can be described as the sum of two components, an average net force and a random force, as follows:

$$\sum_{j \neq i} \vec{F}_{ij}(t) = \left\langle \vec{F} \right\rangle + \vec{X}(t) \tag{1.15}$$

where the random force $\vec{X}$ has an average value of zero. The average net force is responsible for the ballistic, center of mass or ***convective*** motion of the molecules, while the random force causes a random motion giving rise to diffusion.

Convection: Mechanism by which matter is transported as a result of the collective motion of the molecules.

If the magnitude of the average net force is much larger than the magnitude of the random force contribution, the diffusive behavior becomes unimportant.

As a consequence of random molecular motion, heat can also be transferred by a process called *conduction*, which is equivalent to the transfer of mass by diffusion. This equivalence was recognized by Fick [13] who set the quantitative basis of diffusion by adopting the mathematical treatment of heat conduction previously derived by Fourier [14].

Isotropism: In isotropic substances, the structure and properties in the neighborhood of any point are identical.

For ***isotropic*** substances, the motion of a single molecule has no preferential direction. If we consider a system with two zones of different concentration of a certain type of molecules, the total number of molecules diffusing from one zone to the other at a given instant will be proportional to the number of molecules present at each side of the interface. It is, therefore, evident that there will be a *net flux* of molecules from the zone of higher molecular concentration to the zone of lower concentration, and this flux will be proportional to the difference in concentration [15]. It is important noticing that diffusion is the result of random molecular motion, and that the gradient in concentration determines the net flux of molecules in purely diffusive systems. For isotropic substances, the net flux of molecules is proportional to the concentration gradient and the proportionality constant is the *diffusion coefficient* or *diffusivity* of the molecules in the system (**Fick's first law of diffusion**):

$$\vec{J} = -D \cdot \vec{\nabla} C \tag{1.16}$$

where $\vec{J}$ is the rate of transfer of molecules per unit area of a section (net flux), ∇C is the concentration gradient of the diffusing substance, and D is the diffusion coefficient. The negative sign arises because the net flux of molecules takes place from the higher to the lower concentration region. The magnitude of the diffusion coefficient will depend on the geometry (size and shape) of the molecules and on the intermolecular forces present in the system. For an anisotropic medium, the diffusion properties also depend on the direction in which they are measured, and the diffusion coefficient is actually a function of the local spatial composition around the diffusing molecule. Some common examples of anisotropic media are crystals, textile fibers, and polymer films in which the molecules have a preferential direction of orientation. In these cases, Eq. 1.16 remains valid, but the diffusion coefficient is a tensor of rank 2 and not a scalar. It is possible to relate the rate of change in the concentration of the diffusing substance with the net diffusive flux $\vec{J}$, simply from the continuity equation. By making use of the first law, **Fick's second law of diffusion** can be obtained:

$$\frac{\partial C}{\partial t} = -\vec{\nabla} \cdot \vec{J} = D \cdot \nabla^2 C \tag{1.17}$$

If the effect of composition or temperature on the diffusion coefficient cannot be neglected, Fick's laws of diffusion become:

$$\vec{J} = -\vec{\nabla}\left[D(C,T) \cdot C\right] \tag{1.18}$$

$$\frac{\partial C}{\partial t} = \nabla^2 \left[D(C,T) \cdot C\right] \tag{1.19}$$

For nonzero average net interaction forces, the net flux of molecules depends on the ***chemical potential*** rather than on the molecular concentration. The chemical potential or *cohesive energy* (μ) represents the total free energy of a molecule (cf. Eq. i), and it includes the interaction potentials as well as the contribution associated with its thermal energy.

> ***Chemical potential***: The chemical potential is a measure of the net interaction energy to which an individual molecule is subject when present in a particular environment.

The development of a consistent theory of Brownian motion began with the early contributions of Einstein [16] and Smoluchowski [17]. In their pioneering work, they were able to relate the microscopic Brownian motion observed in colloidal systems with the macroscopic molecular Fickian diffusion coefficient.

There are some important considerations in Einstein's solution to the problem of Brownian motion:

- The motion is caused by the exceedingly frequent impacts on the particle of the incessantly moving molecules of fluid in which it is suspended.
- The motion of these molecules is so complicated that its effect on the particle can only be described probabilistically in terms of exceedingly frequent statistically independent impacts.
- Each individual particle executes a motion, which is independent of the motions of all other particles.
- The movements of a given particle at different time intervals are independent processes.

Einstein's analysis led to the following conclusion:

$$\frac{\partial f(x,t)}{\partial t} = D\frac{\partial^2 f(x,t)}{\partial x^2} \quad (1.20)$$

where $f(x,t)$ is the probability function of finding a particle at position x in time t. This expression is analogous to Fick's second law of diffusion in one dimension (Eq. 1.17), considering that the probability function is proportional to the concentration of the diffusing compound. This partial differential equation can be solved analytically for the case of diffusion from a single point (neglecting the interaction between the diffusing particles). The result is the Gaussian distribution function:

$$f(x,t) = \frac{1}{\sqrt{4\pi D}} \frac{\exp\left(-\frac{x^2}{4Dt}\right)}{\sqrt{t}} \quad (1.21)$$

such that

$$\langle x^2 \rangle = 2Dt \quad (1.22)$$

Equation 1.22 is known as **Einstein's diffusion equation** and is valid for the diffusion in one dimension. The generalization to three dimensions is simply:

$$\langle r^2 \rangle = 6Dt \quad (1.23)$$

where r is the distance to the initial position of the molecules.

Please notice that the molecular diffusion coefficient derived by Einstein and the macroscopic diffusion coefficient proposed by Fick are proportional but not necessarily exactly the same. This is particularly true for multicomponent systems [18].

The diffusion equations derived from Smoluchowski's treatment are mathematically identical to those obtained by Einstein. However, Smoluchowski considered a concentration-dependent diffusion coefficient, while Einstein's equation defines a constant diffusion coefficient.

A couple of years after Einstein's original derivation, Langevin presented an alternative method, which was quite different from Einstein's and, according to him, "infinitely more simple." **Langevin's equation for Brownian motion** can be expressed as [19]:

$$m\frac{d^2x}{dt^2} = -6\pi\eta a\frac{dx}{dt} + X \tag{1.24}$$

where m is the mass of the Brownian entity (particle or molecule), a is its hydrodynamic radius, x is its position at a given time t, η is the viscosity of the continuous phase, and X is a random fluctuating force, which is the result of the collisions of the Brownian entity with the surrounding molecules of the continuous phase.

The basic assumptions in Langevin's approach are as follow:

- There are two forces acting on the particle: a viscous drag and a fluctuating force X, which represents the incessant impacts of the molecules of the liquid on the Brownian particle. The fluctuating forces should be positive and negative with equal probability.
- The mean kinetic energy of the Brownian particle in equilibrium should reach, in one dimension, a value of:

$$\left\langle \tfrac{1}{2}mv^2 \right\rangle = \tfrac{1}{2}k_\mathrm{B}T \tag{1.25}$$

- The random force has an average mean value of zero and is independent of its previous values. That is, $\langle X \rangle = 0$ and $\langle X(t)X(t') \rangle = \Gamma\delta(t - t')$, where $\Gamma = 12\pi\eta a k_\mathrm{B}T/\Delta t$ and δ is Dirac's delta function.

Langevin solved Eq. 1.24 and found that:

$$\frac{d\left(x^2\right)}{dt} = \frac{k_\mathrm{B}T}{3\pi\eta a} + C\exp(-6\pi\eta at/m) \tag{1.26}$$

where C is an arbitrary constant. Langevin estimated that the decaying exponential approaches zero with a time constant of the order of 10^{-8} s, which for any practical observation at that time could be neglected. Integrating the last equation, it is found that

$$\left\langle x^2 \right\rangle = \frac{k_B T}{3\pi\eta a} t \tag{1.27}$$

which corresponds to the expression deduced by Einstein, provided that

$$D = \frac{k_B T}{6\pi\eta a} \tag{1.28}$$

This last expression is usually denoted as the **Stokes–Einstein equation**.

1.3.2 Brownian Dynamics (BD) Simulation

The numerical solution to Langevin's equation for Brownian motion (Eq. 1.24) is known as **Langevin Dynamics** simulation. If the system is assumed to relax completely, the solution to the equations of motion corresponds to the method of **Brownian Dynamics** (**BD**) simulation. There are several techniques for the numerical solution to Brownian motion [20]. One of the most representative methods is the **Monte Carlo random flight** (**MCRF**) algorithm [21, 22]. A flowchart for the MCRF algorithm for BD simulation is presented in Fig. 1.5.

Brownian entity: Any individual material entity, either individual molecules, fragments of molecules (such as radicals or ions), or molecules aggregated as a particle, for which Brownian motion is relevant. Brownian entities have sizes usually up to 1 μm. Larger entities are practically insensitive to Brownian motion.

The simulation is restricted to a small cell (usually cubic) containing a given number of ***Brownian entities***. Periodic boundary conditions are applied to take into account the effect of large-scale systems. In the MCRF method, the diffusive displacement on each direction for each molecule or particle at each time-step dt is obtained from a normal Gaussian distribution with mean zero and variance $\sqrt{2Ddt}$.

As the first step, the initial positions of the Brownian entities are randomly determined considering a uniform probability distribution in the simulation cell, always checking that particle superposition is not taking place. The Brownian motion of the particles is then simulated in time-steps given by

$$dt = \max\left(\alpha \frac{d_{\min}^2}{D}, \frac{mD}{k_B T}\right) \tag{1.29}$$

where d_{min} is the minimum separation between the surface of two particles, D is the diffusion coefficient of the particles, α is a "damping" factor, which is selected based on the probability of collision during the simulated time-step dt, m is the mass of the Brownian entities, T is the temperature of the system, and k_B is the Boltzmann constant.

The term mD/k_BT expresses the relaxation time for the motion of the smallest Brownian entity, which is the time resolution of the MCRF simulation method and is obtained from Eqs. 1.26 and 1.28. When the separation between entities is large, the computation efficiency can be improved by increasing the time-step of the simulation proportionally to d_{min}^2/D [22]. This means that the simulated time-step will be proportional to the squared distance between Brownian entities, but the minimum time-step considered will be that of the corresponding relaxation time of the momentum of the smallest Brownian entities. The probability of collision between Brownian entities is related to the damping factor α, according to the expression $\alpha = (2Z^2)^{-1}$, where Z is the inverse of the normal cumulative probability of collision and corresponds to the closest distance between entities expressed in the number of standard deviations. Thus, in order to obtain a collision probability of the order of 10^{-7} for time-steps larger than the relaxation time, a value of α = 0.01852 is used [22]. This value guarantees that practically every collision will occur with a resolution corresponding to the momentum relaxation time. At each time-step, the movement of the entities in each direction (in rectangular coordinates) is calculated as:

$$dx = \xi_G \sqrt{2D_r dt} \tag{1.30}$$

ξ_G is a random number obtained from a Gaussian distribution with mean 0 and variance 1.

Several important applications of the interaction between Brownian entities (such as radicals, macromolecules, or colloidal particles) can be found in the literature [21, 23]. BD simulation is perhaps the most suitable method for the simulation of systems at the colloidal scale, and it is also useful for the simulation of molecules in solution.

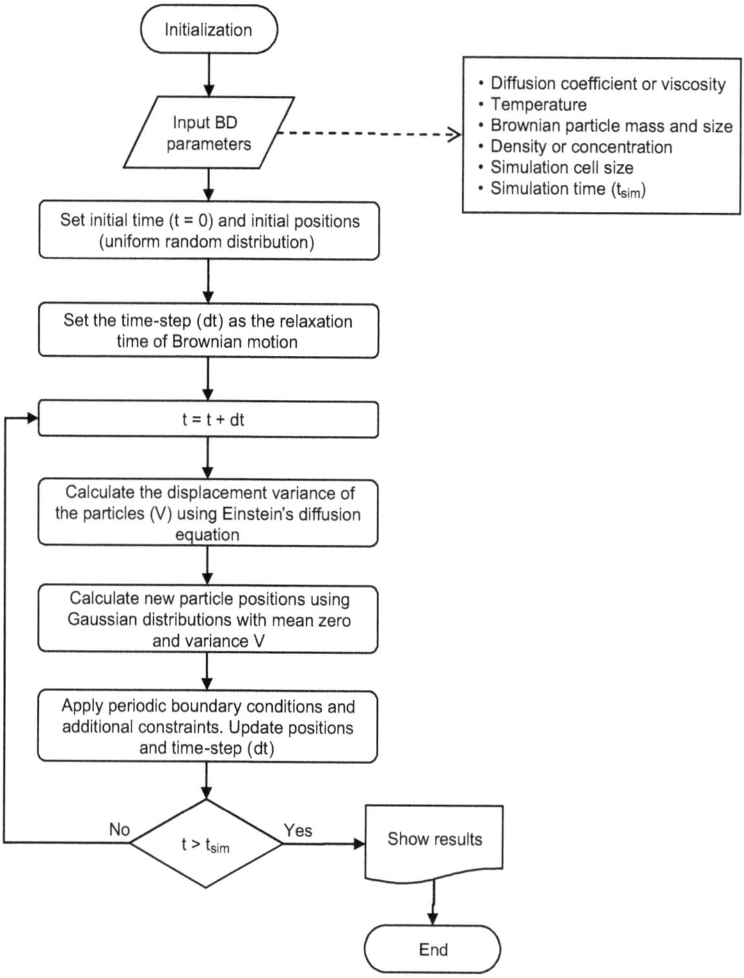

Figure 1.5 Monte Carlo random flight (MCRF) algorithm.

1.3.3 Macromolecular Diffusion

Molecular diffusion is one of the most important processes taking place in heterophase polymerization. The net molecular flux by diffusion is determined by the difference in chemical potential between two different regions, which in turn depends on the intermolecular potentials. At steady state, there is no chemical potential difference and the net molecular flux becomes zero. It is by molecular diffusion that active molecules, monomer molecules, and oligomers present in the continuous phase can reach and enter the polymer particles. Similarly, it is by molecular diffusion in the polymer phase that molecules present inside polymer particles can reach the surface and desorb. Bimolecular reactions, such as propagation and termination for radical polymerization, can also occur only if the reactants (and the reactive sites) approach each other by diffusion. For example, termination requires as the first step the approach of the polymer chains by molecular diffusion and as the second step the segmental diffusion of the active chain ends. If the diffusion process is slow compared to the rate of reaction, the reaction becomes diffusion-controlled.

The diffusivity of a system depends on the nature and strength of the intermolecular interactions, on the kinetic energy of the molecules (associated to the temperature of the system), and on the size and concentration of the molecules. In some systems, the diffusion coefficients are extremely sensitive to the process conditions. A particular and relevant example of this situation is the diffusion in polymer media. The diffusion coefficients for small molecules (solvent or monomer) through polymer solutions in the vicinity of the glass transition are known to change by as much as six orders of magnitude with only a small change in polymer concentration [24]. A mathematical expression relating the diffusion coefficient of a small molecule in a polymer was obtained by Vrentas and Duda [25] using the ***free-volume theory*** developed by Fujita [26]. The free-volume theory is by far the most widely used theory for predicting diffusion coefficients in polymers and polymer solutions. Free volume exists in a system because of geometrical restrictions, random thermal motion, and intermolecular repulsive potentials.

> ***Free-volume theory***: The diffusion of a molecule is only possible if there is enough free space or free volume surrounding the diffusing molecule to accommodate it.

Even if an individual hole may not be large enough to accommodate a diffusing molecule, the cooperative motion of several neighboring molecules may allow two or more holes to merge into a single hole large enough for diffusion to occur. As the free volume of a system is reduced, the energy required by the diffusing molecule to accommodate in a new position increases, and therefore, the diffusion coefficient will decrease exponentially. According to the free-volume theory, the diffusion coefficient of a small molecule in a polymer can be obtained from the following equation:

$$D_{\mathrm{p}} = D_{\mathrm{p01}} e^{-E/k_{\mathrm{B}}T} \exp\left(-\gamma \frac{(1-w_{\mathrm{p}})V_1^* + w_{\mathrm{p}}\xi_{\mathrm{v}}V_2^*}{V_{\mathrm{FH}}}\right) \tag{1.31}$$

where the subscripts 1 and 2 represent the small molecule and the polymer, respectively; E is the attractive energy between the small molecule and its polymer neighbors; D_{p01} is a pre-exponential factor; γ is a correction factor for the free-volume overlap; w_{p} is the weight fraction of polymer; V_i^* is the specific critical hole free volume of the component i; ξ_{v} is the ratio of the critical molar volume of the small molecule jumping unit to that of the polymer jumping unit; and

$$V_{\mathrm{FH}} = (1-w_{\mathrm{p}})K_{11}(K_{21} + T - T_{\mathrm{g1}}) + w_{\mathrm{p}}K_{12}(K_{22} + T - T_{\mathrm{g2}}) \tag{1.32}$$

where V_{FH} is the average size of the free volume per gram of mixture, K_{ij} are parameters of the system, and $T_{\mathrm{g}i}$ is the glass transition temperature for each component.

The free-volume theory of diffusion is a macroscopic semi-empirical theory based on the motion and interaction forces between the molecules. However, if the trajectories and velocities of all the molecules in the system over a certain period of time are known, the diffusion coefficients can be directly obtained simply by using Einstein's description of Brownian motion (Eq. 1.23).

Molecular modeling simulation methods as well as the model of Vrentas and Duda [25] are perhaps the most relevant approaches for the estimation of diffusion coefficients in polymers because they consider explicitly the intermolecular forces between the polymer and the diffusing molecule.

A different expression for molecular diffusion in polymers was obtained by Yasuda et al. [27], who found that the diffusion coefficient can be expressed as

$$D_{\mathrm{p}} = D_{\mathrm{p}}^0 \exp\left(-\frac{\beta x_{\mathrm{v}}(1-\alpha)}{1+\alpha x_{\mathrm{v}}}\right) \tag{1.33}$$

where D_{p}^0 is the self-diffusion coefficient of the small molecule, $\beta = V^*/V_{\mathrm{Fm}}$, $\alpha = V_{\mathrm{Fp}}/V_{\mathrm{Fm}}$, $x_{\mathrm{v}} = \varphi/(1-\varphi)$, V^* is the critical free-volume fraction necessary for diffusion to take place, V_{Fm} and V_{Fp} are the free-volume fractions of small molecules and polymer, and φ is the volume fraction of small molecules.

A simplified approach was presented by Chang et al. [28]. They expressed the diffusion coefficient of a small molecule inside the polymer particle as

$$D_\mathrm{p} = K(\varphi - \varphi_0)^2 \tag{1.34}$$

where K is a constant and φ_0 is a parameter defined as a function of T_g. φ_0 is zero at about 20–30°C above T_g, positive at lower temperatures, and negative at higher temperatures.

An alternative for the estimation of diffusion coefficients, valid only for low polymer concentrations, is the expression for high-viscosity liquids obtained by Hiss and Cussler [29] based on Eyring's theory of absolute reaction rates. In this case,

$$D_\mathrm{p} = \frac{k_\mathrm{B}T}{\eta^{2/3}h^{1/3}}\exp\left(-\frac{\Delta G^{\neq}}{3k_\mathrm{B}T}\right) \tag{1.35}$$

where h is the Planck constant, η is the viscosity of the mixture, and $\Delta G^{\neq}$ is the activation free energy for the diffusion of the molecule.

An additional problem arises when the growth of oligomers is considered, because the diffusion coefficient is reduced by the incorporation of additional monomer units in the chain. In this case, some empirical equations have been developed for the estimation of diffusion coefficients as a function of the oligomer size and monomer fraction in the particles [24, 30]. For example, Griffiths et al. [30] obtained an empirical equation for estimating the diffusion coefficients of oligomers of chain length i, inside monomer-swollen polymer particles as a function of the degree of polymerization of the oligomer and the weight fraction of polymer in the particle:

$$D_{\mathrm{p}i}(w_\mathrm{p}) = \frac{D_{\mathrm{p}1}(w_\mathrm{p})}{i^{0.664+2.02w_\mathrm{p}}} \tag{1.36}$$

This expression is valid for a variety of monomers when the concentration of polymer in the particles is above a critical concentration at which polymer chains start to overlap.

A different approach for estimating the diffusion coefficients of small molecules in polymer particles is the use of molecular modeling simulation methods, such as molecular dynamics, coarse-grained, or Monte Carlo methods [31–33]. In this approach, the molecular motion of a molecule or the oligomer of a given size in a polymer phase of a given composition is simulated taking into account the different intermolecular forces exerted by the neighboring molecules (polymer chains, monomer, etc.). A suitable autocorrelation function is then used to calculate the diffusion coefficient of the small molecule from the trajectories obtained. Additional information about the modeling of diffusion in polymers can be found in the review paper of Masaro and Zhu [34].

Molecular modeling simulation methods as well as the model of Vrentas and Duda are perhaps the most relevant approaches for the estimation of diffusion coefficients in polymers because they explicitly consider the intermolecular forces between the polymer and the diffusing molecule.

1.3.4 Molecular Diffusion by Molecular Dynamics Simulation

Molecular dynamics (MD) simulation is a method used to follow the trajectories and velocities of an ensemble of atoms or molecules subject to interatomic or intermolecular forces for a certain period of time. Atoms and molecules, which are large assemblies of a huge number of quantum particles, can be satisfactorily described by Newton's classical equations of motion [31, 35]:

$$\frac{d\vec{v}_i}{dt} = \frac{1}{m_i}\sum_{j \neq i} \vec{F}_{ij} \tag{1.37}$$

$$\frac{d\vec{x}_i}{dt} = \vec{v}_i \tag{1.38}$$

where $\vec{v}_i$ is the velocity, m_i is the mass and $\vec{x}_i$ is the position of the i-th molecule, $\vec{F}_{ij}$ is the interaction force between the i-th and j-th molecules, and t is the time. Additional external or internal (mean field) forces can also be considered.

By means of MD simulation, different equilibrium and non-equilibrium properties of a system can be determined. Some relevant examples include: the conformation of a molecule in a certain medium, the calculation of transport properties of the system (viscosity, thermal conductivity, and diffusion), and the estimation of the total energy of the system (potential energy + kinetic energy).

Different types of ensembles can be simulated in MD. Perhaps the most commonly used ensembles are the canonical or *NVT* ensemble (constant number of molecules N, volume V, and temperature T) and the microcanonical or NVE ensemble (constant number of molecules N, volume V, and energy E). However, it is also possible to simulate grand-canonical or μVT (constant chemical potential μ, volume, and temperature), isothermal–isobaric or NpT (constant number of molecules, pressure p, and temperature), and isoenthalpic–isobaric or NpH ensembles (constant number of molecules, pressure, and enthalpy H), by using adequate algorithms for solving the equations of motion.

As long as reliable expressions for calculating intermolecular forces are available, the diffusion coefficient of a system can be calculated from MD simulation as follows:

$$D = \frac{1}{2n_{\mathrm{D}}} \lim_{t\to\infty} \frac{<(r(t)-r(0))^2>}{t} = \frac{1}{2Nn_{\mathrm{D}}} \lim_{t\to\infty} \frac{\sum_{i=1}^{N}\sum_{j=1}^{n_D}\left(r_{ij}(t)-r_{ij}(0)\right)^2}{t} \quad (1.39)$$

where N is the number of molecules used for the determination of the diffusivity, n_{D} is the number of spatial dimensions considered, and r_{ij} is the position of the center of mass of the molecule i in the j-th direction. Equation 1.39 is usually referred to as the Wiener–Einstein equation.

An alternative formulation of Einstein's equation was developed by Green and Kubo [36–38], where the diffusivity can be calculated as an autocorrelation function of the velocities of the molecules:

$$D = \lim_{t\to\infty} \int_0^t <\vec{v}(t)\cdot\vec{v}(0)> dt = \frac{1}{N} \lim_{t\to\infty} \int_0^t \sum_{i=1}^{N} \vec{v}_i(t)\cdot\vec{v}_i(0) dt \quad (1.40)$$

Both approaches are equivalent and equally valid. As mentioned before, the use of MD simulation for the determination of diffusion coefficients relies on accurate models for describing the interaction potential between the molecules. Some of the most common models for nonpolar molecules are the Lennard–Jones and Buckingham potentials (see Section 1.2). The interaction parameters of these models can be obtained from physicochemical experimental data (e.g., phase-change enthalpies), but it is also possible to estimate the parameters using semi-empirical or *ab initio* molecular modeling methods [39].

In order to show the dependence of the diffusion coefficient on the intermolecular forces in the system, the Buckingham interaction potential (Eq. 1.12) was used to calculate the molecular self-diffusion coefficients from the Wiener–Einstein expression (Eq. 1.39) using an MD simulation of an *NVT* ensemble. In order to reduce the numerical error during the calculations, the following reduced dimensions were employed:

$$t^* = \sigma\sqrt{\frac{m}{\varepsilon}} \tag{1.41}$$

$$D^* = \frac{D}{\sigma}\sqrt{\frac{m}{\varepsilon}} \tag{1.42}$$

$$T^* = \frac{k_B T}{\varepsilon} \tag{1.43}$$

$$\rho^* = \frac{\rho\sigma^3}{m} \tag{1.44}$$

$$\alpha^* = \mathrm{a} \tag{1.45}$$

$$m^* = \sigma^* = \varepsilon^* = 1 \tag{1.46}$$

where m is the mass of a single molecule, and σ, ε, and α are the interaction parameters of the Buckingham interaction potential (Eq. 1.12).

The values of α^* in the range between 10 and 15 were used, according to the values reported for several interaction parameters obtained using force field methods [40]. The effect of the repulsion steepness of the Buckingham interaction potential (α^*) on the reduced diffusion coefficient is presented in Fig. 1.6. It is observed that the reduced diffusivity is very sensitive to small variations in the value of α^*. However, there is a clear trend for decreasing the values of the reduced diffusion coefficient as α^* is increased. Although the functional dependence of D^* on α^* may be highly nonlinear, for the range of values considered, it can be approximated by a linear function.

Using MD simulation, it is possible to obtain an approximate expression for the diffusion coefficient as a function of the system conditions and interaction parameters [8]:

$$D(T,P,\alpha,m,\sigma) \approx A_1\sigma\sqrt{\frac{k_B T}{m}}\exp\left(-\frac{A_2 N_A \sigma^3}{V_m(T,P)}\right)(A_3-\alpha) \quad (1.47)$$

where $A_1 = 0.093$, $A_2 = 8.6404$, $A_3 = 16.9684$, and $V_m = N_A m/\rho$ is the molar volume. Notice that although the potential well depth (ε) determines the strength of the interaction forces, the net effect on the diffusion coefficient of the molecules is apparently negligible.

The expression in the exponential of Eq. 1.47 contains the effect of the free volume on the diffusion coefficient. The diffusion coefficient is, therefore, mainly determined by the free volume of the system, the temperature of the system, the size and mass of the molecules, and the steepness of the intermolecular repulsion.

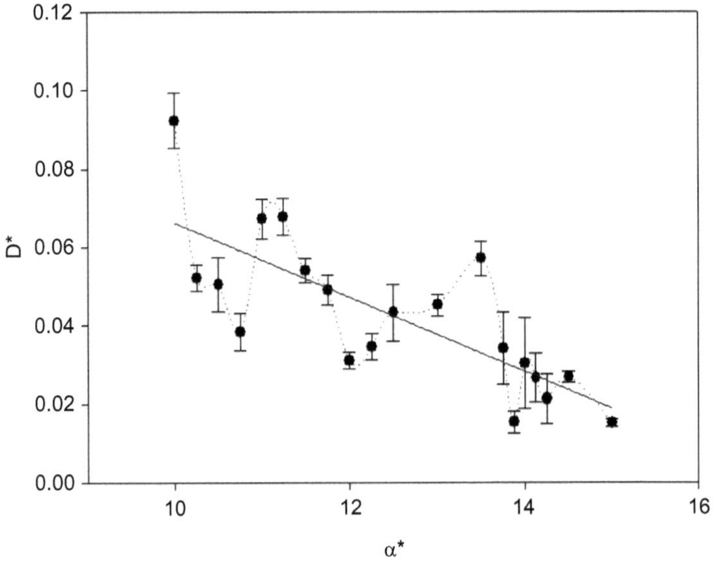

Figure 1.6 Effect of repulsion steepness on reduced diffusion coefficient. $T^* = 2$, $\rho^* = 0.3$. Data points: MD simulation. Error bars: Estimation error using MD simulation. Dotted line (red): smoothed curve. Solid line (blue): best linear fit.

1.4 Molecular Phases

In many cases, the net time-averaged force exerted by different neighboring molecules as well as any other external force on a particular molecule is nonzero, and this results in the displacement of the molecule until it reaches a place where the external forces are finally balanced (on a long-time average basis). A continuous region composed of molecules with an almost zero net time-averaged external force is denoted as a "**phase**." It is important to notice that a phase can contain different types of molecules, or in other words, a phase can be a multicomponent mixture.

All molecules are driven by external forces to the most stable region where external forces are balanced. Let us consider different possible scenarios for any given molecule:

- A molecule present in an unstable region (nonzero net average force) will move until a balance is reached (in a different phase), either by entering a different phase or by aggregating with other molecules to form a new phase (cf. Fig. 1.7A). Thus, a phase change occurs. This is the case of coalescence of droplets or particles of immiscible substances in the absence of stirring, since there is a difference in the gravitational force exerted on molecules with different density forcing the heavier molecules to go the bottom of the vessel. It is important to notice that the difference in gravitational forces is only important if its magnitude is much higher than the magnitude of the local and temporal force fluctuations.
- A molecule present in a stable region can aggregate with other molecules of the same stable region forming a new and more stable phase when environmental conditions change (cf. Fig. 1.7B). This phenomenon also represents a phase change, and it can be driven by changes in environmental conditions (pressure, temperature) and composition. An example is the crystallization of a liquid phase (assemble on top is also possible depending on the density difference between phases). The molecules in the liquid phase are in a stable system until energy is removed from the system. The molecules with lower energy find a new equilibrium point at closer distances, forming regular arrays of molecules known as crystals. Under certain environmental conditions, some systems may present *miscibility gaps*, leading to the coexistence of two or more different stable phases.

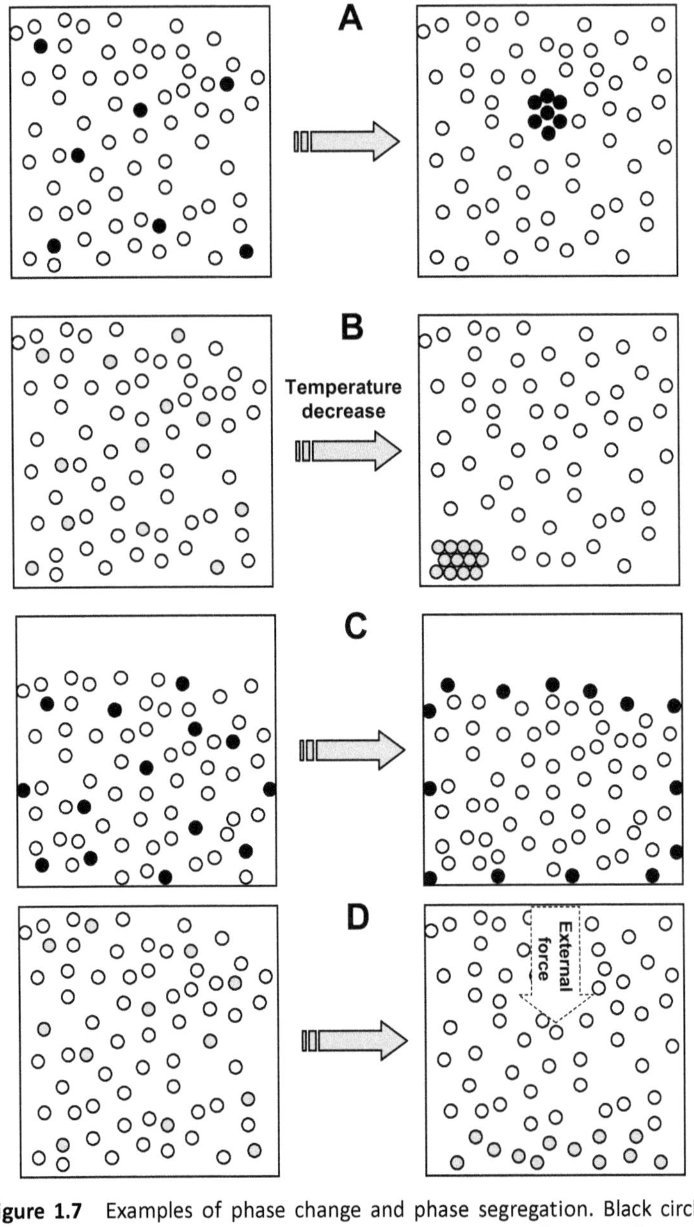

Figure 1.7 Examples of phase change and phase segregation. Black circles represent molecules with nonzero net average interaction forces in the initial state. Gray circles represent molecules with zero net average interaction forces in the initial state but forced to different phases by external changes. White circles represent all other molecules with zero net average interaction forces.

- A molecule is present in an unstable region, and it is not able to find balance inside a phase, but only at interfaces (cf. Fig. 1.7C). In some special cases (e.g., surfactants), molecules can only find balance between two different phases, as they are not fully compatible with only one of them. By staying at the interface, the molecules find their balance in the interfacial force, and at the same time, they provide stability to the whole molecular domain since the interfacial forces are nonzero net forces present at the interfaces.
- A molecule that is already in a stable region with time-averaged balanced external forces tends to remain in its current phase, unless an external force is exerted, which forces the molecule to a different phase (cf. Fig. 1.7D). This is the case of mixing in a reactor, where the external force exerted by the stirrer can segregate a phase composed of a large number of molecules into many phases (e.g., immiscible droplets) composed of relatively few molecules. Sometimes a molecule that is already in a stable region can go to a different phase as a result of local and temporal fluctuations in the external forces. This behavior is also observed during evaporation and condensation processes. Depending on the strength of external forces (as well as cohesive forces), a molecule can either abandon or become part of a condensed phase.

A system composed of a single phase is denoted as homogeneous, whereas a system containing two or more phases is denoted as heterogeneous. Even though this is a very clear definition, the identification of different phases is rather subjective in the sense that the detection of heterogeneity strongly depends on the sensitivity of the observation technique. As it was mentioned earlier, a phase can be identified by a nonzero time-averaged net force acting on the bulk of the molecules composing the phase; however, this definition is dependent on the time scale considered and subsequently, on the length scale considered. From a practical point of view, the determination of time-averaged net forces is not straightforward, and therefore, the identification of phases is done based on the observation of physicochemical properties, e.g., appearance, density, refractive index, etc.

The influence of the observation length scale (or magnification ability of the observer) on the definition of a homogeneous phase is exemplified in Fig. 1.8. A system observed as homogeneous from a macroscopic point of view (Fig. 1.8A) can be shown to be actually heterogeneous at smaller scales (Figs. 1.8B and 1.8C). When both the inner squares in Fig. 1.8A are compared, no differences in their properties are observed and, therefore, the system is perceived as homogeneous. The same result is obtained when the inner squares of Fig. 1.8B are compared. No significant differences in their properties are found even though a heterogeneous internal structure can be observed. Finally, the inner squares of Fig. 1.8C clearly show different (instantaneous) properties, demonstrating the heterogeneity of the system at smaller length scales. Please notice that observation of different regions of small-scale systems for longer times will again result in similar properties; thus, the time scale of the observation is also important.

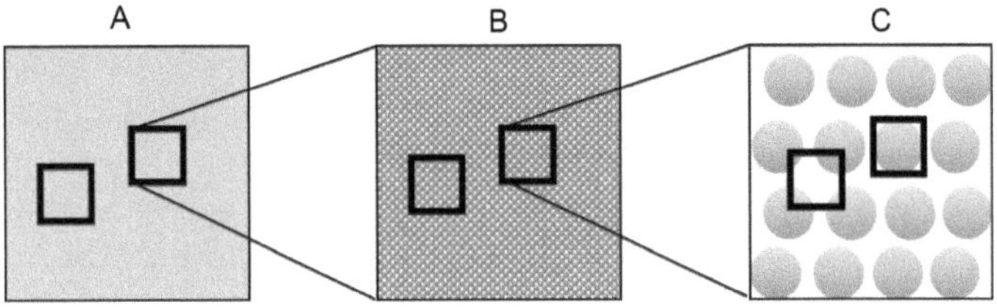

Figure 1.8 Illustration of the effect of observation length scale on the heterogeneity of a system. In sketch (A), the system is observed from a macroscopic perspective. Sketch (B) corresponds to a zoom on a section of system (A). Sketch (C) corresponds to a zoom on a section of system (B). In each sketch, the outer and inner squares represent different observation volumes.

Since chemical systems are composed of individual molecules, it is possible to state that *all chemical systems are heterogeneous at the molecular level, even though they can be perceived as homogeneous at the macroscopic level*. Homogeneity is actually an apparent condition of a system, when it is observed at certain particular time and length scales. Thus, ***homogeneous systems are only particular cases of the more general heterogeneous systems***. This is the reason why *homogeneous polymerization* can be considered a particular case of the more general *heterophase polymerization* processes.

Molecular heterogeneity principle: All chemical systems are heterogeneous at the molecular level, even though they can be perceived as homogeneous at the macroscopic level.

Quantification of homogeneity and heterogeneity is possible. *Dispersity* is a quantitative measure for heterogeneity that characterizes the degree of division of the heterophase system. It is defined as the reciprocal average characteristic length scale of the dispersed phase. A chemical system where all components are of similar size and perfectly miscible with each other appears homogeneous for any analytical technique at any moment of time that probes spatial dimensions, which are much larger than the molecular volume of the components (cf. Fig. 1.8A). If, however, reactants participate with molecular dimensions that differ by orders of magnitude like during polymerization, an apparently homogeneous system might appear heterogeneous (cf. Fig. 1.8C).

Spatial inhomogeneity of a macroscopically homogeneous volume element where chemical reactions take place will strongly depend on the *reaction-diffusion kinetics* at the molecular scale. Under such conditions, homogeneity is a function of characteristic times (defined by kinetics) and volumes (defined by molecular dimensions). In other words, the important issue here is how fast concentration differences can be re-equilibrated by molecular diffusion after a particular reaction has occurred. Thus, molecular diffusion and kinetics compete in this consideration, particularly by their corresponding characteristic time scales.

Molecular phases are very important for chemical reactions since chemical forces act locally at very short distances. ***A single molecule can only react with its neighboring molecules and not with the same probability with any molecule in the system as it is assumed in ideal chemical kinetic models***. Thus, the local environment around a given molecule determines its reaction kinetics; and since the local molecular environment is different for each phase present in the system, different reaction rates should be expected for molecules of the same chemical species, which are present in different phases. An overall picture of the kinetics of a heterogeneous system should take into account not only the kinetics for each phase, but also the molecular partitioning among all phases, and the molecular exchange between phases.

1.5 Molecular Behavior at Interfaces

The boundary between two different adjacent phases is known as an *interface*. At the macroscopic scale, this is a clear and straightforward definition. However, at the molecular scale, considering the molecular heterogeneity of matter, the concept of an interface becomes fuzzy. Molecules are in permanent motion, and they may keep transferring from one phase to another (where those phases are defined at particular time and length scales). The exact place and moment when a molecule abandons a phase and becomes part of another is practically impossible to determine, because by doing so, we would be changing the time and length scales of observation, thus disturbing the definition of the original phases. This is somehow similar to Heisenberg's *uncertainty principle*: by trying to measure a property, the original system is disturbed leading to incorrect results. Therefore, the interface between two different phases is an ideal and quite easy concept for geometrically clearly defined contact areas but with no clear representation at the molecular scale. However, the ideal concept of a molecular interface can be used to further understand, in principle, the behavior of adjacent phases.

Let us consider the behavior of a molecule or molecular aggregate at the interface between two different phases. In the bulk of each phase, such molecule or aggregate has a corresponding diffusion coefficient, but at the interface, the behavior may be completely different because the molecules at the interface are subjected to different forces and thus they have different mobility compared to the molecules in the bulk of each phase. For strongly adsorbed molecules (surfactants below its saturation concentration), diffusion at interfaces happens mainly in two dimensions. The cooperative motion of assemblies of such molecules resembles rafting. The situation depicted in Fig. 1.9 is a different one as no third phase molecules (surfactants) are considered to be present. Blue spheres (left) represent molecules belonging to a certain phase, whereas yellow spheres (right) represent molecules of an adjacent phase. Green spheres (center) represent the molecules at the ideal flat interface between those two phases. Black arrows indicate the net force vector acting on each molecule. According to the definition of a phase given in Section 1.4, the net force acting on the molecules of each phase is zero. ***However, the net force acting on the molecules***

at the interface cannot be zero because their surroundings have different properties.

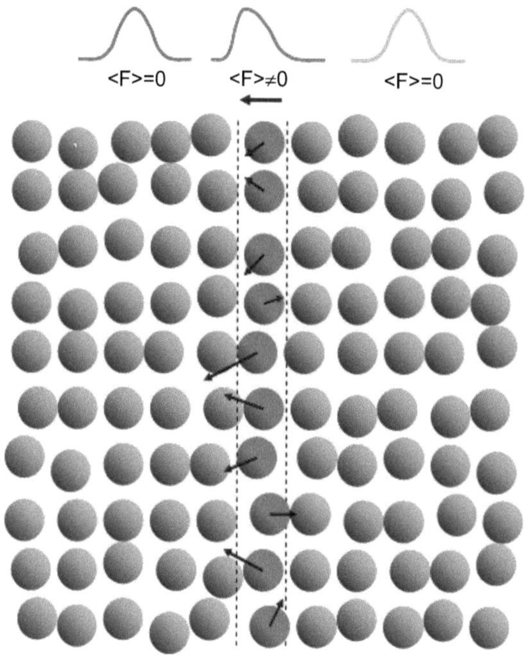

Figure 1.9 Diffusion of molecules at a flat interface between two phases.

According to Langevin's equation (Eq. 1.24), external force fields and interaction potentials are neglected in the description of molecular Brownian motion, although the interaction potentials are implicitly considered in both the drag force and the random force through the viscosity term. Thus, it is possible to explicitly consider additional nonzero net forces in Eq. 1.24 and perform BD simulation assuming a net drift in the system as a result of these forces. Assuming negligible differences in viscosity between both phases, the equation of motion in the direction normal to the interface ($\vec{x}$) can be expressed as:

$$m\frac{d^2\vec{x}}{dt^2} = -6\pi\eta a\frac{d\vec{x}}{dt} + \vec{F} \tag{1.48}$$

where $\vec{F}$ is the force acting on the molecule at the interface, and η is the bulk viscosity of the phase in the direction of the motion of the molecule.

Averaging Eq. 1.48 and using Eq. 1.28:

$$\frac{d\langle\vec{v}\rangle}{dt}+\frac{k_{\mathrm{B}}T}{mD}\langle\vec{v}\rangle=\frac{\langle\vec{F}\rangle}{m} \tag{1.49}$$

where $\vec{v}$ is the molecular velocity, and D is the effective diffusion coefficient of the molecule. Solving Eq. 1.49, the following expression for the average velocity of the molecules at the interface is obtained:

$$\langle\vec{v}\rangle=\frac{D}{kT}\left(\langle\vec{F}\rangle-Cme^{-\frac{k_{\mathrm{B}}T}{mD}t}\right) \tag{1.50}$$

where C is an integration constant. When times longer than the relaxation time are considered, the average drift experienced by the molecules is obtained:

$$\langle\vec{v}\rangle=\frac{d\langle\vec{x}\rangle}{dt}=\frac{D\langle\vec{F}\rangle}{k_{\mathrm{B}}T} \tag{1.51}$$

The molecular velocity can be calculated as the sum of two contributions, the drift at the interface and Brownian motion:

$$\vec{v}=\langle\vec{v}\rangle+\vec{v}_{\mathrm{BM}} \tag{1.52}$$

where $\vec{v}_{\mathrm{BM}}$ is described by a Gaussian distribution function with mean zero and standard deviation $\sqrt{2D/\Delta t}$.

Integrating Eq. 1.52 results in

$$\Delta\vec{x}=\frac{D\langle\vec{F}\rangle}{k_{\mathrm{B}}T}\Delta t+\xi_{\mathrm{G}}\sqrt{2D\Delta t}\,\hat{u}_x \tag{1.53}$$

where ξ_{G} is a random number obtained from a Gaussian distribution of mean zero and standard deviation 1, and $\hat{u}_x$ is the unit vector in the positive direction. If the average net force acting on the molecules at the interface is zero, the behavior predicted by pure Brownian motion is obtained. In this case, the probability for a single molecule to move in the positive ($\hat{u}_x$) or negative ($-\hat{u}_x$) direction is 1/2. If the average net force is not zero, the molecules will have a higher probability of moving in the direction of the net force.

Let us now consider an *average* net force acting on the molecules at the interface in the negative direction: $\vec{F}=-F\hat{u}_x$, being $F>0$. Please notice that it is a time-average force, and not an instantaneous force. Thus, it may eventually act on the positive direction resulting in the molecule also moving in the positive direction. The probability of the molecule moving in the positive direction will be given by

$$P(\Delta x > 0) = 1 - P\left(\xi_G \le \frac{F}{2k_B T}\sqrt{2D\Delta t}\right) = \frac{1 - \text{erf}\left(\frac{F}{2k_B T}\sqrt{2D\Delta t}\right)}{2} \tag{1.54}$$

The probability of a molecule moving in the positive direction subjected to an average net force in the negative direction relative to the probability under pure Brownian motion represents the efficiency factor in one dimension:

$$f = \frac{P(\Delta x > 0)}{1/2} = 1 - \text{erf}\left(\frac{F}{2k_B T}\sqrt{2D\Delta t}\right) \tag{1.55}$$

For relatively low drift values, the term $\sqrt{2D\Delta t}$ can be regarded as the net displacement of the molecule and, therefore,

$$F\sqrt{2D\Delta t} \approx F\Delta x = E_{\text{transfer}} \tag{1.56}$$

where E_{transfer} is the energy of phase transfer of the molecule and corresponds to the work made by the system to displace the molecule a distance Δx by means of the force F. Therefore,

$$f \approx 1 - \text{erf}\left(\frac{E_{\text{transfer}}}{2k_B T}\right) \approx \frac{2}{\sqrt{\pi}}\exp\left(-\frac{E_{\text{transfer}}}{k_B T}\right) \tag{1.57}$$

which is similar to the usual expression used for the description of the efficiency of a process limited by an energy barrier. Equation 1.57, valid for only one dimension, implies that the energy barrier against phase transfer can be supplied by the kinetic energy of the molecules. Since thermal motion takes place in three dimensions, the corrected expression for the ***efficiency factor*** is

$$f_{3D} \approx \frac{2}{\sqrt{\pi}}\exp\left(-\frac{E_{\text{transfer}}}{3k_B T}\right) \tag{1.58}$$

Equations 1.57 and 1.58 also indicate that ***the activation energy for the transfer of a molecule across an interface is proportional to the average net force opposing to the transfer acting at the interface***. At a molecular scale, this force can be obtained from the interaction potentials between the molecules at the interface and their surrounding molecules. It is also important to notice that the mobility of the molecules is reduced at the interface as a result of the energy barrier for phase transfer.

Phase transfer activation energy: Work required for displacing a molecule across the interface between two phases. It is proportional to the average net force opposing transfer.

The dynamic unbalance of forces at the interface between different phases leads to the phenomenon of ***interfacial tension***. These forces at the interface, responsible for the energy barrier to the transfer of molecules from one phase to the other, are also responsible for the different rates of molecular transfer between phases, depending on the nature of the molecule transferred. It also represents a mechanical barrier, the strength of which depends on the particular conditions. The magnitude of the interfacial tension is extremely important because it determines how intensely both phases interact. The higher the interfacial tension (say about 70 mN/m), the interfacial area tends to minimize. On the contrary, at very low interfacial tension (as low as 0.001 mN/m) triggered by high concentration of surfactant and a special combination of surfactant/co-surfactant, a bicontinuous phase is formed possessing a huge interfacial area. In this case, one cannot distinguish between water-in-oil (W/O) or oil-in-water (O/W) emulsion; in fact it is a *microemulsion*.

Interfacial tension: It is the net total energy per unit area of the interface, resulting from the forces that pull molecules at the interface toward the bulk of each of the adjacent phases.

1.6 Polymerization

Polymerization is a process leading to the synthesis of large molecules (*polymers* or *macromolecules*) as a result of the binding of molecular building blocks called ***monomers***. Molecules obtained by binding only a relatively small number of monomer units are usually denoted as ***oligomers***. Particularly for heterophase polymerization, one must consider the fact that the properties of oligomers greatly differ from that of polymers. Important differences exist with respect to their physical properties (aggregation state, mobility, solubility, association, and swelling behavior) and all mechanical properties. In addition, the shorter the chain length, the stronger the influence of the end groups on these properties. The bonds between monomer units can be chemical (e.g., covalent bonds) or physical (e.g., hydrogen bonding, metal coordination). When monomer units are physically bonded, the resulting macromolecules are denoted as *supramolecular polymers* [41, 42].

Monomer: The term "*monomer*" is derived from the Greek words *mono* (one) and *meros* (part).

Oligomer: Short polymers are usually denoted as *oligomers* (some parts), while large macromolecules are designated as *polymers* (many parts). Oligomer comprises between 2 and 10–100 monomer units. Even though there is not a clear limiting size between oligomers and polymers, the key difference is that for an oligomer, its properties significantly change when one monomer unit is added or removed.

Carothers [43] classified polymerization processes, depending on the final products obtained, into two main groups: *addition polymerization* and *condensation polymerization*. The basic difference between these two groups is that the mass of a macromolecule formed by addition polymerization is exactly the sum of the molecular masses of all the monomers used in its synthesis. On the contrary, the molecular mass of a macromolecule formed by condensation is less than the sum of its components because during the incorporation of a monomer into the chain, a small by-product molecule is formed. The common feature is that the monomers must be at least bifunctional in order to carry out a polymerization reaction.

Flory, on the other hand, classified polymerization processes depending on their mechanism of reaction into *step-growth polymerization* and *chain-growth polymerization* [3]. The main difference between these two mechanisms is that in step-growth polymerization, the reacting sites (*functional groups*) are consumed after each polymerization step; whereas for chain-growth polymerization, the reacting sites (*active chemical sites* such as *free radicals* and *ions*) are not necessarily consumed after each chain-growth event. As a result, chain-growth polymerization is faster than step-growth polymerization, leading to larger polymers formed in shorter times, but also leaving unreacted monomer until the end of the polymerization.

Ring-opening polymerization is a particular case of chain-growth polymerization where an active site breaks a bond in a cyclic monomer and incorporates the resulting acyclic structure into the backbone of the polymer. In the case of supramolecular polymerization, a *ring-chain polymerization* mechanism is possible, but the ring opening takes place by physical forces rather than by an active chemical site.

In general, supramolecular polymerization consists on addition polymerization processes taking place via the step-growth polymerization mechanism (e.g., *isodesmic* and ring-chain polymerization). There is another particular type of supramolecular polymerization known as *cooperative polymerization*, which resembles the behavior of chain-growth polymerization (long chains are quickly formed), even though it still operates by step-growth. The reason for this behavior is that monomers and long polymers are thermodynamically more stable than certain oligomers (*nuclei*) [44].

According to the chemical mechanism of monomer incorporation (type of active site involved), chain-growth polymerization processes can be classified into *free-radical polymerization, ionic polymerization*, and *coordination polymerization*. In free-radical polymerization, the growing chain contains at least one unpaired electron (free radical), which reacts readily with a molecule with at least one unsaturated bond, leading to chain growth. In free-radical polymerization, the radicals can be generated in very different ways. The simplest ways are naturally given, including thermal and radiative radical generation. Radiative initiation can be caused by light or background radiation on earth. It is also possible to generate radicals from the decomposition of sensitive molecules called *initiators*. Some initiators decompose with temperature (*thermal initiation*), some others under the effect of light (*photoinitiators*), and others generate radicals after an electron transfer reaction (*redox initiators*). The radicals generated by the decomposition of an initiator are denoted as *primary free radicals*.

Free radicals can react in many different ways in a typical polymerization process (Fig. 1.10). Some of the most relevant reactions involving radicals include:

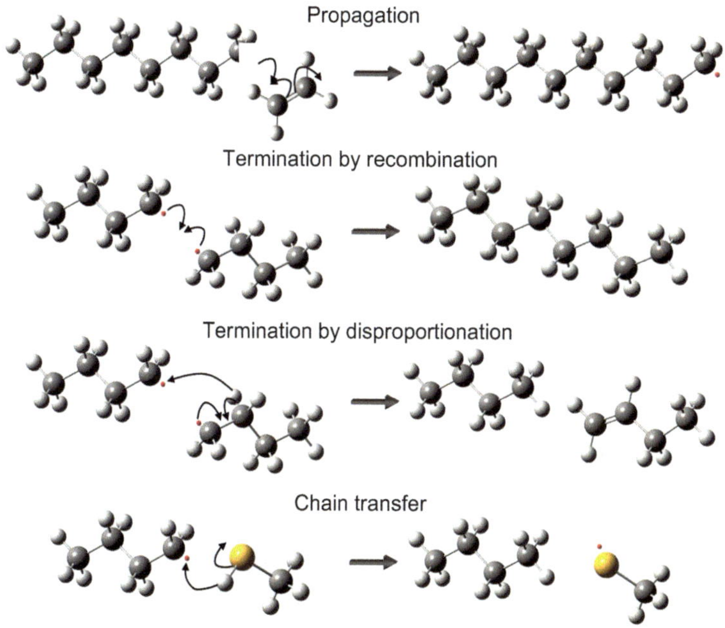

Figure 1.10 Typical reactions of free radicals in a polymerization process. White: hydrogen atoms, Gray: carbon atoms, Yellow: sulfur atoms (a mercaptan as example for chain transfer agent), Red: unpaired electrons.

- **Addition to carbon–carbon double bonds:** Given the high electron density and relative weakness of a carbon–carbon double (or triple) bond in an unsaturated molecule, the unpaired electron of the radical easily breaks one of the bonds and adds covalently to one of the carbon atoms. After this addition, the atom at the opposite side of the double bond ends with an unpaired electron due to the bond breakage. By means of this mechanism, both the free radical and the unsaturated molecule become covalently bonded, and the new molecule is also a free radical, but now the unpaired electron belongs to a different atom. In polymerization, if the original radical is a primary radical, this reaction is known as *initiation*; otherwise, it is known as *propagation*. Radicals are highly reactive species; therefore, they are able to react with almost any molecule in its vicinity. A radical may add to the double bond of a monomer but also may abstract a labile atom from other molecules (e.g., solvents) or even from its own polymer chain (in the case of growing radicals, e.g. so-called back-biting during radical polymerization of butyl acrylate).

- **Termination:** Termination is the reaction between a pair of radicals. As a result of this reaction, both radicals are consumed. There are two different types of termination reactions depending on the products obtained: *recombination* and *disproportionation*. In termination by recombination, a new covalent bond is formed between both unpaired electrons. The final product is, therefore, a single molecule. On the other hand, in termination by disproportionation, one of the unpaired electrons is located preferentially at one of the radical centers, and also resulting in a hydrogen-transfer event. The hydrogen-donor molecule forms a double bond, while the other molecule becomes saturated. Also by means of disproportionation, both radicals disappear, but one of the reacting molecules (the former hydrogen donor) can react again with a free radical via a propagation reaction.
- **Chain transfer:** Chain transfer is basically another hydrogen-transfer reaction. If the free radical is in the vicinity of a molecule with a weakly bonded hydrogen atom, the hydrogen atom is easily abstracted by the radical and forms a bond with the unpaired electron, whereas the broken bond results in an unpaired electron transferring the radical to the second molecule. When the new radical formed is stable (i.e., no or low reactivity), this process is known as *inhibition* or *retardation*. There are different types of chain transfer reaction depending on the nature of the hydrogen donor, which can be: a solvent molecule (e.g., water in emulsion polymerization), a monomer molecule, a polymer molecule (leading to branching or grafting), another part of the chain itself (back-biting), a surfactant molecule, an initiator molecule (*iniferter*), an inhibitor or any molecule deliberately incorporated to promote chain transfer reactions (e.g., mercaptans), allowing a certain control over the molecular mass of the chains. The molecules added for this purpose are denoted as *chain transfer agents*.

In ionic polymerization, the growing chain contains a strong nucleophilic or electrophilic ionic end group, which is also capable of reacting with an unsaturated bond or with a cyclic compound. However, ionic polymerization is very sensitive to the presence of other ions or strongly polar molecules (such as water); therefore, it is not well suited for aqueous polymerization processes. Coordination polymerization is a special type of ionic polymerization characterized by the use of a transition metal compound (*coordination initiator*), which strongly interacts with the double bond of a monomer. This interaction is stereo-selective and is referred to as *coordination*. Due to the high electronic density of the transition metal, the molecular orbitals of the monomer are strongly perturbed, and the double bond can be easily broken. Compared to the other polymerization mechanisms, the coordination polymerization is peculiar with respect to the growth mechanism because the new monomer is inserted between the catalyst (coordination initiator) and the polymer chain. The other polymerization mechanisms are different because the new monomer is added far away from the initiator residue.

Even though there are different chemical reactions specific for each polymerization mechanism, different polymerization reactions and mechanisms may take place simultaneously (competitively or not) in a single polymerization system. And this is particularly true for heterophase polymerization systems where different sets of reactions and mechanisms may take place at each particular phase of the system.

1.7 Heterophase Polymerization

From a physical point of view, polymerization processes can be classified into *homogeneous* and *heterogeneous* depending on the physical state of the reaction mixture. If the monomer molecules as well as the polymer obtained are ***soluble*** in the medium, the process is said to be homogeneous. Typical examples of homogeneous polymerization processes are *bulk* (when the monomer is the medium and the polymer formed is soluble in it) and *solution* polymerization (when a solvent for monomer, polymer, and all other ingredients is used). Otherwise, the process is designated as heterogeneous or *heterophase polymerization* because more than one phase may be present at some moment during the polymerization (i.e., one or more phases dispersed in a continuous phase). In fact, if the interaction between the polymerization system and the atmosphere (e.g., monomer evaporation, oxygen diffusion, etc.) is considered, polymerization systems previously considered homogeneous now become heterogeneous since another phase is involved in the process. Other phenomena, such as wall fouling for example, also involve the occurrence of additional phases in the system.

> ***Solubility***: A chemical compound is considered to be soluble in a solvent when it coexists in a single phase with the solvent. The solubility is quantified as the maximum amount of material that apparently remains soluble in a solvent. These are macroscopic definitions, which depend on the observation and interpretation of a homogeneous phase (see Section 1.4).

When the final polymeric material is distributed in a fluid medium forming *stable* individual particles, it is called a *polymer dispersion*. Please notice that a polymeric dispersion seems homogeneous to the naked eye (see for example Fig. 1.1), but under a strong-enough light microscope, it appears heterogeneous and, provided the particles are large enough and the dispersion is properly diluted, the Brownian movement of the particles can be observed. Although any fluid can be used as the dispersion medium as long as it is not a solvent for the dispersed polymer, water is usually the preferred continuous phase due to safety, economic, and environmental reasons. Aqueous polymer dispersions are also known as *polymer latexes* or *latices*. In recent years, aqueous heterophase polymerization processes have become increasingly important technologically and commercially, not only because of the production of high-performance polymer materials, but also for being environmentally friendlier [45–48].

Another interesting fact about polymer dispersions is that they, based on polymer mass, are manufactured in a range between tons (commodities) and grams (very specific applications), respectively, covering a price range between a few parts of a cent up to hundreds of Euros per gram of polymer.

Size and size distribution are important properties of polymer dispersions, not only for the polymerization kinetics and application fields but also for their stability during polymerization, storage, and application. The stability of the dispersed phase is achieved by using ***amphiphilic molecules***, which are composed of one moiety soluble in the continuous phase and the other soluble in the dispersed phase. These amphiphilic molecules are also called *stabilizers* or *surface active agents* (*surfactants*). The stabilizers can be *ionic* (anionic, cationic, or *zwitterionic*) or *nonionic* (block copolymers, graft copolymers). The mechanism of stabilization using amphiphiles can be *electrostatic, steric*, or *electrosteric*. The final (stable) size of the dispersed phase strongly depends on the amount and nature of the stabilizer used. The higher the stabilizer concentration, the larger the interfacial area that can be stabilized, and thus, finally smaller particles can be obtained (or the maximum of the particles size distribution is shifted toward lower values).

Amphiphilic molecules: Molecules with both lyophilic (affinity for the dispersion medium) and lyophobic (no affinity for the dispersion medium) properties. They are also known as *amphipathic* molecules or simply as *amphiphiles*.

Depending on the size and composition of the different phases formed during the polymerization, heterophase polymerization processes in dispersed emulsion phases can be classified into *precipitation, suspension, micro-suspension, dispersion, emulsion (macroemulsion), inverse emulsion, miniemulsion, inverse miniemulsion, microemulsion*, and *inverse microemulsion,* among others [7, 46, 48, 49]. It is important to notice that these names, although recommended by the IUPAC, are not systematic and can be misleading. Sometimes they designate the initial condition of the system (e.g., emulsion polymerization), whereas in others they indicate the final state of the system (e.g., precipitation polymerization). The most relevant characteristics of the different types of homogeneous and heterogeneous polymerization processes are compared in Table 1.2.

There is another view on the taxonomy of heterophase polymerization based on the ratio of the necessary recipe components. A possibility to distinguish the different heterophase polymerization techniques is the consideration of L_M/L_W and L_S/L_M where L_M, L_W, and L_S are the reactor loads with monomer, water, and stabilizer

Table 1.2 Types of polymerization processes

Type of polymerization	Continuous phase	Initiator	Stabilizer	Monomer solubility	Particle size	Special features
Bulk	Monomer	Lyophilic	None	Soluble	–	
Solution	Any	Lyophilic	None	Soluble	–	
Precipitation	Any	Any	None	Soluble	>1 mm	
Suspension	Any	Lyophobic	Polymeric or colloid	Low	10–500 μm	
Micro-suspension	Any	Lyophobic	Polymeric + surfactant	Low	1–10 μm	
Dispersion	Any	Lyophilic	Polymeric	Low-soluble	1–20 μm	
Emulsion	Water	Any	Any type (low amounts)	Low	5 nm – 10 μm	a, b
Inverse emulsion	Organic	Any	Any type (low amounts)	Low	5 nm – 10 μm	a, b
Miniemulsion	Water	Any	Any (high amounts)	Insoluble	50–500 nm	a, b, c
Inverse miniemulsion	Organic	Any	Any (high amounts)	Insoluble	50–500 nm	a, b, c
Microemulsion	Water	Any	Any (very high amounts)	Low*	30–100 nm	c, d
Inverse microemulsion	Organic	Any	Any (very high amounts)	Low*	30–100 nm	c, d

Notes: * The monomer solubility is originally low. After the addition of the stabilizer, solubilization of the monomer as nanodroplets in the continuous phase is achieved.

a — high polymerization rate, b — high molecular mass, c — high energy input required for emulsification, d — no energy required for emulsification in mass units, respectively. Applying these ratios to typical values employed in the different variants of heterophase polymerizations leads to a kind of phase diagram (Fig. 1.11) showing the operational space with respect to the reactor loads with monomer, water, and stabilizer. Figure 1.11 elucidates that emulsion polymerization is the most versatile and flexible heterophase polymerization technique with respect to stabilizer concentration and solid content. Moreover, it is necessary to emphasize that among all heterophase polymerization techniques, emulsion polymerization is the only one that can be carried out stabilizer free.

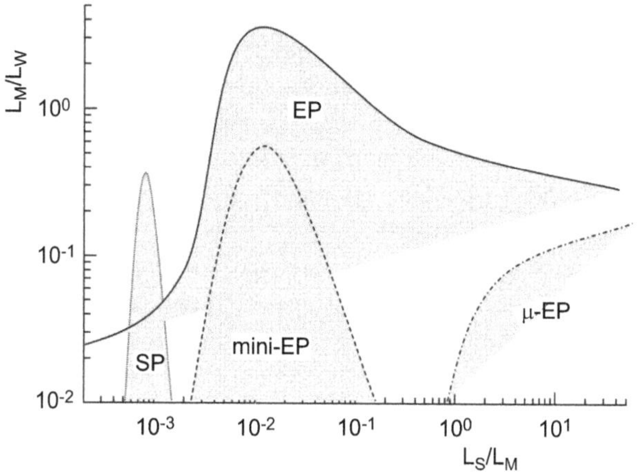

Figure 1.11 Phase diagram of the different heterophase polymerizations showing the operational space with respect to the reactor load with water (L_W), monomer (L_M), and stabilizer (L_S), SP — suspension polymerization, EP — emulsion polymerization, mini-EP — miniemulsion polymerization, μ-EP — microemulsion polymerization; note the logarithmic scale of both axes and the fact that for surfactant-free emulsion polymerization, the ratio L_S/L_M is zero.

Precipitation polymerization is characterized by the large size (>1 mm) of the particles obtained, although smaller particles are possible, particularly when crosslinkers are used. The monomer is completely soluble in the continuous phase, but the polymer chains are insoluble. Particles are initially formed by the collapse of larger oligomers when they reach the critical solubility. Since no stabilizer is used, particles grow mainly by aggregation although some monomer might be absorbed and polymerized inside the particles. As particles grow, the ratio between the surface cohesive forces and the particle mass becomes weaker, and they eventually stop growing when equilibrium between random forces and cohesive forces is achieved. The continuous phase must not necessarily be liquid. It can also be a gas, as in the case of gas-phase polyolefin polymerization.

In *suspension polymerization*, the monomer has a limited solubility in the continuous phase; therefore, continuous stirring is required for promoting mass transfer between the dispersed monomer phase and the continuous phase. Also, a stabilizer is used, initially to hinder coalescence of monomer droplets, and later for stabilizing the polymer particles, leading to smaller particles (10–500 μm) when compared to precipitation polymerization. The polymerization is expected to begin in the monomer phase, by using lyophobic initiators. However, polymerization in the continuous phase also takes place due to thermodynamic reasons: equilibration of the chemical potential of each species in each phase. *Micro-suspension polymerization* is similar to suspension polymerization, but more efficient surfactants are used as stabilizers, leading to smaller particles (1–10 μm).

In *dispersion polymerization*, the monomer can be completely or partially soluble in the continuous phase, and the initiator is lyophilic. The polymer, of course, is insoluble in the continuous phase. This polymerization is similar to precipitation polymerization, but since a polymeric stabilizer is used, smaller particles (1–20 μm) are obtained.

Particularly for *emulsion polymerization*, it is assumed that the continuous phase is water, with the monomer being only partially soluble in the aqueous phase. The term emulsion polymerization is due to the fact that the monomer is initially present in the form of an oil-in-water emulsion (or water-in-oil emulsion with a subsequent phase inversion).

Since monomer emulsions are unstable (at least for technically/commercially relevant conditions), continuous stirring is required for maintaining the system emulsified. One particular feature of emulsion polymerization is the segregation of active species involved in the polymerization reaction. This allows both, increasing the polymerization rate and the molecular mass of the polymer chains. The production of water soluble polymers requires an organic continuous phase offering only limited solubility for the corresponding monomers. Such polymerization is denoted as *inverse emulsion polymerization.*

Miniemulsion polymerization is a special variant of emulsion polymerization, where the initial monomer emulsion appears to be stable (frequently called kinetic stability), i.e., it will remain relatively stable without macroscopic phase separation for longer times (ideally, long enough for carrying out the polymerization). Molecular systems are highly dynamic and continuously changing; thus stability is a macroscopic, not a molecular, concept. Such *kinetic* stability (referred to the slow rate of degradation of the miniemulsion in relation to the rate of polymerization) is achieved by the reduction in the monomer droplets size (with the use of high-power ultrasound or high-pressure homogenizers), and by the use of a hydrophobic additive (so-called hydrophobe) inside the droplets, which provides attractive forces that help prevent monomer diffusion toward the continuous phase. In fact, thermodynamically speaking, a *miniemulsion* is unstable. However, when it is assumed that the hydrophobic co-stabilizer is almost completely insoluble in the continuous phase, the situation is comparable with swollen polymer particles, which means that the individual droplets cannot vanish by leaking out completely due to the Laplace pressure, but only by coalescence/coagulation caused by colloidal instability.

Drops of different sizes have different chemical potentials. Thus, there are two ways to increase the stability of an emulsion. First, in order to reach thermodynamical stability, the drop size distribution of an emulsion must be monodisperse. Second, the amount of *hydrophobe* per drop must be adjusted according to the drop size. However, both ways are practically impossible to realize. Miniemulsion polymerization is also favored when the monomers have a very low solubility in the aqueous phase. This way, polymerization is expected to occur ***mainly*** inside the original monomer droplets, without the formation of new particles in the continuous phase.

Finally, *microemulsion polymerization*, which is another special case of emulsion polymerization, involves polymerizing a thermodynamically stable monomer emulsion (i.e., *microemulsion*). Thermodynamic stability of microemulsions is achieved by using significant proportions of stabilizers (surfactants and co-surfactants) in order to obtain ultralow interfacial tensions.

The consequences of ultralow interfacial tension are significant for the shape and structure of microemulsions, which is fundamentally different compared with all other emulsions discussed above.

Following P. A. Winsor, microemulsion phases can be distinguished as either of type Winsor I (dispersion of oil in water), Winsor II (dispersion of water in oil), and Winsor III (bicontinuous mixture of oil and water) [50]. Typically, Winsor I/II coexist with excess oil and water, respectively, and Winsor III with excess oil up and excess water down. In contrast, Winsor IV type stands for a single phase of water, oil, and surfactant without excess phases. Note that all these microemulsion types exist in the absence of an external emulsification force (absence of mixing). In all cases, the emulsified oil and water are "*mesophases, having dimensions on an intermediate scale, e.g., 50 nm. On the macroscopic level the microemulsion behaves as a stable phase subject to the ordinary laws of thermodynamics.*" [51].

Phase diagrams of microemulsions are very complex and dependent, besides temperature, on the specific properties and concentrations of all components, including electrolytes (or salinity) [52]. Known morphologies, beside oil drops in water or water drops in oil, are sponge-like phases where the surfactants form an extended bilayer separating the oil and water phase, or bicontinuous phases where surfactant monolayers separate water from oil, and which are located in the phase diagrams in proximity to lamellar phase regions.

Considering microemulsions only as ordinary emulsions with just smaller drops is surely not correct. There might be good reasons to consider microemulsions in particular corners of the phase diagram as normal emulsions with much smaller drop sizes, but this is by far not the general case. It might be true for oil-in-water or water-in-oil systems at very low oil and water volume fractions, respectively, so-called micellar solutions. However, even "*In the case of thermodynamically stable micellar solutions the term 'microemulsion' is a misnomer and should not be used in this connection*" [53].

Due to the extremely low interfacial free energy (which is almost in the order of thermal fluctuations), microemulsions are dynamic systems experiencing spontaneous fluctuations leading to structural changes within milliseconds. In a seminal paper, the structure of microemulsions was investigated and explained by Helfrich using a free energy equation where the free energy is strongly associated with curvature deformations in the surfactant (bi)layer [54].

Typically, polymerizations in microemulsions lead also to latexes consisting of spherical particles but, according to the much higher surfactant concentrations and considerably lower monomer volume fractions, of much smaller sizes than normally obtained in ordinary emulsion polymerizations [55]. A comprehensive study of aqueous microemulsion polymerization of butyl acrylate with a sodium dodecylsulfate/AOT sodium salt of sulfosuccinic acid bis(2-ethylhexyl)ester surfactant mixture [recipe: surfactant ratio 3:1 by mass varied from overall 5 g over 7.5 g to 10 g, 2 g of monomer, 18 mg of potassium peroxodisulfate as initiator, 100 g – (total surfactant mass in g) gives the mass of water] by means of reaction calorimetry, dynamic light scattering, and surface tension measurements revealed unexpected results regarding the polymerization kinetics and colloidal properties of the final latexes [56]. With increasing surfactant concentrations, the overall rate of polymerization decreases, the time increases before the maximum in the reaction rate profile is reached, and the average particle size in the final latexes increases as well. An attempt is made to explain the experimental results consistently with a particle nucleation mechanism based on the classical nucleation theory. The surface tension measurements at the liquid–vapor interface (γ_{lv}) of the neat surfactant solution, the initial microemulsion, and the final latex revealed other interesting results. With increasing surfactant concentration, the γ_{lv} values of the surfactant mixture decrease from 31.5 to 29.9 mN/m. Upon adding the monomer, the surface tension decreases to values of some 27 mN/m almost independent of the surfactant concentration. Although this decrease is small, it is significant and beyond the reproducibility of the measurement. This value is close to γ_{lv} for butyl acrylate (which is 26 mN/m [57]) and thus, indicating a monomer layer on top of the microemulsion. The surface tensions of the final latexes are almost exactly the same as those measured for the neat surfactant solutions proving, first, almost complete conversion of the monomer and, second, the complete coverage of the particle–water interface with surfactant and the existence of empty micelles. The coexistence of empty micelles and completely covered particles is typical for final latexes of microemulsion polymerizations.

Due to the high surfactant concentrations at the end of the batch polymerization, a semi-batch polymerization variant with monomer feed seems straightforwardly possible. In this case, particle nucleation occurs during the whole batch and semi-batch polymerization periods also in the absence of micelles. During the post-addition batch polymerization period, coalescence greatly contributes to particle growth. The particular method of semi-batch microemulsion polymerization allows obtaining latexes with particle sizes below 40 nm (dynamic light scattering intensity-weighted diameter) at polymer concentrations near 30% wt. and stabilizer amounts between 5% and 6% relative to the polymer mass [58].

Note also that microemulsions of aqueous monomer solutions (such as acrylamide) in organic continuous phases have been subjected to polymerization (inverse microemulsion polymerization) [59]. However, using surfmers (polymerizable surfactants) to create bicontinuous microemulsions, it is possible to fix these structures via polymerization and thus, to obtain the corresponding nanoporous materials [55].

For spherical microemulsions, the monomer droplets can be so small that the microemulsions might appear translucent. The formation of the microemulsion is spontaneous, and thus it does not require high energy inputs as in the case of miniemulsion polymerization. Since the ultralow interfacial tension facilitates molecular transfer, active species are no longer segregated, and thus polymerization rates and molecular masses are lower.

Although the physical appearance of the reaction mixture and the physical and chemical properties of the final product obtained are different for each type of polymerization process, the physical and chemical mechanisms involved, as well as the stability criteria, in all cases are in principle the same.

Depending on the initial state of the system, heterophase polymerization can be classified into *ab initio* (no polymer particles present at the beginning of the process) and *seeded* polymerization (previously prepared polymer particles are used). Regarding the amount of monomer present in the system, heterophase polymerization processes can be classified into *flooded* (free-monomer phase, monomer drops are present) or *starved* (monomer feed is controlled in a way that the monomer concentration is below its solubility limit to avoid a free-monomer phase and hence monomer drops). Furthermore, heterophase polymerization can be carried out in batch, semi-batch, or continuous operation, as any other chemical reaction. However, not all heterophase polymerization reactions and recipes are suitable for continuous operation because to obtain a steady-state operating point at high conversion, a special rate of polymerization profile in combination with a suitably residence time is needed. This is particularly true for *ab initio* conditions, because of the sensitive nature of particle nucleation, which causes, even under tightly controlled conditions, the occurrence of sustained oscillation of all polymerization and product properties (such as latex surface tension, average particle size, particle size distribution, solids content, and conversion) [60–64]. Although most initial efforts to perform continuous heterophase polymerizations, and particularly continuous emulsion polymerizations, were oriented toward using continuous stirred tank reactors (CSTRs) [65–68], other types of reactors and configurations have been used, including:

- **CSTR trains:** The presence of multiple heterophase polymerization reactors in series can dampen oscillations [69]. Even a two-reactor configuration is possible, where particle nucleation takes place only in the first (small) reactor [70].
- **Tubular reactors:** A direct extension of a series of CSTR reactors. The main disadvantage of tubular reactors for continuous heterophase polymerization is the risk of reactor fouling and plugging. Different configurations have been proposed, including: conventional plug-flow reactors [71], *loop reactors* [72, 73] (which are tubular reactors with partial recycle), and pulsed tubular reactors [74].
- **Pulsed packed columns:** An extension of pulsed tubular reactors, increasing the reactor diameter and including packing for improving mixing [75–77]. This type of reactor can also reduce oscillations in particle size and also minimize fluctuations in polymerization rate.
- **Couette–Taylor vortex flow reactors:** This is a modification of a cylindrical column reactor, where an internal rotating cylinder is placed inside the reactor instead of the packing. The clearance between the rotating cylinder and the reactor wall is designed to achieve a particular flow pattern [78–80]. This approach allows obtaining a narrow particle size distribution and is quite useful for systems sensitive to mechanical shear.

Continuous heterophase polymerization is a key technology for the industrial-scale production of polymer commodities. Additional challenges faced by continuous heterophase polymerization include increasing the concentration of the polymer dispersion (because the final concentration obtained by continuous processes is usually lower than for batch or semi-batch processes) and improving sensors and online measurements for process control. The concentration of polymer dispersions, along with the recovery of surfactants and continuous phase, can be done using ultrafiltration technologies [81–83], for example. Online sensors available for continuous heterophase polymerization include density, surface tension, monomer composition, particle size distribution, molecular mass distribution, and many other [84–87].

The most general type of heterophase polymerization is perhaps the *ab initio* semi-batch polymerization, which will be described in detail. This polymerization is considered the most general case, because it involves the phenomenon of particle formation and particle growth; however, both temporally separated. Moreover, the polymerization conditions permanently change. Also, the *ab initio* semi-batch polymerization starts as batch *ab initio* polymerization, i.e., the monomer feed is started only after particle nucleation.

Generally, any semi-batch polymerization, whether seed latex is applied or not, has to start, formally considered, with the formation of the reaction loci even if it happens timely and spatially separated from the main reactor as it is the case for seeded polymerizations. Producing the seed latex in a separate reactor allows a much better control of particle nucleation. If such seed latex is produced in larger amounts, it can be applied in different seeded semi-batch polymerizations even with different recipes, since its amount in the final latex particles, assuming a particles' growth factor of 5, is only less than 1% by mass.

A semi-batch polymerization may also start as a batch *ab initio* process, where a desired initial amount of monomer, the continuous phase, and optionally some additives (e.g., stabilizer, initiator, etc.) are added to the reactor. If the monomer is not completely soluble in the continuous phase, then a liquid/liquid dispersion will be formed by mixing. If the dispersed phase is in the colloidal range (around 1 nm to 1 μm), such dispersion is called an *emulsion*. Normally, such liquid/liquid dispersion is not stable (neither kinetically nor thermodynamically), and it will lead to phase separation after stopping stirring. However, if a suitable stabilizer is added to the system (or if the monomer has amphiphilic properties), the stability of the dispersion can be improved. Furthermore, by increasing the stabilizer content (by about a factor of 10 and additionally applying a co-surfactant), it is also possible to carry out the batch polymerization under conditions favoring microemulsion formation, which subsequently leads to semi-batch microemulsion polymerization [88] (cf. above).

A stabilized dispersion remains apparently stable (without macroscopic phase separation) during the time scale of the polymerization process even though it might be thermodynamically unstable, particularly considering the common practice of permanently using mechanical stirring during polymerization. Depending on the relative amount of each component and on the relative affinity of the stabilizer to each phase, a dispersion of monomer in the continuous phase (M/C) or a dispersion of the continuous phase in the monomer (C/M) can be obtained. For emulsion polymerization processes, these dispersions will be denoted as O/W and W/O, respectively. For inverse emulsion polymerizations (organic continuous phase and liquid hydrophilic monomer or an aqueous solution thereof), they will correspond to W/O and O/W, respectively.

Hydrophilic–lipophilic balance (HLB): Describes the ratio between the hydrophilic (lipophobic) moiety and the lipophilic (hydrophobic) moiety of an amphiphilic molecule. The HLB value can be obtained either considering the molecular mass of each moiety or by group contribution methods.

The relative affinity of the stabilizer to both phases is usually quantified using the ***HLB (hydrophilic–lipophilic balance)*** parameter. The formation of O/W dispersions is favored by high-HLB (>8) surfactants, which are more easily dissolved in the aqueous phase (for example, the most prominent surfactant sodium dodecyl sulfate has an HLB value of 36 [89]), while the formation of W/O dispersions is preferred by low-HLB (<8) surfactants, which dissolve more easily in the organic phase.

When liquid/liquid dispersions are formed, the size of the discontinuous phase will depend basically on the type and amount of amphiphile used (if any), and on the mechanical energy applied to the system (e.g., stirring rate, ultrasound power, etc.), which has a direct effect on breakage and coalescence phenomena. After stirring for a certain period of time, whose length depends on stabilizer concentration, viscosity of both phases, stirrer type, and stirrer intensity, a dynamic equilibrium between droplet coalescence (leading to drop size increase, farther away from the stirrer) and re-breaking (causing decrease in drop size in the vicinity of the stirrer) is established. Consequently, even for stable systems (without macroscopic phase separation), the size and shape of each droplet of discontinuous phase are actually in permanent change, not only as a result of those breakage and coalescence processes but also as a result of the transfer of molecules between phases by diffusion. Only under certain conditions, it is possible to obtain a relatively steady system, as it is the case for *miniemulsions* [90]. However, from an industrial point of view, meeting and keeping such conditions during polymerization are pretty challenging, if achievable at all [91].

Once the reactants are added to the system, it is only a matter of time before reactions take place as a result of molecular collisions. Those reactions can take place in any phase of the reacting system, as long as the reactants are present with enough kinetic energy to overcome reaction energy barriers. Particularly, polymerization reactions lead to the formation of macromolecular chains in the system with different sizes, giving rise to a molecular mass distribution of the macromolecules. If the attractive forces between a macromolecule and the molecules in the continuous phase are stronger than the attractive forces of the macromolecule with molecules at a discontinuous phase, with other macromolecules or with other parts of the same macromolecule, it will likely remain solubilized in the continuous phase. On the other hand, the fate of the macromolecule will be determined by the stronger attractive forces found in its particular path. Thus, macromolecules may form aggregates with other chains or with other species in the system forming a separate new phase, or they can transfer into a discontinuous phase, or they can even collapse forming a single chain or a *pauci chain* separate phase. In any case, whenever a discontinuous phase acquires its first macromolecular chain, it starts to be considered a polymer particle. This phenomenon corresponds to a *nucleation*-type first-order transition known as *particle formation* [92]. Polymer particles can be formed in the absence as well as in the presence of stabilizer.

Stabilizer molecules have different effects on heterophase polymerization [93]. When stabilizers are present in the system, they promote the presence of a larger number of smaller entities of the discontinuous phases, increasing the probability of particle formation. For chain-growth polymerization reactions, the presence of smaller entities of the discontinuous phases has the additional effect of causing a segregation effect of the active sites (e.g., isolated radicals or ions in individual particles), which increases the rate of polymerization and the molecular mass of the chains.

Stabilizers are also important for controlling the number of particles produced and their final particle size as a result of the balance between aggregation and stabilization of polymer particles. Stabilizers should be carefully chosen according to their effects on the product characteristics and on the final application properties. In most practical cases, the stabilizer remains in the final dispersions and thus it influences application properties. However, for certain dispersions made by continuous emulsion polymerization, which require quite a high surfactant concentration (up to 5 wt.% based on monomer or even higher) during the polymerization process, caused by the peculiarities of the continuous process, strategies have been developed to reduce the surfactant amount before the application, whereby the removed surfactant solution can be re-used in another polymerization. Some of these properties may include the wettability of the dispersion and adhesion of the polymer to certain surfaces for coating and adhesive applications, the rheological behavior of the dispersion, or the compatibility with other materials (fillers, pigments, etc.) [94]. Most commercial recipes contain mixtures of stabilizers and additional ingredients depending on the specific requirements of the final application or the specific conditions needed during polymerization [46].

The adsorption energy of a particular surfactant determines its tendency to form micelles and to adsorb at interfaces as well as the strength of micelles. The adsorption energy of a surfactant is critical for its behavior during emulsion polymerization. The stronger the adsorption, the lower the critical micelle concentration (CMC) and the larger the particle number for a given recipe (cf. Refs. [95, 96] and references therein). Note that the adsorption strength of a surfactant at the polymer particle–water interface depends on the compatibility with the polymer (particularly on the polarity of the polymer). For an alkyl chain surfactant, the following relations exist. The higher the polarity of the interface, the larger the area covered by a single surfactant molecule meaning that the lower the number of stabilizing groups per unit interfacial area, the higher the polarity of the interface. In the adsorption layer, the alkyl chain of an adsorbed surfactant molecule interacts with the polymer of the particle and with the alkyl chains of the neighboring surfactant molecules. Clearly, this is a competitive situation and controlled by the magnitude of both interaction energies. Provided the interaction between the surfactant chains is energetically favored compared to the interaction with the polymer, the consequences for the emulsion polymerization can be disastrous. If this happens, the CMC should be an extraordinary point in a plot of the particle number in the final latex versus the applied surfactant concentration. Indeed, such behavior was experimentally observed (results were presented at an international meeting of the International Polymer Colloids Group in Irsee, Germany, 2002, by K. Tauer) during the emulsion polymerization of styrene with the sodium salt of perfluorooctanoic acid (PFOA) as stabilizer and PEGA200, a water-soluble poly(ethylene glycol)–azo initiator with a molecular weight of the poly(ethylene glycol) chain of 200 g/mol, as uncharged initiator (cf. Fig. 1.12).

Figure 1.12 shows the development of the final particle number with increasing surfactant concentration in comparison with sodium dodecylsulfate (SDS), the "standard surfactant" for studies of emulsion polymerization. The log N–log C_S plot shows for both surfactants a linear increase until the surfactant concentration is close to the CMC. For SDS above the CMC, the particle number continuously increases, although with decreasing slope. On the contrary, for PFOA, beginning at the CMC, the final particle number decreases sharply, and the latex coagulates. The fluorosurfactant likes itself much more than the polystyrene (fluorophobic effect), and hence the micelle formation is energetically favored. Consequently, the PFOA molecules desorb; they do not stabilize the particles any longer, and the colloidal system collapses. For SDS, the decline in the increase in the particle number indicates also slightly decreasing stabilizing ability of the surfactant due to the increasing interaction of the alkyl chains. But obviously, the mutual interaction of the alkyl chains is, in relation to the interaction with polystyrene, not strong enough so that complete desorption as for PFOA does not take place.

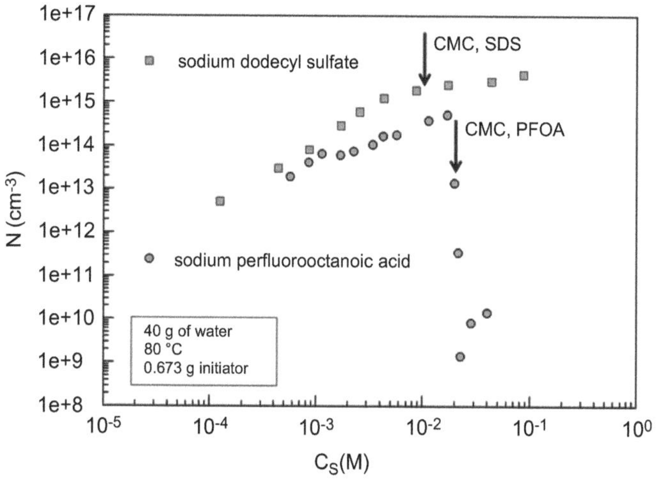

Figure 1.12 Final particle number (N) versus surfactant concentration (C_S) for the emulsion polymerization of styrene with sodium salt of perfluorooctanoic acid (PFOA) and sodium dodecylsulfate as stabilizer and PEGA200, a water-soluble poly(ethylene glycol)–azo initiator with a molecular weight of the poly(ethylene glycol) chain of 200 g/mol. The arrows mark the position of the CMC for the surfactants.

A general picture of chain-growth heterophase polymerization is presented in Fig. 1.13. Generally, in step-growth heterophase polymerization, initiator or active site molecules must not necessarily be present. The chain-growth heterophase polymerization scheme includes the discontinuous phase (e.g., monomer) droplets with varying sizes, molecularly dissolved monomer (may represent different monomer species), initiator molecules, living polymers, dead polymers, monomer–polymer aggregates (surfactant-free polymer particles), stabilizer molecules (free or adsorbed), stabilizer aggregates (micelles), monomer–stabilizer aggregates (monomer droplets), monomer–polymer–surfactant aggregates (stabilized polymer particles), and general polymer particles (polymer chains present in a discontinuous phase), all immersed in a continuous phase of solvent molecules. As it can be seen, any combination of molecular aggregates is possible. Of course, its probability of occurrence will depend on kinetic and thermodynamic factors.

The size of the entities involved in a typical heterophase polymerization process ranges from molecular species of less than 1 nm in size, to colloidal entities with a size ranging from a few nanometers up to hundreds of micrometers, to almost millimeter-sized discontinuous phase droplets. For a certain amount of a given component, the number concentration of the corresponding entities is strongly dependent on its size distribution. The size and concentration of the different entities involved in heterophase polymerization system can be determined using a combination of different experimental techniques, including electron microscopy, light scattering, light microscopy, ultracentrifugation, spectroscopy, chromatography, and many others.

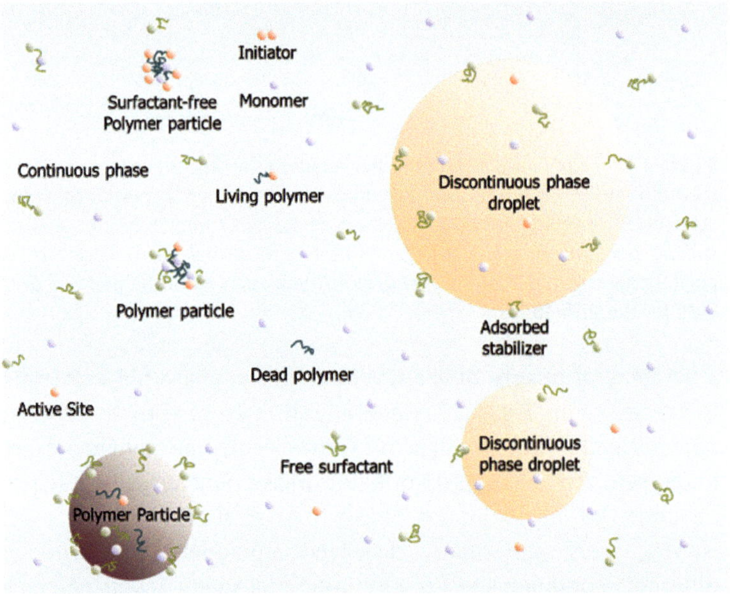

Figure 1.13 Illustrative representation of chain-growth heterophase polymerization (sizes of molecules and aggregates, as well as proportions of free and adsorbed surfactant molecules are not to exact scale).

The reactions taking place in all phases of the system will continue until their rate of reaction is significantly reduced. This may occur when some reactants are eventually depleted (e.g., monomers, active sites, functional groups, etc.), their kinetic energy is not enough to react anymore (e.g., cooling the system), or there are mass transfer limitations to the reaction (e.g., gelation/solidification of polymer particles). If the polymer particles obtained at the end of the polymerization are stabilized in the continuous phase, the final product is known as polymer dispersion. If the particles are not stable, they will more or less loosely aggregate and eventually precipitate from the continuous phase, but they can be re-dispersed by stirring if needed, provided the glass transition temperature of the polymers is high enough to prevent fusion of the polymer chunks.

The initiator used in an industrial emulsion polymerization is typically thermally initiating and water soluble, despite the fact that also oil-soluble initiators can be applied (cf. Section IV) and photoinitiation as well [97, 98]. The various initiation systems applicable in emulsion polymerization might support the idea that the role of the initiator is mainly limited to starting the chain growth. However, due to the complexity of the reactions involved and interaction between colloid chemistry and polymerization kinetics, the role of the initiator goes much more beyond. In an experimental study of emulsion polymerization, it was proven that all components of the initiator–emulsifier system (i.e., anionic, cationic, and neutral) and their particular combination influence the whole process and all latex and polymer properties [99]. Exemplarily, for the combination AOT (sodium salt of sulfosuccinic acid bis(2-ethylhexyl)ester) as emulsifier and peroxodisulfate as initiator anion, even the kind of counter ion (ammonium or potassium) has a profound influence on the particle size of the latex particles.

As a final word, before discussing in detail the fundamental mechanisms of heterophase polymerization in the next chapter, it is important mentioning some factors that have motivated the industrial development of polymer dispersions by heterogeneous polymerization [1, 2, 47, 100–102]:

- The possibility of simultaneously obtaining polymers of very high molecular mass at high polymerization rates, which are required for certain high-performance applications (segregation effect).
- The technology of feed procedures that allows carrying out, on the one hand, polymerization at the maximum rate and, on the other hand, producing particles with the desired morphology and composition on nanometer-size scales.

- The easy processability of the high molecular mass material, due to the low viscosity of the dispersion.
- The increased safety and productivity of the reaction, as the continuous phase works also as a sink for the energy liberated during the polymerization reaction, allowing a better temperature control and a reduced risk of thermal runaway at high polymerization rates.
- The wide range of products with different physical, chemical, and performance properties that can be obtained by heterophase polymerization.

References

1. Tauer, K. (2002). Heterophase polymerization. In: *Encyclopedia of Polymer Science and Technology*, Vol. 6 (John Wiley & Sons), pp. 410–527.
2. Antonietti, M. and Tauer, K. (2003). 90 years of polymer latexes and heterophase polymerization: More vital than ever, *Macromolecular Chemistry and Physics*, **204**, pp. 207–219.
3. Flory, P. J. (1953). *Principles of Polymer Chemistry* (Cornell University Press, USA).
4. Hohenstein, W. P. and Mark, H. (1946). Polymerization of olefins and diolefins in suspension and emulsion. Part II, *Journal of Polymer Science*, **1**(6), pp. 549–580.
5. Frenkel, S. Y. (1963). On the theory of heterophase polymerization—I. Molecular weight distribution in the suspension polymerization of a water-soluble monomer, *Polymer Science USSR*, **4**(1), pp. 136–146.
6. Mandal, B. M. (2013). *Fundamentals of Polymerization* (World Scientific Publishing Co., Singapore). p. 407.
7. Slomkowski, S., Alemán, J. V., Gilbert, R. G., Hess, M., Horie, K., Jones, R. G., and Stepto, R. F. (2011). Terminology of polymers and polymerization processes in dispersed systems (IUPAC Recommendations 2011), *Pure and Applied Chemistry*, **83**(12), pp. 2229–2259.
8. Hernandez, H. (2008). *Multiscale Simulation of Heterophase Polymerization: Application to the Synthesis of Multicomponent Colloidal Polymer Particles*. Doctoral Dissertation (MPI of Colloids and Interfaces/Universität Potsdam, Germany).
9. Israelachvili, J. N. (2011). *Intermolecular and Surface Forces*, 3rd Ed. (Academic press, USA).

10. Butt, H. J. and Kappl, M. (2010). *Surface and Interfacial Forces* (Weinheim: Wiley-VCH, Germany).
11. Kaplan, I. G. (2006). *Intermolecular Interactions: Physical Picture, Computational Methods and Model Potentials* (John Wiley & Sons, USA).
12. Prausnitz, J. M., Lichtenthaler, R. N., and de Azevedo, E. G. (1998). *Molecular Thermodynamics of Fluid-Phase Equilibria*, 3rd Ed. (Prentice Hall PTR, USA).
13. Fick, A. (1855). Ueber diffusion, *Annalen der Physik*, **170**(1), pp. 59–86.
14. Fourier, J. (1822). *Théorie Analytique de la Chaleur.* (Éditions Jacques Gabay, France).
15. Hernandez, H. (2019). Calculation of molecular fluxes and equivalent pressure in ideal gases, *ForsChem Research Reports*, **4**, 2019–03. doi: 10.13140/RG.2.2.35898.44483.
16. Einstein, A. (1906). Zur Theorie der Brownschen Bewegung, *Annalen der Physik*, **19**, pp. 371–381.
17. von Smoluchowski, M. (1906). Zur kinetischen Theorie der Brownschen Molekularbewegung und der Suspensionen, *Annalen der Physik*, **21**, pp. 756–780.
18. Hernandez, H. (2019). On the relationship between molecular and macroscopic diffusion in ideal gases, *ForsChem Research Reports*, **4**, 2019–07. doi: 10.13140/RG.2.2.30954.98242.
19. Lemons, D. S. and Gythiel, A. (1997). Paul Langevin's 1908 paper "On the theory of Brownian motion"["Sur la théorie du mouvement brownien," CR Acad. Sci.(Paris) 146, 530–533 (1908)], *American Journal of Physics*, **65**(11), pp. 1079–1081.
20. Bluett, V. M. and Green, N. J. B. (2006). On the competition between scavenging and recombination in solutions of macromolecules, *The Journal of Physical Chemistry. A*, **110**(18), pp. 6112–6121.
21. Bolton, C. E., Green, N. J. B., Harris, R., and Pimblott, S. M. (1998). Competition between geminate recombination and reaction with a macromolecule, *The Journal of Physical Chemistry A*, **102**(4), pp. 730–739.
22. Hernandez, H. F. and Tauer, K. (2007). Brownian dynamics simulation of the capture of primary radicals in dispersions of colloidal polymer particles, *Industrial & Engineering Chemistry Research*, **46**, pp. 4480–4485.
23. Adamczyk, Z., Jaszczółt, K., Michna, A., Siwek, B., Szyk-Warszyńska, L., and Zembala, M. (2005). Irreversible adsorption of particles on heterogeneous surfaces, *Advances in Colloid and Interface Science*, **118**(1–3), pp. 25–42.

24. Karlsson, O. J., Stubbs, J. M., Karlsson, L. E., and Sundberg, D. C. (2001). Estimating diffusion coefficients for small molecules in polymers and polymer solutions, *Polymer*, **42**, pp. 4915–4923.
25. Vrentas, J. S. and Duda, J. L. (1976). Diffusion of small molecules in amorphous polymers, *Macromolecules*, **9**(5), pp. 785–790.
26. Fujita, H. (1961). Diffusion in polymer-diluent systems, *Advances in Polymer Science*, **3**, pp. 1–47.
27. Yasuda, H., Lamaze, C. E., and Peterlin, A. (1971). Diffusive and hydraulic permeabilities of water in water-swollen polymer membranes, *Journal of Polymer Science Part A-2*, **9**, pp. 1117–1131.
28. Chang, K. H. S., Litt, M. H., and Nomura, M. (1981). The reinvestigation of vinyl acetate emulsion polymerization (I): The rate of polymerization. In: El-Aasser, M. S. and Vanderhoff, J. W. (Eds.) *Emulsion Polymerization of Vinyl Acetate*. (Applied Science Publishers, USA).
29. Hiss, T. G. and Cussler, E. L. (1973). Diffusion in high viscosity liquids, *AIChE Journal*, **19**(4), pp. 698–703.
30. Griffiths, M. C., Strauch, J., Monteiro, M. J., and Gilbert, R. G. (1998). Measurement of diffusion coefficients of oligomeric penetrants in rubbery polymer matrixes, *Macromolecules*, **31**, pp. 7835–7844.
31. Frenkel, D. and Smit, B. (2002). *Understanding Molecular Simulation* (Academic Press, USA).
32. Hofmann, D., Fritz, L., Ulbrich, J., and Paul, D. (2000). Molecular simulation of small molecule diffusion and solution in dense amorphous polysiloxanes and polyimides, *Computational and Theoretical Polymer Science*, **10**, pp. 419–436.
33. Harmandaris, V. A., Adhikari, N. P., van der Vegt, N. F., Kremer, K., Mann, B. A., Voelkel, R., Weiss, H., and Liew, C. (2007). Ethylbenzene diffusion in polystyrene: United atom atomistic/coarse grained simulations and experiments, *Macromolecules*, **40**(19), pp. 7026–7035.
34. Masaro, L. and Zhu, X. X. (1999). Physical models of diffusion for polymer solutions, gels and solids, *Progress in Polymer Science*, **24**, pp. 731–775.
35. Berendsen, H. J. C. (2007). *Simulating the Physical World* (Cambridge University Press, UK).
36. Green, M. S. (1952). Markoff random processes and the statistical mechanics of time-dependent phenomena, *The Journal of Chemical Physics*, **20**(8), pp. 1281–1295.
37. Green, M. S. (1954). Markoff random processes and the statistical mechanics of time-dependent phenomena. II. Irreversible processes in fluids, *The Journal of Chemical Physics*, **22**(3), pp. 398–413.

38. Kubo, R. (1957). Statistical-mechanical theory of irreversible processes. I. General theory and simple applications to magnetic and conduction problems, *Journal of the Physical Society of Japan*, **12**(6), pp. 570–586.

39. Jensen, F. (2007). *Introduction to Computational Chemistry*, 2nd Ed. (John Wiley & Sons, Ltd., USA).

40. Svedung, H., Nordholm, S., and Nyeland, C. (2001). Simple analysis of intermolecular potentials: The mBq pair potential for collisions and energy transfer, *Physical Chemistry Chemical Physics*, **3**(12), pp. 2209–2215.

41. Yang, L., Tan, X., Wang, Z., and Zhang, X. (2015). Supramolecular polymers: Historical development, preparation, characterization, and functions, *Chemical Reviews*, **115**(15), pp. 7196–7239.

42. Hartlieb, M., Mansfield, E. D., and Perrier, S. (2020). A guide to supramolecular polymerizations, *Polymer Chemistry*, **11**(6), pp. 1083–1110.

43. Carothers, W. H. (1929). Studies on polymerization and ring formation. I. An introduction to the general theory of condensation polymers, *Journal of the American Chemical Society*, **51**(8), pp. 2548–2559.

44. De Greef, T. F., Smulders, M. M., Wolffs, M., Schenning, A. P., Sijbesma, R. P., and Meijer, E. W. (2009). Supramolecular polymerization, *Chemical Reviews*, **109**(11), pp. 5687–5754.

45. Asua, J. M. (1997). *Polymeric Dispersions: Principles and Applications* (Kluwer Academic Publishers, The Netherlands).

46. Urban, D. and Takamura, K. (2002). *Polymer Dispersions and Their Industrial Applications* (Wiley-VCH, Germany).

47. Blackley, D. C. (1997). *Polymer Latices*, Vol. 1 (Chapman & Hall, UK).

48. Tauer, K. (2004). *Latex Particles*. In: Caruso, F. (Ed.) *Colloids and Colloid Assemblies* (Wiley-VCH, Germany), pp. 1–51.

49. Hunkeler, D., Candau, F., Pichot, C., Hemielec, A. E., Xie, T. Y., Barton, J., Vaskova, V., Guillot, J., Dimonie, M. V., and Reichert, K. H. (1994). Heterophase polymerizations: A physical and kinetic comparison and categorization, *Advances in Polymer Science*, **112**, pp. 115–133.

50. Winsor, P. A. (1948). Hydrotropy, solubilisation and related emulsification processes, *Transactions of the Faraday Society*, **44**, pp. 376–398.

51. Reiss, H., Ellerby, H. M., and Manzanares, J. A. (1993). Configurational entropy of microemulsions: The fundamental length scale, *The Journal of Chemical Physics*, **99**(12), pp. 9930–9937.

52. Strey, R. (1994). Microemulsion microstructure and interfacial curvature, *Colloid and Polymer Science*, **272**(8), pp. 1005–1019.

53. Ekwall, P., Mandell, L., and Fontell, K. (1970). Some observations on binary and ternary aerosol OT systems, *Journal of Colloid and Interface Science*, **33**(2), pp. 215–235.

54. Helfrich, W. (1973). Elastic properties of lipid bilayers: Theory and possible experiments, *Z Naturforsch C*, **28**, pp. 693–703.

55. Yong, C. P. and Gan, L. M. (2005). Microemulsion polymerizations and reactions, *Advances in Polymer Science*, **175**, pp. 257–298.

56. Tauer, K., Ramírez, A. G., and López, R. G. (2003). Effect of the surfactant concentration on the kinetics of oil in water microemulsion polymerization: A case study with butyl acrylate, *Comptes Rendus Chimie*, **6**(11–12), pp. 1245–1266.

57. Daubert, T. E., Danner, R. P., Sibul, H. M., and Stebbins, C. C. (1998). *Physical and Thermodynamic Properties of Pure Chemicals: Data Compilation* (Taylor & Francis, USA).

58. Ramírez, A. G., López, R. G., and Tauer, K. (2004). Studies on semibatch microemulsion polymerization of butyl acrylate: Influence of the potassium peroxodisulfate concentration, *Macromolecules*, **37**(8), pp. 2738–2747.

59. Leong, Y. S. and Candau, F. (1982). Inverse microemulsion polymerization, *The Journal of Physical Chemistry*, **86**(13), pp. 2269–2271.

60. Jacobi, B. (1952). Zur kolloidchemie der emulsionspolymerisation, *Angewandte Chemie*, **64**(19–20), pp. 539–543.

61. Brooks, B. W. (1973). Particle nucleation rates in continuous emulsion polymerisation reactors, *British Polymer Journal*, **5**(3), pp. 199–211.

62. Senrui, S., Kodama, A., and Takehisa, M. (1974). Radiation-induced emulsion polymerization of ethylene. VI. Continuous flow system, *Journal of Polymer Science: Polymer Chemistry Edition*, **12**(10), pp. 2403–2417.

63. Kiparissides, C., MacGregor, J. F., and Hamielec, A. E. (1979). Continuous emulsion polymerization. Modeling oscillations in vinyl acetate polymerization, *Journal of Applied Polymer Science*, **23**(2), pp. 401–418.

64. Tauer, K. and Müller, I. (1993). Modeling sustained oscillations in continuous emulsion polymerization of vinyl chloride, *Acta Polymerica*, **44**(6), pp. 285–293.

65. Owen, J. J., Steele, C. T., Parker, P. T., and Carrier, E. W. (1947). Continuous preparation of butadiene-styrene copolymer, *Industrial & Engineering Chemistry*, **39**(1), pp. 110–113.

66. Poehlein, G. W. and Dougherty, D. J. (1977). Continuous emulsion polymerization, *Rubber Chemistry and Technology*, **50**(3), pp. 601–638.

67. Brooks, B. W., Kropholler, H. W., and Purt, S. N. (1978). Emulsion polymerization of styrene in a continuous stirred reactor, *Polymer*, **19**(2), pp. 193–196.

68. Rawlings, J. B. and Ray, W. H. (1988). The modeling of batch and continuous emulsion polymerization reactors. Part II: Comparison with experimental data from continuous stirred tank reactors, *Polymer Engineering & Science*, **28**(5), pp. 257–274.

69. Gerrens, H. and Kuchner, K. (1970). Continuous emulsion polymerisation of styrene and methyl acrylate, *British Polymer Journal*, **2**(1), pp. 18–24.

70. Penlidis, A., MacGregor, J. F., and Hamielec, A. E. (1989). Continuous emulsion polymerization: Design and control of CSTR trains. *Chemical Engineering Science*, **44**(2), pp. 273–281.

71. Ueda, T., Omi, S., and Kubota, H. (1971). Experimental study of continuous emulsion polymerization of styrene, *Journal of Chemical Engineering of Japan*, **4**(1), pp. 50–54.

72. Rollin, A. L., Patterson, I., Bataille, P., and Huneault, R. (1977). The effect of flow regime on the continuous emulsion polymerization of styrene in a tubular reactor, *The Canadian Journal of Chemical Engineering*, **55**(5), pp. 565–571.

73. Araújo, P. H., Abad, C., de la Cal, J. C., Pinto, J. C., and Asua, J. M. (1999). Emulsion polymerization in a loop reactor: Effect of the operation conditions, *Polymer Reaction Engineering*, **7**(3), pp. 303–326.

74. Paquet Jr, D. A. and Ray, W. H. (1994). Tubular reactors for emulsion polymerization: I. Experimental investigation, *AIChE Journal*, **40**(1), pp. 73–87.

75. Hoedemakers, G. F. M. and Thoenes, D. (1990). Continuous emulsion polymerisation in a pulsed packed column. In: Lemstra, P. J. and Kleintjens, L. A. (Eds.) *Integration of Fundamental Polymer Science and Technology—4* (Springer, The Netherlands), pp. 182–193.

76. Meuldijk, J., van Strien, C. J. G., van Doormalen, F. A. H. C., and Thoenes, D. (1992). A novel reactor for continuous emulsion polymerization, *Chemical Engineering Science*, **47**(9–11), pp. 2603–2608.

77. Mayer, M. J. J., Meuldijk, J., and Thoenes, D. (1996). Application of the plug flow with axial dispersion model for continuous emulsion polymerization in a pulsed packed column, *Chemical Engineering Science*, **51**(13), pp. 3441–3448.

78. Imamura, T., Saito, K., Ishikura, S., and Nomura, M. (1993). A new approach to continuous emulsion polymerization, *Polymer International*, **30**(2), pp. 203–206.

79. Kataoka, K., Ohmura, N., Kouzu, M., Simamura, Y., and Okubo, M. (1995). Emulsion polymerization of styrene in a continuous Taylor vortex flow reactor, *Chemical Engineering Science*, **50**(9), pp. 1409–1416.

80. Wei, X., Takahashi, H., Sato, S., and Nomura, M. (2001). Continuous emulsion polymerization of styrene in a single Couette–Taylor vortex flow reactor, *Journal of Applied Polymer Science*, **80**(11), pp. 1931–1942.

81. DelPico, J. (1979). Ultrafiltration process for the concentration of polymeric latices. U.S. Patent No. 4,160,726.

82. Huddleston Jr, G. R. and Turner, J. W. (1982). Ultrafiltration of vinyl resin latices and reuse of permeate in emulsion polymerization. U.S. Patent No. 4,340,702.

83. Shen, J. J. and Mir, L. (1982). Reducing energy requirement in latex concentration by ultrafiltration, *Industrial & Engineering Chemistry Product Research and Development*, **21**(1), pp. 63–68.

84. Kiparissides, C., MacGregor, J. F., Singh, S., and Hamielec, A. E. (1980). Continuous emulsion polymerization of vinyl acetate. Part III: Detection of reactor performance by turbidity-spectra and liquid exclusion chromatography, *The Canadian Journal of Chemical Engineering*, **58**(1), pp. 65–71.

85. Schork, F. J. and Ray, W. H. (1983). On-line measurement of surface tension and density with applications to emulsion polymerization, *Journal of Applied Polymer Science*, **28**(1), pp. 407–430.

86. Chien, D. C. H. and Penlidis, A. (1990). On-line sensors for polymerization reactors, *Journal of Macromolecular Science. Part C: Polymer Reviews*, **30**(1), pp. 1–42.

87. Richards, J. R. and Congalidis, J. P. (2006). Measurement and control of polymerization reactors, *Computers & Chemical Engineering*, **30**(10–12), pp. 1447–1463.

88. Puig, J. E. and Rabelero, M. (2016). Semicontinuous microemulsion polymerization, *Current Opinion in Colloid & Interface Science*, **25**, pp. 83–88.

89. Kruglyakov, P. M. (2000). *Hydrophile–Lipophile Balance of Surfactants and Solid Particles: Physicochemical Aspects and Applications* (Elsevier, Amsterdam), p. 178.

90. Landfester, K., Bechthold, N., Tiarks, F., and Antonietti, M. (1999). Miniemulsion polymerization with cationic and nonionic surfactants: A very efficient use of surfactants for heterophase polymerization, *Macromolecules*, **32**(8), pp. 2679–2683.

91. Asua, J. M. (2014). Challenges for industrialization of miniemulsion polymerization, *Progress in Polymer Science*, **39**(10), pp. 1797–1826.

92. Tauer, K. and Kühn, I. (1995). Modeling particle formation in emulsion polymerization: An approach by means of the classical nucleation theory, *Macromolecules*, **28**, pp. 2236–2239.

93. Puig, J. E., Mendizábal, E., López-Serrano, F., and López, R. G. (2012). Surfactant assisted polymerization methods. In: Somasundaran, P. (Ed.) *Encyclopedia of Surface and Colloids Science* (Taylor & Francis, USA).

94. Tauer, K. (2002). Surface chemistry in the polymerization of emulsion. In: Holmberg, K., Shah, D. O., and Schwuger, M. J. (Eds.) *Handbook of Applied Surface and Colloid Chemistry*, Vol. 1 (John Wiley & Sons, USA), pp. 175–200.

95. Tauer, K. (2001). Emulsion polymerization. In: Texter, J. (Ed.) *Reactions and Synthesis in Surfactant Systems* (Marcel-Dekker, USA), pp. 429–453.

96. Tauer, K. (2003). The role of emulsifiers in the kinetics and mechanisms of emulsion polymerization. In: Karsa, D. R. (Ed.) *Surfactants in Polymers, Coatings, Inks & Adhesives* (Blackwell Publishing, UK), pp. 32–70.

97. Krüger, K., Tauer, K., Yagci, Y., and Moszner, N. (2011). Photoinitiated bulk and emulsion polymerization of styrene–evidence for photo-controlled radical polymerization, *Macromolecules*, **44**(24), pp. 9539–9549.

98. Laurino, P., Hernandez, H. F., Bräuer, J., Krüger, K., Grützmacher, H., Tauer, K., and Seeberger, P. H. (2012). Snowballing radical generation leads to ultrahigh molecular weight polymers, *Macromolecular Rapid Communications*, **33**(20), pp. 1770–1774.

99. Tauer, K. and Müller, H. (2003). On the role of initiator in emulsion polymerization, *Colloid and Polymer Science*, **281**(1), pp. 52–65.

100. Daniel, J.-C. (2003). A long history with many challenges to meet in the future: Free-radical emulsion polymerization and aqueous polymer

dispersions. In: Elaissari, A. (Ed.) *Colloidal Polymers* (Marcel Dekker, USA), pp. 1–23.

101. Pich, A. and Richtering, W. (2010). Microgels by precipitation polymerization: Synthesis, characterization, and functionalization, *Advances in Polymer Science*, **234**, pp. 1–37.

102. Richez, A. P., Yow, H. N., Biggs, S., and Cayre, O. J. (2013). Dispersion polymerization in non-polar solvent: Evolution toward emerging applications, *Progress in Polymer Science*, **38**(6), pp. 897–931.

Chapter 2

Mechanisms of Heterophase Polymerization

Heterophase polymerization is the result of the complex interaction between several different individual physical and chemical processes, including molecular diffusion (absorption, adsorption, and desorption processes), phase changes, chemical reactions, and many other processes. Heterophase polymerization is a prominent example of how (molecular) chemistry and colloid chemistry mutually interact. This means for chemical reactions, size effects of both the reacting species and the reaction locus strongly influence the process. On the other hand, colloid chemical processes or effects such as adsorption, interfacial tension, and coalescence/coagulation strongly depend on the chemical changes during the process. Understanding the mechanism of heterophase polymerization requires understanding of the polymerization mechanism and the colloid chemistry as well as the interactions between them.

In this chapter, the molecular mechanisms of the most important processes taking place in heterophase polymerization are discussed in detail, aiming at providing a clear molecular picture of heterophase polymerization as a whole. These processes are:

- Particle formation (Section 2.1), responsible for the transformation or emergence of discontinuous phases, i.e., particles as main polymerization loci, containing polymer chains.

Heterophase Polymerization: Basic Concepts and Principles
Hugo Hernandez and Klaus Tauer

ISBN 978-981-4877-32-9 (Hardcover), 978-1-003-11929-6 (eBook)
www.jennystanford.com

- Molecular transfer (Section 2.2), i.e., the motion/diffusion of all types of molecules through the system, responsible in a large extent for the characteristic kinetic behavior of heterophase polymerization. The most important events considered by molecular transfer are absorption by polymer particles (Section 2.2.1) and desorption from polymer particles to the continuous phase (Section 2.2.2). Molecular transfer equilibrium (Section 2.2.3), from which monomer equilibrium swelling is a very important particular case, determines the distribution of monomer and relevance of the polymerization loci.
- Polymerization (Section 2.3). Polymerization reactions lead to polymer chains in any system, either homogeneous or heterogeneous. It is, therefore, important to understand its dynamics, quantitatively determined by the rate of polymerization. Other concepts usually related to the kinetics of polymerization involve the degree of polymerization, the reaction conversion, and the distributions of polymer chain lengths.
- Particle dynamics (aggregation and breakage of polymer particles and monomer droplets) (Section 2.4), determines the size distribution and concentration of polymer particles in the system and influences the rate of heterophase polymerization in a very complex way by affecting the number of polymerization loci, the concentration of monomer in each particle, and the number of active sites per particle.

Most of the references mentioned in this chapter are related to the mechanism of emulsion polymerization because it has traditionally attracted the attention of researchers more than any other type of heterophase polymerization. This is probably because emulsion polymerization has been the most relevant heterophase polymerization process used by industry.

2.1 Particle Formation

Although particle formation (or *particle nucleation*) is, at least for emulsion polymerization, the most important reaction step as it generates the reaction loci where the formation of high-molecular-weight polymers takes place, it lasts in an *ab initio* batch process typically only a few minutes and happens at very low conversions. Thus, the experimental investigation is extremely challenging and requires sophisticated techniques.

Multi-stage heterophase polymerization: Type of heterophase polymerization where the reaction conditions (e.g., feed composition, feed rate, temperature, etc.) are drastically changed at least once during the polymerization process.

In ***multi-stage heterophase polymerizations***, particle formation may or may not occur repeatedly under the different conditions, and if it does, then a bimodal or multimodal particle size distribution is obtained. For some applications, a multimodal particle size distribution is desirable because these latexes, at given solids content, exhibit lower viscosity than their monomodal counterparts. However, for special applications, extremely monodisperse latex particles are required, instead. In both cases, particle nucleation must be controlled in different ways.

Generally, understanding the nucleation of a new phase is of great importance for both natural processes taking place on earth, such as weather phenomena or mineral formation, and many industrial processes, such as crystallization of organic or inorganic materials and silicon wafer production. There have been many approaches for the description of nucleation processes for more than 100 years [1]. The most successful and general approach based on thermodynamic considerations is known as the classical nucleation theory (CNT) and can be used to scientifically understand nucleation independent of the particular system.

A fundamental understanding of the thermodynamics, kinetics, and mechanisms of nucleation processes is of great importance for the control of particle size distribution and particle morphology. The most important question that must be addressed before the investigation of the kinetics of particle formation is the definition of a ***polymer particle***.

Polymer particle: From a kinetic point of view, a polymer particle is a molecular aggregate of at least one polymer chain and other molecular species (monomers, solvent, additives, etc.), which can effectively isolate polymerization reactive sites from reactants in the continuous phase.

From a purely kinetic point of view, a polymer particle can be interpreted as a molecular aggregate of one or more polymer chains with other chemical species such that the polymerization reactive sites (functional groups or chemical active sites) in the aggregate can be effectively segregated from the polymerization reactive sites either in the continuous phase or in any other discontinuous phase present in the system. Those reactive sites may be exposed to local monomer concentration different from that in the continuous phase, significantly affecting the polymerization rate and the molecular mass distribution (MMD) of the chains formed. For example, if the segregation of an active site in chain-growth polymerization is achieved and the local monomer concentration inside the particles is higher than in the continuous phase, then higher molecular mass polymer chains can be obtained at higher polymerization rates.

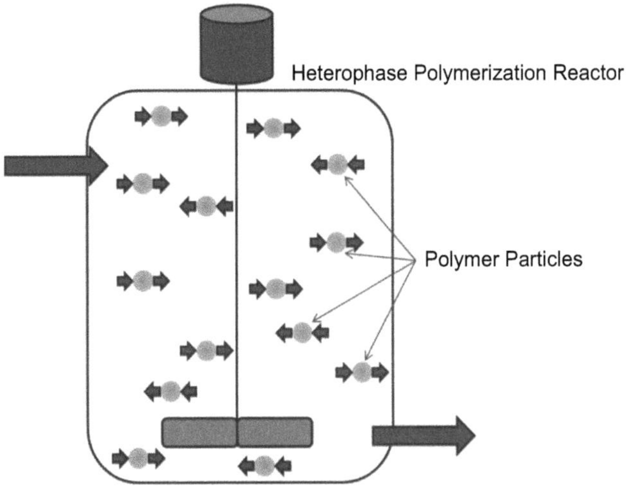

Figure 2.1 Schematic representation of polymer particles during heterophase polymerization as dispersed micro-reactors in a macro-reactor. Each particle exchanges mass and energy with the continuous phase. Mixing inside the particle is achieved by diffusion (equilibrating the chemical potential).

A polymer particle, which is a dispersed phase where polymerization takes place, can then be considered a micro-reactor dispersed in the macro-reactor (see Fig. 2.1). The particles, with respect to monomer, can be homogeneous across the entire volume or possess a shell (monomer-rich)–core (polymer-rich) structure. The particular distribution of the monomer depends on the Thiele modulus (cf. Chapter 0). On the contrary, with respect to radicals of any chain length, the particle's structure is not homogeneous.

The interaction between molecules inside the polymer particle and molecules present in any other phase of the system is possible but only if molecular transfer processes across the particle interface take place. Those transfer processes are represented in Fig. 2.1 as input and output flows to/from the dispersed micro-reactors.

Interfacial reactions pose a challenge to the previous kinetic definition of a polymer particle because they correspond to reactions between molecules in different phases. However, as it was already mentioned in the previous chapter, a clear distinction of the interface is not possible at a molecular level, and for that reason, interfacial reactions might be considered taking place either inside the particle or in the continuous phase. Furthermore, when interfacial reactions are sporadic compared to reactions taking place inside the particles, their occurrence can be safely neglected.

Polymer Particle: From a thermodynamic point of view, a polymer particle is a thermodynamically stable phase separated from the continuous phase as a result of the favorable reduction in Gibbs' free energy.

From a thermodynamic point of view, a ***polymer particle*** can be defined as a stable new phase in the system, which grows spontaneously, as long as enough building blocks are supplied, as a result of the favorable reduction in Gibbs free energy. This definition, while conceptually clear, is not so easily used in practice because of the difficulty in determining Gibbs free energy for individual segregated phases. Furthermore, the above-mentioned interaction between colloid chemistry and chemical reactions poses an additional challenge for such determination.

The formation of polymer particles is perhaps the most difficult event to be investigated in heterophase polymerization because its characteristic time and length scales are below the sensitivity of most experimental methods available.

Different models have been proposed to explain the formation of polymer particles, including self-aggregation (precipitation), cooperative aggregation, and phase transfer (Fig. 2.2).

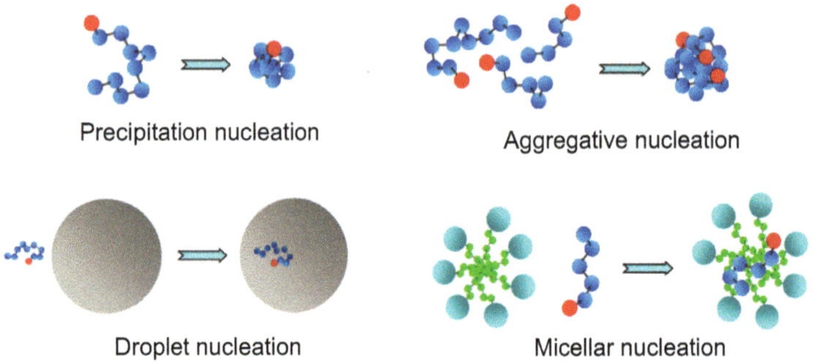

Figure 2.2 Models of particle formation as discussed for various techniques of heterophase polymerization. Blue spheres: monomer units; red spheres: reactive sites; cyan spheres: lyophilic moiety of the surfactant; green spheres: lyophobic unit of the surfactant; gray spheres: monomer droplets.

Homogeneous nucleation: Nucleation event taking place without being influenced by interfaces, i.e., involving only the macromolecular chains dissolved in the original homogeneous starting phase in this case. Since macromolecules have quite a high activation free energy, homogeneous nucleation becomes a highly unlikely event.

The nucleation of polymer particles can be ***homogeneous***, if only macromolecular chains in the continuous phase are involved in the transition, or ***heterogeneous***, if an interface of any species (surfactant, monomer, etc.) present in a different phase (surfactant micelle, monomer droplet, etc.) is involved in the transition. Therefore, the presence of interfaces is a requisite for the formation of particles by heterogeneous nucleation. Again, homogeneous nucleation has higher activation energy than heterogeneous nucleation, which is a generally valid statement and holds within the frame of the CNT for any nucleation event. The most prominent example is the formation of rain drops, which happens at the surface of nuclei (dust, salt, bacteria, pollen, etc.) floating in the atmosphere

[2]. The CNT has also been used to explain particle formation in dispersion polymerization in organic media [3].

Heterogeneous nucleation: Nucleation event involving additional species other than the nucleating species (for example, monomer, surfactant, initiator, etc.) present in a different phase and possessing an interface. The active presence of an interface reduces the free activation energy of the nucleation process.

In a typical heterophase polymerization, the conditions are favorable for heterogeneous nucleation [2]. Before the polymerization is started, the reaction system consists of at least monomer droplets, single surfactant molecules, and surfactant aggregates with aggregation numbers smaller than the critical micelle concentration (CMC). All colloidal objects that have a different chemical composition than the molecules going to be nucleated can be considered foreign objects. Within the frame of CNT, the interaction of the nucleating molecules with the foreign objects is crucial, determining how strong the activation free energy of nucleation will be reduced (cf. Fig. 2.3). Heterogeneous nucleation is a situation where at least three different components participate: the continuous phase (or mother phase in crystallization experiments), the aggregates of the nucleating species, and the foreign particle (or substrate). In any case, the nucleation happens at lower supersaturation and takes place at the interface of the foreign objects. The interaction between the three components is described by the contact angle between the aggregate of the nucleating species and the foreign particle, which is determined by the interfacial tensions between the three components.

The relation between the activation free energies for homogeneous (ΔG_{hom}) and heterogeneous (ΔG_{het}) nucleation is given by Eq. 2.1, as a function of the relativizing factor, $f(m)$ (Eq. 2.2), where m (Eq. 2.3) is the cosine of the contact angle (Θ), which is determined by the three interfacial tensions σ_{21}, σ_{31}, and σ_{32} (Eq. 2.4).

$$\Delta G_{\text{het}} = f(m)\Delta G_{\text{hom}} \tag{2.1}$$

$$f(m) = \frac{1}{4}(2+m)(1-m)^2 \tag{2.2}$$

$$m = \cos \Theta \tag{2.3}$$

$$\Theta = \frac{\sigma_{31} - \sigma_{32}}{\sigma_{21}} \tag{2.4}$$

The compatibility between the nucleating species and the substrate determines the contact angle, which can take values between 0° and 180°. Θ = 180° means no interaction at all or complete incompatibility, and Θ = 0° stands for complete compatibility. Thus, when $\Theta = 180° \rightarrow m = -1 \rightarrow f(m) = 1$. On the other hand, when $\Theta = 0° \rightarrow m = 1 \rightarrow f(m) = 0$. The former case means that the foreign particle does not interact with the nucleating species and, hence, has no influence on the nucleation and the nucleation process is homogeneous. If the contact angle is zero, substrate and nucleating species attract each other and, because the activation free energy is zero, the nucleating species precipitates to the substrate as soon as the solubility is reached.

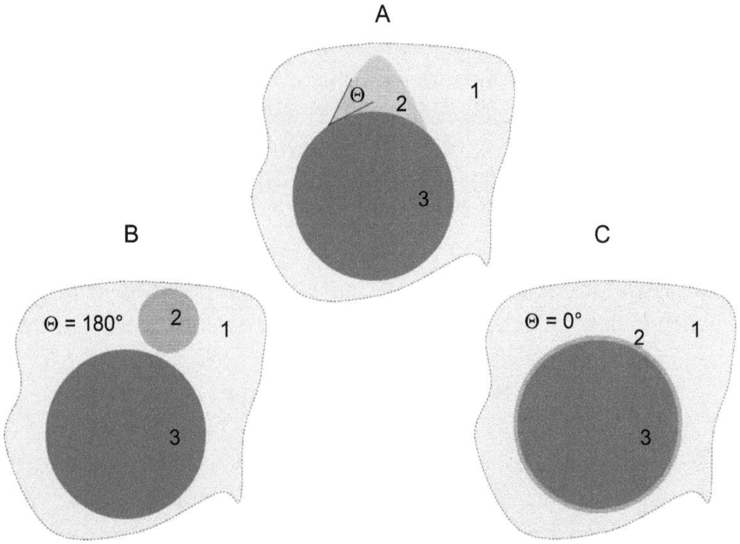

Figure 2.3 Schematic drawing illustrating heterogeneous nucleation; 1 – continuous phase, 2 – nucleating species, 3 – foreign phase; Θ – contact angle; A, B, C – different interaction strengths between the nucleating species and the foreign particles; A – intermediate interaction strength or compatibility, B – no interaction at all or complete incompatibility, C – strong interaction or complete compatibility.

When the conditions during a typical heterophase polymerization are considered, a scenario with no mutual interaction between the nucleating species and the foreign objects (monomer droplets and small surfactant aggregates) is rather unlikely. Most of the polymers are soluble in their monomers (the exception is polyacrylonitrile), and the interactions with suitable surfactants are obligatory. Consequently, both the activation free energies of nucleation and the supersaturation in heterophase polymerization are quite low.

2.1.1 Precipitation Nucleation

The first possible mechanism of particle formation is the collapse of a single polymer chain initially dissolved in the continuous phase. This is a straightforward assumption based on a simple picture. Consider a monomer with certain solubility in the continuous phase in which also the polymerization is started. With increasing degree of polymerization, the solubility of the growing chain decreases stepwise until the chain precipitates, and hence a *pauci* chain particle is formed. A polymer chain may be present in its coiled or dissolved state when the attraction of the molecules in the continuous phase is stronger than its own intramolecular attraction. At a certain temperature, the intramolecular forces will dominate and the chain segments will attract each other in such a way that the polymer collapses. If the temperature of the system is kept constant, but the composition of the continuous phase is modified reducing the interaction between the macromolecule and the continuous phase to the point that the intramolecular interactions become stronger than the repulsion originated from thermal motion, the polymer will also collapse. Additionally, for a growing polymer, the addition of a new monomer unit to the chain will increase the intramolecular attraction, and unless the solvent–segment interaction is stronger than the segment–segment attraction, the chain will reach a certain length at which the intramolecular attraction overcomes the thermal repulsion forces and the chain collapses. In this case, the size of the nuclei scales with the third root of the chain length of the collapsing macromolecule. Since thermal motion exerts a fluctuating force on the chain, the length at which the chain collapses may vary around a certain average critical chain length. This type of single-chain particle formation mechanism is usually called *precipitation particle formation* and was suggested by Fitch and Tsai [4] and improved some years later by Ugelstad and Hansen [5].

In principle, the Hansen–Ugelstad–Fitch–Tsai (HUFT) model states that the rate of particle formation by precipitation corresponds to the rate of formation of polymer chains with the critical chain length. Then, each critical-length polymer chain becomes an independent polymer particle. Even though the original idea of precipitation nucleation was developed for radical emulsion polymerization, the general concept is valid for any other type of heterophase polymerization. This can be summarized using the following expression:

$$\frac{dN_{\mathrm{p}}}{dt} = \frac{N_{\mathrm{A}}}{2} \sum_{i=1}^{j_{\mathrm{cr}}-1} k_{p,i,j_{\mathrm{cr}}-i} \left[P_i\right]_c \left[P_{j_{\mathrm{cr}}-i}\right]_c \tag{2.5}$$

where N_{p} is the number concentration of particles in the system, N_{A} is the Avogadro constant, $[P_i]_{\mathrm{c}}$ represents the concentration of polymer chains of length i in the continuous phase, $k_{p,i,j_{\mathrm{cr}}-i}$ represents the rate of polymerization between chains of length i and j_{cr}-i, and P_1 represents a monomer molecule. Actually, the complexity of Eq. 2.5 resides in the fact that the concentrations of the monomer and of each polymer with sub-critical chain length are represented by their own ordinary differential equations, which must be solved simultaneously. For certain cases, Eq. 2.5 can be transformed into a simplified expression. For example, according to the HUFT theory, for radical chain-growth emulsion polymerization, the rate of particle formation can be determined by

$$\frac{dN_{\mathrm{p}}}{dt} = 2fk_{\mathrm{d}}\left[I\right]\left(1 + \frac{k_{\mathrm{tw}}\left[R\right]_{\mathrm{w}}}{k_{\mathrm{p}}\left[M\right]_{\mathrm{w}}} + \frac{k_{\mathrm{c}}N_{\mathrm{p}}}{k_{\mathrm{p}}\left[M\right]_{\mathrm{w}}}\right)^{1-j_{\mathrm{cr}}} \tag{2.6}$$

where f is the initiation efficiency factor, k_{d} is the rate coefficient of initiator decomposition, k_{tw} is the rate coefficient of radical (R) termination in water, k_{p} is the rate coefficient of radical propagation, k_{c} is the rate coefficient of radical capture by particles, N_{p} is the total number concentration of particles, and j_{cr} is the critical chain length for precipitation.

Even though the HUFT theory is widely accepted and obligatory part in newer textbooks, some critical comments are necessary:

(1) An expression for the size of the particles just after nucleation (i.e., size of nucleus) is missing, (2) the model requires precise numerical values of kinetic parameters, which are not directly and independently accessible; and (3) j_{cr} data have to be experimentally determined, which is extremely challenging.

2.1.2 Aggregative Nucleation

An alternative mechanism of homogeneous particle formation consists in the simultaneous interaction of different oligomers present in the continuous phase causing a multi-chain collapse. This mechanism is known as *aggregative nucleation*. It is evident that the rate of nucleation in this case will depend on the amount and length of the polymer chains present in the continuous phase relative to their solubility, that is, on the ***supersaturation*** of the polymer. This mechanism can be described by the theory of homogeneous nucleation developed by Becker and Döring [1], presently known as the ***classical nucleation theory*** (CNT). From a thermodynamic point of view, nucleation is characterized by the necessity to overcome a certain energy barrier by means of thermal fluctuations. This energy barrier or activation energy represents the free energy of formation of a critical particle and strongly depends on the specific conditions of the system. In 1995, a modeling framework for aggregative nucleation in emulsion polymerization processes based on CNT was published [6]. The model allows the estimation of the chain length of the nucleating oligomers, the number of chains per nucleus, the diameter of the nucleus, the total number of nuclei formed, and the rate of nucleation, using the activation energy of nucleation as the only adjustable parameter. The general mechanism for heterophase polymerization is described next.

Degree of supersaturation (S): Ratio between the concentration of a solute and its solubility in the continuous phase.

Classical nucleation theory: Theory based on the fact that nucleation takes place after thermal fluctuations overcome the energy barrier of forming a critical particle.

According to the CNT, the rate of particle formation by the aggregation of oligomers of chain length j can be calculated as [6]:

$$\frac{dN_{\mathrm{p},j}}{dt} = \frac{2D_{\mathrm{c},j}}{\upsilon_\mathrm{c}}\left(\frac{\pi\rho_\mathrm{c}N_\mathrm{A}}{6M_\mathrm{c}}\right)^{2/3}\exp\left(-\frac{\Delta G_{\max}}{k_\mathrm{B}T}\right) \tag{2.7}$$

$$\Delta G_{\max} = \frac{\pi\sigma^3}{3}\left(\frac{4jM_\mathrm{m}}{\rho_\mathrm{p}N_\mathrm{A}k_\mathrm{B}T\ln S_j}\right)^2 \tag{2.8}$$

where $D_{c,j}$ is the diffusion coefficient of the oligomers of chain length j in the continuous phase, υ_c is the average molar volume of the continuous phase, ρ_c is the density of the continuous phase, M_c is the average molecular mass of the continuous phase, N_A is the Avogadro constant, k_B is the Boltzmann constant, T is the temperature of the system, σ is the interfacial tension between the nucleus and the continuous phase, M_m is the molecular mass of a monomer unit, ρ_p is the density of the polymer, and S_j is the supersaturation of the oligomer of length j, given by the ratio of concentration (C_j) to solubility, which in turn is obtained using the Flory–Huggins interaction parameter (χ):

$$S_j = \frac{M_\mathrm{m}C_j}{\rho_\mathrm{p}}\left(\frac{j^2(\chi-1)+2j}{j(1-\chi)-1}\right) \tag{2.9}$$

It is also possible to consider particle formation as the result of the aggregation of oligomers of different chain lengths. In this case, the total supersaturation of the oligomers forming the aggregate can be expressed as:

$$S = \prod_{j=1}^{J} S_j^{\,m_j} \tag{2.10}$$

where m_j is the number of oligomers of length j forming the aggregate. According to the CNT, the free energy of formation for this cluster is

$$\Delta G = -k_{\mathrm{B}}T\sum_{j=1}^{J} m_j \ln(S_j) + \sigma(4\pi)^{1/3}\left(\frac{3M_{\mathrm{m}}}{\rho_{\mathrm{p}}N_{\mathrm{A}}}\right)^{2/3}\left(\sum_{j=1}^{J} jm_j\right)^{2/3} \tag{2.11}$$

Therefore, the change in free energy by adding an additional oligomer of size *i* is

$$\left(\frac{\partial G}{\partial m_i}\right)_{j\neq i} = -k_{\mathrm{B}}T\ln(S_i) + \frac{2i\sigma}{3}(4\pi)^{1/3}\left(\frac{3M_{\mathrm{m}}}{\rho_{\mathrm{p}}N_{\mathrm{A}}}\right)^{2/3}\left(\sum_{j=1}^{J} jm_j\right)^{-1/3} \tag{2.12}$$

The concentration of oligomers in the aqueous phase can be obtained from the kinetics of polymerization in the continuous phase. A new stable particle is formed as soon as

$$\left(\frac{\partial G}{\partial m_i}\right)_{j\neq i} \leq 0 \tag{2.13}$$

i.e., when

$$\frac{1}{i}\ln(S_i) \geq \frac{2\sigma}{3k_{\mathrm{B}}T}(4\pi)^{1/3}\left(\frac{3M_{\mathrm{m}}}{\rho_{\mathrm{p}}N_{\mathrm{A}}}\right)^{2/3}\left(\sum_{j=1}^{J} jm_j\right)^{-1/3} \tag{2.14}$$

Equation 2.14 implies that the energy of activation for the nucleation of a particle will strongly depend on the size and amount of oligomers forming the precursor and the size and concentration of the oligomer, which reaches the critical nucleus size. Therefore, there will not be a single unique value of the activation energy of particle nucleation. The noise and variability of the nucleation process is increased by this effect.

The precipitated polymer aggregate can grow by subsequent absorption and polymerization of monomer or by the incorporation of additional chains or clusters of chains into the particle. The aggregation of polymer clusters proceeds until the particles are sufficiently stabilized by the adsorbed surfactant (from solution, monomer droplets, or micelles) or by ionic groups incorporated into the chains. It is important in this context to realize that the growth of polymer particles via aggregation of polymer clusters or precursors has been called *coagulative nucleation* [7], despite that it is not a true nucleation process. In fact, the so-called "coagulative nucleation" is not a mechanism of particle formation, but a colloidal aggregation process determining the size and number of particles in the system.

The nucleation model based on the CNT is fundamentally different compared to the precipitation nucleation model. CNT is a multi-chain process mainly controlled by thermodynamics and modulated by the polymerization kinetics in the continuous phase.

On the other hand, the application of CNT for heterophase polymerization is a nice example for the special situation when applying colloid chemistry to polymerization. Compared to nucleation of salt crystals or rain drops where the nucleating species do not change, during heterophase polymerization the chain length of the oligomers grows and therewith the physical properties as well, particularly, solubility decreases and supersaturation increases.

2.1.3 Heterogeneous Nucleation

In heterophase polymerization, the presence of a wide variety of molecular species in the system usually results in the appearance of different interfaces. Even if all components are completely soluble in the reaction medium, the inevitable presence of small contaminants, or even the interfaces with the reactor walls, provides enough interfacial area for reducing the activation free energy for the nucleation of the nucleating species, and thus heterogeneous nucleation should necessarily take place.

Perhaps the most important chemical species involved in heterogeneous nucleation are amphiphilic molecules because they contain a lyophobic moiety that interacts strongly with the polymer chains, while their lyophilic moiety interacts with the continuous phase reducing the interfacial tension of the particle.

Surfactants facilitate nucleation by lowering both the interfacial tension and the free-energy barrier, thus leading to faster rates of particle formation. The effect of the amphiphile on the nucleation process will depend also on the concentration of surfactant in the continuous phase. The lyophobic interaction of surfactant molecules is so strong that they are subject to their own phase separation process, which is called ***micellization***.

Micellization: Denotes the process of formation of micelles in a solution of amphiphiles. Micellization is also a nucleation process.

Micelles (or surfactant aggregates) are formed after reaching certain saturation in the continuous phase denoted as the ***critical micelle concentration*** (CMC). It should be noted that the CMC depends on temperature as well as the concentration of electrolytes and other components (also monomer) in the continuous phase. Thus, when applying CMC values to evaluate emulsion polymerization, one should use experimental data obtained under conditions as close as possible to those under polymerization. If a macromolecule is captured by a micelle, segregation of the polymer takes place and thus, a new polymer particle is formed. This mechanism of particle formation is called *micellar nucleation* [8]. Compare in this context the critical discussion regarding the coexistence of monomer drops and swollen micelles in Section II.

Micelles: Supramolecular structures formed as a result of the aggregation of amphiphiles in solution above a certain critical concentration (critical micelle concentration).

Not only separate amphiphile aggregates but also separate aggregates of monomer molecules (monomer droplets) can be involved in the formation of polymer particles via capture (bulk absorption or interfacial adsorption) of macromolecules, oligomers, or primary radicals from the continuous phase. This capture process, which is in principle the aggregation of macromolecular chains and/or active sites to monomer droplets, is usually referred to as droplet nucleation [9]. Droplet nucleation is considered to be more relevant when using monomers with very low solubility in the continuous phase, for example in miniemulsion polymerization [10]. A particular case of droplet nucleation is the association of macromolecular chains and/or active sites with very small aggregates of monomer molecules (nanodroplets). Such monomer nanodroplets are usually formed by a process known as spontaneous emulsification [11]. Nanodroplet nucleation resulting from spontaneous emulsification is probably one of the most important mechanisms of particle formation in heterophase polymerization and especially in emulsion polymerization. ***Spontaneous emulsification*** has been observed in oil–water systems in the absence of any stabilizer compound and in the absence of any type of mechanical mixing.

Critical micelle concentration: Concentration of amphiphiles above which micelles are spontaneously formed. Below CMC, amphiphilic molecules are not able to form stable aggregates but remain dissolved in the continuous phase.

Spontaneous emulsification: Consists in the formation of relatively stable molecular aggregates (nanodroplets) when two immiscible liquids are in contact, even in the absence of stabilizers and without mechanical energy input. Spontaneous emulsification is the result of molecular diffusion at the interface of the immiscible phases.

At first sight, the determining step for the formation of new particles by heterogeneous nucleation is the capture (or aggregation) of a polymer chain (dead or alive) or an active molecule by an insoluble molecular aggregate. Active molecules can be primary radicals in the case of radical chain-growth polymerization, or any other molecule capable of initiating a polymerization reaction inside the aggregate. For example, initiation by radiation and heat can also occur inside monomer droplets. Polymer chains and active molecules containing lyophilic groups, when captured by the molecular aggregate, will remain very close to the interface instead of being buried in the bulk of the particle. Some of those groups may also help stabilize the growing molecular aggregate.

However, there is another important effect that must be considered, which is the energy (or enthalpy) change during polymerization. When a small monomer aggregate becomes a polymer particle by the capture of an active molecule or a growing chain, polymerization reactions may take place inside the aggregate. During exothermic polymerization reactions (for example, typical chain-growth polymerization systems present highly exothermic propagation reactions), the energy released during the reaction will be quickly dissipated to the surrounding molecules, causing an important increase in molecular speeds (increase in local temperature over molecular distances [12]). If the reaction takes place close to the interface of a monomer drop, the energy released during polymerization may result in desorption of the active molecule or the growing chain, abandoning its status of polymer particle. Furthermore, if the molecular aggregate is too small, the energy release may result in the breakage or destruction of the particle.

According to the initial results of Fitch and Tsai [13], the critical chain length for chain precipitation in the emulsion polymerization of methyl methacrylate (MMA) with sulfate or sulfonate end groups is in the order of 65 to 75 units, which is a very reasonable result for a polymer collapse process. However, the critical chain length for particle formation estimated from experimental kinetic data of the emulsion polymerization of MMA at 80°C has been found to be in the order of 10 units [8]. Following the CNT for aggregative nucleation, the critical chain length for MMA emulsion polymerization at 50°C is 11 [6]. In the case of styrene, at 70°C the experimental critical chain length was found to be 5 [8], and according to the CNT, the critical value for styrene at 50°C is 6 [6]. For vinyl acetate at 80°C, the experimental critical chain length is 20 [14], while the calculated value at 50°C using the CNT framework is 22.

In general, the experimental critical values observed are too low for a polymer precipitation or collapse process. Let us consider, for example, the critical chain length of styrene: At 70°C, the critical length is only 5 monomer units. For this chain length, there is no physically possible single-chain conformation that can completely isolate the active site from the continuous phase, which is required by the definition of a particle presented above. On the other hand, if the chain is surrounded by other chains or by individual monomer molecules, the segregation of the active site is possible. In addition, the probability of finding a particle conformed only of polymer chains is extremely low compared to the probability of finding a particle formed by monomer molecules and polymer chains. For these reasons, the idea of a droplet nucleation mechanism controlled by the enthalpy of propagation seems to be quite reasonable. According to this mechanism, the size of the droplet also plays an important role especially for smaller droplets because the energy barrier of desorption is reduced with decreasing size due to the positive contribution of the Laplace pressure [15] to molecular desorption:

$$E_{\mathrm{des}} = E_{\mathrm{nc}} + \Delta\mu + \frac{2V_{\mathrm{m}}\sigma}{r_{\mathrm{d}}} \tag{2.15}$$

where E_{nc} is the contribution of nonconservative forces, $\Delta\mu$ is the difference in chemical potential of the oligomer between the aqueous phase and the droplet or particle, V_m is the molecular volume of the oligomer, σ is the interfacial tension between the droplet or particle and water, and r_d is the radius of the droplet or particle.

In the case of small droplets, the minimum chain length required for particle formation may be some units longer than for larger droplets. Even if the radical is not close to the surface, the dissipation of the released energy through the droplet or particle may lead to desorption of some monomer units, and if the droplet is small enough, this could result in the complete disintegration of the droplet.

A similar argument can be proposed for the formation of particles by micellar nucleation. What is called monomer-swollen micelle is actually a very small droplet stabilized by a surfactant layer. Even though the surfactant layer can reduce the Laplace pressure effect and can also increase the energy barrier for desorption, the propagation of a growing chain inside the micelle might still lead to desorption of the chain (especially for smaller chains). This was evidenced experimentally by Harkins [16] using X-ray diffraction. In these experiments, Harkins found that the micelles absorbed styrene before polymerization, increasing the thickness of the micelles by 7.2 Å. Just after polymerization of the monomer by the action of the X-rays, the thickness of the original micelle was restored indicating that the polymer formed abandoned the micelle and appeared in the aqueous phase. Once the radical is desorbed, it can more easily swell molecular monomer or associate with small monomer droplets and adsorb free surfactant molecules rather than enter the micelle again, depending on the relative concentration of monomer and surfactant in the system. As a result of this, it is possible to state that micellar nucleation is not a significant mechanism of particle formation in heterophase polymerization, as it has been demonstrated experimentally using online conductivity measurements [17].

Using the CNT approach, the first "*stable*" molecular clusters of monomer are formed at supersaturation values between 5 and 6, containing between 60 and 80 molecules per cluster and with a diameter of about 2.75–3 nm. Even larger molecular aggregates are formed close to the source of monomer, where the relative supersaturation values are even higher. The concept of *stability* in this case is referred to the stability to cluster dissociation, and not to cluster aggregation. These droplets formed by the association of molecularly dissolved monomer units are much smaller than those formed by *comminution* techniques such as mechanical stirring or the application of ultrasound and can grow further by the incorporation of additional monomer units present in the continuous phase or by coalescence with other droplets. This process, as it was previously mentioned, is known as *spontaneous emulsification.*

In conclusion, different mechanisms of particle formation can take place simultaneously in a heterophase polymerization process. The relative relevance of each mechanism will depend strongly on the particular composition and conditions of the system.

2.2 Molecular Transfer

Polymer particles are not static entities. On the contrary, they are dynamic aggregates of molecules that are in permanent motion and transformation. Any molecule present in a polymer particle, either a small molecule or a growing chain of any length, continuously moves and collides with neighboring molecules. As a result of such motion, molecules can abandon the aggregate and return to the continuous phase (desorption). Similarly, any molecule in the continuous phase can eventually collide with the aggregate and become a part of it (capture). All those molecular transfer events ultimately have a significant effect on the kinetics of heterophase polymerization, as well as on the MMD and particle size distribution of the polymer dispersion obtained.

When the molecular species transferred is a solvent for the polymer chains, the capture process of molecules and molecular aggregates of any size by particles is called *swelling*, and desorption is called *deswelling*. The dynamics of swelling in polymer particles will be considered specifically in the last part of this section. Even though the principles of molecular transfer are the same, for those particular cases when a large number of molecules are present in the system, the behavior observed macroscopically can be described differently.

The phase transfer process takes place when a molecule or a molecular aggregate reaches the interface between the polymer particle and the continuous phase with a kinetic energy high enough to overcome the resistance to transfer at the interface. The resistance to transfer is not necessarily the same in both directions through the interface. Some transfer processes may be additionally facilitated by intermolecular attractive forces, as observed during swelling studies [11].

Given the experimental difficulties for the precise observation of molecular transfer, a different approach to investigate transfer dynamics can be used based on the simulation of the Brownian motion of individual molecules or molecular aggregates either in the continuous phase or inside the polymer particles. One main advantage of the simulation of Brownian motion (*Brownian dynamics (BD) simulation*) is that the capture and desorption rate coefficients can be easily determined based on the well-established mechanistic equations of Brownian motion, and in addition, capture kinetics can be determined without the interference of competitive events (i.e., chemical reactions).

2.2.1 Capture

The mechanism of molecular capture by polymer particles in heterophase polymerization has been a matter of controversy for many years, particularly regarding radical capture in free-radical emulsion polymerization. Different mechanisms have been proposed as rate-determining steps in order to describe the kinetics of molecular capture by polymer particles. In addition, there are plenty of experimental results supporting or contradicting each of the mechanisms proposed, and there is, until now, no agreement regarding which is the "correct" mechanism of capture. The most important mechanisms of molecular capture in heterophase polymerization include:

- **Collisional mechanism:** The limiting step for the capture or absorption of molecules by polymer particles is assumed to be the ballistic collision between molecules in the continuous phase and the polymer particles. In this case, the rate of molecular capture is proportional to the surface area of the polymer particles [4]. This approach is based on fundamental principles assuming a ballistic motion (straight, inertial motion) of the molecules in the continuous phase. The average rate of capture of molecules of species S per particle (ρ_S) in this case is given by

$$\rho_S = \left(\frac{\pi k_B T}{2m}\right)^{1/2} d_p^2 N_A [S]_c \tag{2.16}$$

where k_B is the Boltzmann constant, T is the temperature of the system, m is the mass of the colliding molecule, d_p is the particle diameter, N_A is the Avogadro constant, and $[S]_c$ is the molar concentration of the transferred species S in the continuous phase.

- **Diffusion-controlled mechanism:** While molecules can be considered to present a ballistic behavior in gases and low-density systems, for high-density systems, collisions with other molecules in the continuous phase are inevitable. Those frequent collisions result in a chaotic change in the speed and direction of the molecules. Thus, the overall displacement of the molecules is not ballistic but diffusional. The same principle holds for the polymer particles, although their diffusion is much slower. For diluted dispersions of polymer particles, the rate of molecular collision by diffusion per particle can be calculated using **Smoluchowski's equation** [18]:

$$\rho_S = 2\pi D_{S,c} d_p N_A [S]_c \tag{2.17}$$

where $D_{S,c}$ is the diffusion coefficient of the molecule of species S in the continuous phase. In this case, the rate of molecular capture depends linearly on the particle diameter. Equation 2.17 is the analytical solution of the fundamental Fick's equations applied to a single particle dispersed in an infinite medium. Considering that not every molecule–particle collision leads to an absorption event, because of interfacial resistance, a rate-reduction factor or an absorption efficiency factor (F) is included in the model [4]:

$$\rho_S = 2\pi D_{S,c} d_p N_A [S]_c F_S \tag{2.18}$$

F_S represents the absorption efficiency factor that describes the degree to which absorption is lowered compared to ideal capture, which can be given by

$$F_S = \frac{1}{\left(\frac{D_{S,c}}{K_{S,eq} D_{S,p} (X \coth X - 1)}\right) + \frac{E_{S,p}}{k_B T}} \tag{2.19}$$

where

$$X = \frac{d_\mathrm{p}}{2}\sqrt{\frac{r_\mathrm{S,p}}{D_\mathrm{S,p}N_\mathrm{A}[S]_\mathrm{p}}} \tag{2.20}$$

$K_\mathrm{S,eq}$ is the equilibrium partition coefficient of molecules of species S between the particles and the continuous phase; $E_\mathrm{S,p}$ is the potential energy barrier for transfer of molecules of species S in the direction from the continuous phase to the particle; $D_\mathrm{S,p}$ is the diffusion coefficient for the molecules inside the particle, and $r_\mathrm{S,p}$ is the rate of consumption of molecules of species S inside the particles (leading to irreversible absorption); and $[S]_\mathrm{p}$ is the molar concentration of molecules of species S inside the particle. A different expression for the efficiency factor, considering molecular re-desorption instead of an interfacial energy barrier, was proposed by Nomura et al. [19]:

$$F_\mathrm{S} = \frac{r_\mathrm{S,p}V_\mathrm{p}}{\delta_\mathrm{S} + r_\mathrm{S,p}V_\mathrm{p}} \tag{2.21}$$

where δ_S is the overall molecular desorption rate of molecules of species S per particle, and V_p is the volume of the particle. These equations are valid, as it was previously mentioned, only in the limit of infinite dilution of the particles. That is, they should be applied, if at all, probably only at the beginning of the nucleation process.

- **Colloidal mechanism:** Penboss et al. [20] considered polymer particles to be electrostatically stabilized polymer colloids and used the Deryaguin–Landau–Verwey–Overbeek (DLVO) theory (cf. Section 2.4) to determine the rate of molecular absorption by colloidal aggregation. The resulting dependence of the rate of capture on the size of the particles was found to be approximately linear:

$$\rho_\mathrm{S} = \pi D_\mathrm{S,c} d_\mathrm{S}\left(d_\mathrm{p} + d_\mathrm{S}\right) N_\mathrm{A}[S]_\mathrm{c}\,\kappa \exp\left(-\frac{E_\mathrm{S,p}}{k_\mathrm{B}T}\right) \tag{2.22}$$

where d_S is the molecular diameter for species S, κ is the reciprocal of the Debye length, and $E_\mathrm{S,p}$ is the energy barrier of the capture process caused by electrostatic repulsion. This expression has been obtained also from Fick's equations in infinitely diluted dispersions, but considering stabilization only by electrostatic repulsion. It is, therefore, very similar to the diffusion-controlled mechanism (Eq. 2.18), where

$$F_S = \frac{d_S}{2}\kappa \exp\left(-\frac{E_{S,p}}{k_B T}\right) \tag{2.23}$$

and $d_p \approx d_p + d_S$.

Cheong and Kim [21] performed the simulation of surfactant-free emulsion polymerization over electrostatically stabilized seed particles and showed that when highly charged seed particles are used, the electrostatic repulsion, resulting from the electrical charge of seed particles, reduces the absorption rate of the growing oligomeric chains in agreement with the predictions of the colloidal mechanism. This mechanism is only valid for systems where the size of the polymer particles is in the colloidal range. This would exclude precipitation polymerization and suspension polymerization.

- **Propagation-controlled mechanism (for oligomers):** Maxwell et al. [22] proposed that the rate-determining step for the capture of polymer chains by particles is the growth of the chains in the continuous phase up to a certain critical chain length after which the capture process is imminent. According to this model, only oligomers of a critical chain length can enter the particles. Entry, thus, becomes independent of particle size and charge. The rate of capture per particle of chains of critical length j_{cr}, assuming a propagation-controlled mechanism, is

$$\rho_{j_{cr}} = \frac{N_A}{2N_p}\sum_{i=1}^{j_{cr}-1} k_{p,i,j_{cr}-i}\left[P_i\right]_c\left[P_{j_{cr}-i}\right]_c \tag{2.24}$$

where $k_{p,i,j_{cr}-1}$ is the polymerization rate constant in the continuous phase and $[P_i]_c$ is the monomer concentration in the aqueous phase. The critical chain length for irreversible capture is $j_{cr} \approx 1 + \text{int}(-23/(RT \ln([M_{sat}]_c/[M]_{ref})))$, where R is the ideal gas constant in kJ/molK, T is the absolute temperature in K, $[M_{sat}]_c$ is the saturation concentration of monomer in the continuous phase in mol/L, $[M]_{ref}$ = 1 mol/L is a reference monomer concentration, int is a rounding operator, and ln represents the natural logarithm operator. Please notice that such expression for determining j_{cr} is purely empirical. There are, however, some important unresolved issues regarding this mechanism of radical capture. Generally speaking, a mechanism based on exclusion and allowing only a single species to enter a particle is rather problematic than scientifically convincing. Thus, no single value of j_{cr} can provide an adequate fit to different experimental data [19, 23]. Additionally, it has been experimentally evidenced that any kind of species present in the continuous phase can enter the particles, even primary free radicals [24]. Therefore, not only critical chains can be absorbed by the particles, but also smaller oligomers. Even though the capture of such oligomers with chain length less than j_{cr} may not be thermodynamically favored, their concentration in the continuous phase is greater than that of critical chain length oligomers, and thus the frequency of particle–oligomer collisions is considerable higher [25]. Moreover, due to side reactions in the continuous phase, oligomers may have different end groups corresponding to different thermodynamic properties and, accordingly, the capture conditions should change. For the sake of completeness, this entry model resembles the idea of heterogeneous nucleation with strong interaction (contact angle of 0°) between the growing chain in the continuous phase (nucleating species) and the particles (foreign objects).

- **Stochastic simulation:** BD simulation has also been used to determine capture rate coefficients [26, 27]. In this case, the diffusive behavior of a molecule (or molecular aggregate) is simulated assuming Brownian motion, and the trajectory of the molecule (or aggregate) is followed until a collision with a polymer particle is observed. An example of such trajectory is presented in Fig. 2.4.

  The ratio between the collision rate obtained by BD simulation ($\rho_{S,BD}$) and the ideal capture rate given by Smoluchowski's equation (Eq. 2.17) is defined as the **Smoluchowski number (*Sm*)**:

$$Sm = \frac{\rho_{S,BD}}{2\pi D_{S,c} d_p N_A [S]_c} \tag{2.25}$$

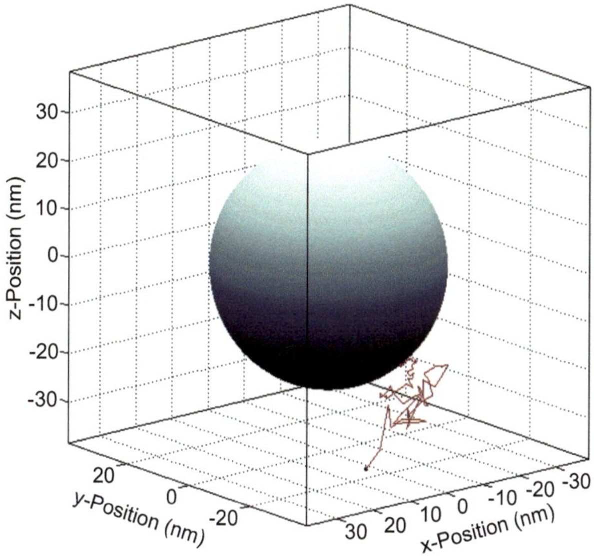

Figure 2.4 Schematic representation of a molecule (black dot) diffusing in the continuous phase, following a random-walk (Brownian motion) trajectory (red line) until hitting a polymer particle. The sphere represents a polymer particle dispersed in the continuous phase. The simulation box corresponds to the average volume around each polymer particle in the dispersion.

At very low volume fractions of polymer particles, BD simulation predicts the collision rate obtained with the Smoluchowski equation ($Sm = 1$), while for concentrated polymer dispersions (volume fractions > 0.1%), the Smoluchowski number and, therefore, the capture rate coefficient present a linear dependence with respect to the volume fraction of polymer particles in the dispersion.

$$Sm = 1 + \upsilon\varphi_{\mathrm{p}} \tag{2.26}$$

where υ is a dimensionless constant obtained for the system under the particular conditions considered in the simulation (~17.98 for the particular case considered in Ref. [26]) and φ_{p} is the volume fraction of particles in the dispersion. Similar conclusions were obtained by Rzepiela et al. [28], who found that the rate of aggregation of colloidal particles obtained by simulation always exceeds the value predicted by the Smoluchowski theory for fast aggregation, and only for volume fractions below about 0.1%, the discrepancy is small. The values of the rate of absorption per particle obtained by BD simulation under a wide range of conditions were used to obtain the following *semi-empirical expression*:

$$\rho = 2\pi D_{\mathrm{S,c}} N_{\mathrm{A}} \left(\frac{\upsilon \pi N_{\mathrm{p}} d_{\mathrm{p}}^{4}}{6} + d_{\mathrm{p}} \right) [S]_{\mathrm{c}} \tag{2.27}$$

This means that the effect of polymer volume fraction on collision kinetics under diffusion-controlled conditions explains the different results obtained during the experimental determination of radical capture kinetics in emulsion polymerization [29].

Interaction forces, interfacial tensions, the presence of stabilizer molecules at the surface of the polymer particles, and many other physical and chemical effects may lead to an increase in free energy during molecular capture, and thus to the existence of an energy barrier for molecular capture. In these cases, only a fraction of the molecules colliding with the particles will be effectively captured (capture efficiency), and the other fraction will just bounce back. When a molecule is not captured by a particle because of such energy barrier, the molecule will remain close to the particle surface; therefore, it will have a very high probability of hitting the same particle again, leading to a series of multiple collisions in a very short time before the molecule finally goes away from the particle surface. BD simulations show that the effect of the magnitude of the energy barrier at different temperatures on the capture efficiency can be described in a general way as [27]:

$$\rho_{\mathrm{S}} = \begin{cases} \rho_{\mathrm{S,0}} & , \quad E < E^{*} \\ \rho_{\mathrm{S,0}} e^{-\frac{(E-E^{*})}{3RT}} & , \quad E \geq E^{*} \end{cases} \tag{2.28}$$

where $\rho_{\mathrm{S,0}}$ is the capture rate in the absence of energy barriers (Eq. 2.27), and E^* is the magnitude of an apparent threshold energy caused by multiple collisions between the molecule and the polymer particle. E^* is expected to be a function of the molecular and particle diameters. It is important to notice that the energy barrier considered may contain the effect of different types of interfacial interactions, being consistent with the colloidal mechanism in surfactant-free systems, but also with surfactant-stabilized systems [30].

Although the functional dependence of the absorption rate coefficient on the particle diameter is very different for the various models discussed, it is quite intriguing that all of them have been successfully validated with experimental data in emulsion polymerization [31–33]. Unfortunately, there are no direct measurements of molecular capture (i.e., data free of model assumptions), so it is not possible to test the capture models unambiguously [34]. Sood and Awasthi [35] demonstrated that depending on the experimental conditions, the particle size dependency of the entry rate coefficient determined using indirect methods might not be reflected on the measured entry rate, and thus erroneous conclusions might be obtained from experimental results. However, the results obtained by stochastic simulation seem to describe the most general mechanism of molecular absorption in heterophase polymerization.

2.2.2 Desorption

The process of molecular desorption can be regarded as the opposite to molecular absorption or molecular capture. However, for most heterophase polymerization processes, provided the conversion is not too low, absorption and desorption differ at least in one crucial point. For entry, the entity coming from the continuous phase (medium with quite low viscosity) enters a polymer particle where the inside viscosity is clearly higher. For an exiting species, the situation is just the opposite. Considering the internal viscosity of the particles, clearly entry and exit of smaller entities is favored. On the other hand, exit and entry strongly depend on conversion. Note that diffusion coefficients are strongly influenced by both the size of the diffusing entity and the viscosity of the medium.

During desorption, a molecule is transferred from the interior of the particles to the continuous phase. The mechanism by which this transfer takes place is also similar: the molecule must reach the inner side of the particle surface and then it must overcome the barrier for desorption exerted by the interface. Desorption causes a decrease in the concentration of molecules in the particles. If the desorbed molecules are involved in polymerization reactions, then the rate of polymerization will necessarily be affected by desorption. For free-radical heterophase polymerization systems, radical exit is perhaps the major cause of loss of free-radical activity inside a particle.

Ugelstad et al. [36] proposed that the rate of molecular desorption should be inversely proportional to the surface of the particle. Nomura and Harada [37], aware of the necessity of quantitatively estimating desorption rate coefficients for the prediction of rates of emulsion polymerization, developed a quantitative model of desorption using a semi-empirical model of interfacial mass transfer. They continued improving this model during more than one decade. Asua et al. [38] complemented the model of desorption by considering the fate of the molecules in the aqueous phase. Further improvements in the semi-empirical model have been proposed by considering the layer of stabilizer around electrosterically stabilized polymer particles as an additional resistance to the desorption process [39]. The semi-empirical macroscopic model of equilibrium desorption developed by Nomura et al. has been widely used to describe molecular desorption in heterophase polymerization. However, the thermodynamic limit assumption involved in the derivation of this model may not be fulfilled by a radical desorption process in a real emulsion polymerization system, because in this case, the radicals are present in a very low concentration inside or around the particles.

The rate of desorption of molecules of species S (δ_S) from a single particle is given by

$$\delta_S = k_S N_A [S]_p V_p \tag{2.29}$$

where k_S is the rate coefficient of molecular desorption, N_A is the Avogadro constant, V_p is the volume of the particle, and $[S]_p$ is the molar concentration of species S inside the particle.

The diffusion-controlled mechanism of molecular desorption is widely accepted right now, but there are minor differences regarding the mathematical treatment of the problem leading to different expressions for the rate of desorption; however, all of them predict the same inverse dependence on the surface area of the particles. One of the reasons for the use of different expressions is caused by different interpretations of the desorption process. In order to avoid confusion, the following types of desorption are considered [40]:

- **Simple molecular desorption:** It is the result of the diffusive motion of the molecules when they are not involved in any reaction inside the polymer particles, and there is no interfacial energy barrier for desorption. The rate coefficient of simple molecular desorption ($k_{\mathrm{S,0}}$) is determined by the average velocity at which the molecules diffuse out of the particle and is a function of the particle size (d_{p}) and the diffusion coefficient of the molecule of species S inside the polymer particle ($D_{\mathrm{S,p}}$):

$$k_{\mathrm{S,0}} = \lambda \frac{D_{\mathrm{S,p}}}{d_{\mathrm{p}}^2} \tag{2.30}$$

  where λ is a constant with a value of 60 for perfectly spherical particles.

- **Equilibrium molecular desorption:** Equilibrium desorption takes into account the different solubility of the molecular species S between the polymer particles and the continuous phase. The rate coefficient of equilibrium desorption ($k_{\mathrm{S,eq}}$) can be related to the simple radical desorption rate coefficient by

$$k_{\mathrm{S,eq}} = \frac{k_{\mathrm{S,0}}}{\left(1 + K_{\mathrm{S,eq}} \dfrac{D_{\mathrm{S,p}}}{D_{\mathrm{S,c}}} \dfrac{s_{\mathrm{c}}}{s_{\mathrm{p}}}\right)} \tag{2.31}$$

  where $D_{\mathrm{S,c}}$ is the diffusion coefficient of the molecules in the continuous phase, $K_{\mathrm{S,eq}}$ is the partition coefficient of species S between the polymer particle and the aqueous phase, s_{c} is the thickness of the stagnant layer in the aqueous phase, and s_{p} is the thickness of the diffusion layer in the polymer phase. The denominator in Eq. 2.31 acts as a correction factor when interfacial barriers for molecular transfer are present. The equilibrium between phases is presented in more detail in Section 2.2.3.

- **Net molecular desorption:** The net desorption of a molecule occurs when it escapes the polymer particle after surviving all competitive reactions that may be taking place inside the particle. The rate coefficient of net molecular desorption ($k_{S,net}$) is determined by

$$k_{S,net} = \left(N_A V_p \sum_i v_{S,i}^{+} r_i + \rho_S \right) \frac{k_{S,eq}[S]_p}{k_{S,eq}[S]_p - V_p \sum_i v_{S,i}^{-} r_i} \tag{2.32}$$

where V_p is the volume of the particle, $v_{S,i}^{+}$ represents the stoichiometric coefficients of only the reactions generating the molecular species S inside the particle (only positive coefficients), $v_{S,i}^{-}$ represents the stoichiometric coefficients of only the reactions consuming molecular species S inside the particle (only negative coefficients), ρ_S is the rate of molecular absorption, and r_i are the corresponding molar reaction rates for each reaction taking place inside the particle.

- **Effective molecular desorption:** A molecule is considered to be effectively desorbed from the particle only after it reacts in the continuous phase. This might be useful from the kinetic perspective but a restriction from the colloid chemistry view. This definition accounts for the fact that desorbed molecules that diffuse through the continuous phase may be reabsorbed by a polymer particle and continue reacting therein, without significantly affecting the kinetics of polymerization. The rate coefficient of effective molecular desorption is ($k_{S,eff}$):

$$k_{S,eff} = N_A V_p \sum_i v_{S,i}^{+} r_i \frac{P_c(1-P_p)}{1-(1-P_w)(1-P_c)} \tag{2.33}$$

where

$$P_p = \frac{1}{1 - \dfrac{k_{S,0}[S]_p}{\sum_i v_{S,i}^{-} r_i}} \tag{2.34}$$

is the probability of reaction of the molecule inside the particles and

$$P_c = \frac{1}{1 - \dfrac{\rho_S N_p}{N_A \sum_j v_{S,j}^{-} r_j}} \tag{2.35}$$

is the probability of reaction of the molecule in the continuous phase. N_p is the concentration of particles in the system, and r_j represents the rate of reactions involving the molecular species S taking place in the continuous phase.

Simple molecular desorption rate coefficients have also been obtained from the simulation of the Brownian motion of radicals inside the particles using a variable time-step Monte Carlo random flight (MCRF) method for BD simulation, similar to the method used for estimating molecular capture by polymer particles. A linear regression of the observed simple molecular rate coefficients yields the expression presented in Eq. 2.30, with a value of $\lambda = 57.14$ [40], in very good agreement with the value of 60 predicted by theory.

Stochastic molecular simulation methods can also be used for predicting desorption rates under nonideal conditions, for example, for non-spherical particles [41], or non-homogeneous particles (such as core–shell or gradient polymer particles) [42]. It is also possible to incorporate the effect of chemical reactions with molecular desorption (effective molecular desorption), by means of a hybrid BD-kinetic Monte Carlo (kMC) simulation. kMC, also known as the stochastic simulation algorithm (SSA) [43], is a powerful tool for modeling competitive chemical and physical processes, particularly for small systems, as is the case of polymer particles.

BD simulation can be used to follow the molecular trajectory inside the particle and to determine the precise moment of desorption, while at the same time kMC simulates the reactions taking place simultaneously both inside the particles and in the continuous phase. In order to test the model for net molecular desorption, let us consider an emulsion polymerization system where monomer-derived radicals are generated by chain transfer to monomer inside the particles. Those primary radicals are assumed to be the only species capable of desorption with zero energy barrier for desorption; larger oligomers are assumed to have an infinite energy barrier for desorption. Inside the particles, the only competitive reaction considered is the propagation of the radicals. There will be no radical reabsorption, since only the net rate coefficient of radical desorption will be determined. All these assumptions are considered only for comparison with the available analytical results. However, the hybrid BD-kMC simulation method can be used also without any of these restrictions.

Figures 2.5 to 2.8 show the results obtained using the hybrid BD-kMC simulation, varying the propagation rate coefficient or the chain transfer to monomer rate coefficient. In both cases, the probability of reaction of the radical inside the polymer particle, P_p, and the net primary radical (R) desorption rate coefficient $k_{R,net}$ obtained by simulation are presented and compared to the corresponding theoretical model (Eqs. 2.34 and 2.32, respectively).

It is observed that the values for the probability of reaction inside the particles and the net radical desorption rate coefficient obtained by the hybrid BD-kMC method are in very good agreement with those predicted by the theoretical models (Eqs. 2.32 and 2.34).

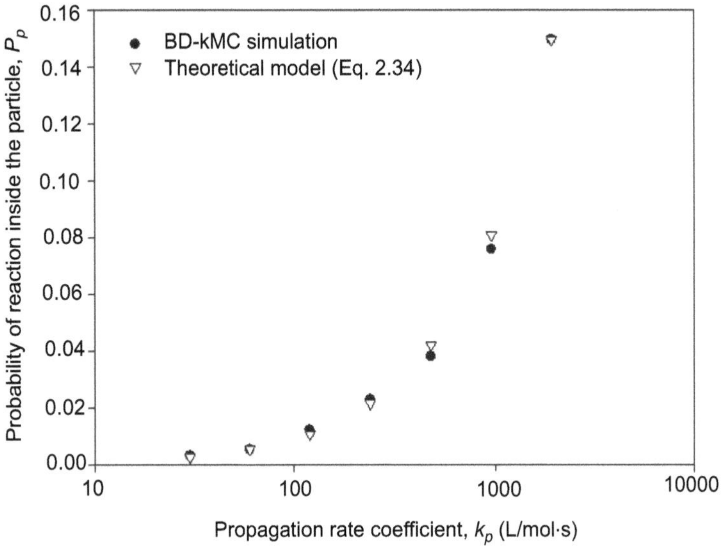

Figure 2.5 Probability of reaction inside the polymer particle (P_p) for different values of the propagation rate coefficient. $D_p = 10^{-11}$ m²/s, $d_p = 100$ nm, $[M]_p = 5.5$ mol/L, $k_{fm} = 0.02$ L/mol·s.

When the rate coefficient for propagation is high, the rate coefficients for net desorption obtained by simulation are slightly higher than those calculated from Eq. 2.32 (Fig. 2.6). This difference can be explained by the fact that the simulations are performed under non-steady-state conditions. That means at the beginning of the simulation, only radicals that desorb faster (i.e., close to the particle surface) are able to leave the particle without propagating, leading to a relatively higher estimate of the desorption rate. In Fig. 2.7, the inherent variability of the stochastic simulation method is evident when compared to the theoretical model (Eq. 2.34), where the chain transfer to monomer rate coefficient has no effect on the probability of reaction inside the particle. The variability of the simulation results can be reduced by increasing the number of radicals simulated, but this will demand more computer time. A trade-off between accuracy and speed is always unavoidable when using computational methods.

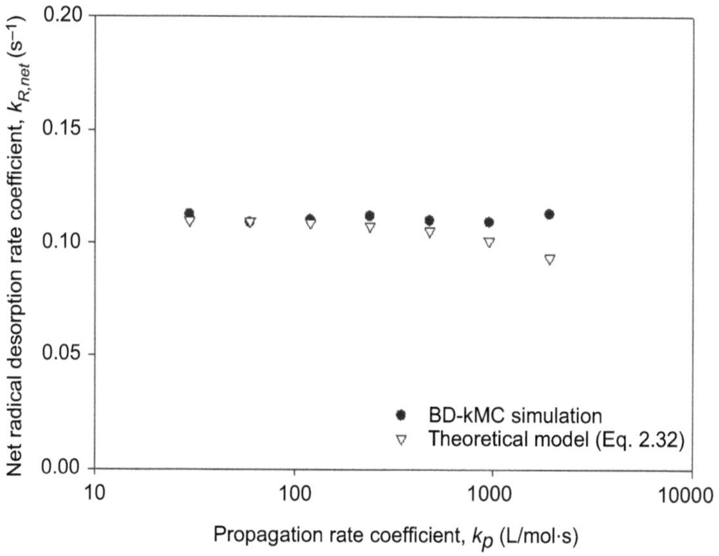

Figure 2.6 Net primary radical desorption rate coefficient ($k_{R,net}$) for different values of the propagation rate coefficient. $D_p = 10^{-11}$ m^2/s, d_p = 100 nm, $[M]_p$ = 5.5 mol/L, k_{fm} = 0.02 L/mol·s.

The use of the hybrid BD-kMC method can be extended to the determination of effective molecular desorption rate coefficients simply by incorporating into the kMC model the reactions in the continuous phase, and including in the BD model the reabsorption of desorbed molecules into the particles. In addition, a higher degree of accuracy can be obtained by considering the energy barrier for absorption or desorption as a function of the molecular size (e.g., chain length). In this case, also the MMD of oligomers inside the particles and in the continuous phase can be obtained.

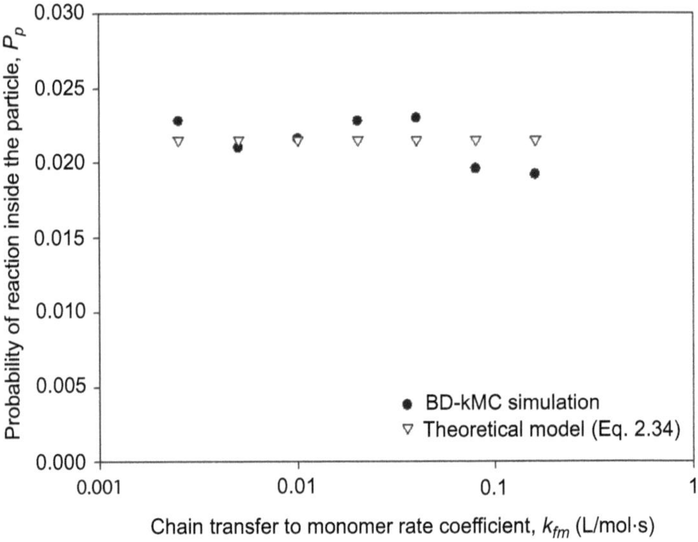

Figure 2.7 Probability of reaction inside the polymer particle (P_p) for different values of the chain transfer to monomer rate coefficient. $D_p = 10^{-11}$ m^2/s, d_p = 100 nm, $[M]_p$ = 5.5 mol/L, k_p = 240 L/mol·s.

Previously, it was mentioned that Asua [39] improved the model of desorption, by considering the effect of a surfactant layer around the particles. He derived an analytical model based on the solution to Fick's equation of mass transfer under steady-state conditions for the calculation of the effective desorption rate. The reabsorption of radicals by the particle was assumed to follow Smoluchowski's equation. He also considered a stagnant film surrounding the surfactant layer. His model can be summarized as follows (the original nomenclature was changed for consistency):

$$k_{\mathrm{R,eff}} = \Psi_{\mathrm{s}} \frac{\Psi_{\mathrm{f}} N_{\mathrm{A}}}{\Psi_{\mathrm{p}} m_{\mathrm{R}}} \left(1 - \frac{\Psi_{\mathrm{s}} N_{\mathrm{p}}}{\Psi_{\mathrm{s}} N_{\mathrm{p}} + k_{\mathrm{p,c}} [M]_{\mathrm{c}} + 2k_{\mathrm{t,c}} [R]_{\mathrm{c}}} \right) \tag{2.36}$$

where

$$\Psi_{\mathrm{s}} = \frac{2\pi D_{\mathrm{R,c}} d_{\mathrm{p}}}{1 + 2\frac{D_{\mathrm{R,c}}}{D_{\mathrm{R,s}}} \frac{s_{\mathrm{s}}}{d_{\mathrm{p}}} + \frac{D_{\mathrm{R,c}}}{D_{\mathrm{R,p}} m_{\mathrm{R}}} \left(\frac{1}{\frac{d_{\mathrm{p}}}{2}\sqrt{\Psi_{\mathrm{p}}} \coth \frac{d_{\mathrm{p}}}{2}\sqrt{\Psi_{\mathrm{p}}} - 1} \right)} \tag{2.37}$$

$$\Psi_{\mathrm{f}} = \frac{k_{\mathrm{fm,p}}[M]_{\mathrm{p}}}{V_{\mathrm{p}} N_{\mathrm{A}} D_{\mathrm{R,p}}} \tag{2.38}$$

$$\Psi_{\mathrm{p}} = \frac{k_{\mathrm{p,p}}[M]_{\mathrm{p}} + 2k_{\mathrm{t,p}} \dfrac{n-1}{V_{\mathrm{p}} N_{\mathrm{A}}}}{D_{\mathrm{R,p}}} \tag{2.39}$$

Here m_{R} is the mass of the radical, s_{s} is the thickness of the surfactant layer, $D_{\mathrm{R,s}}$ is the diffusion coefficient of the radicals in the surfactant layer, and n is the number of radicals in the particle.

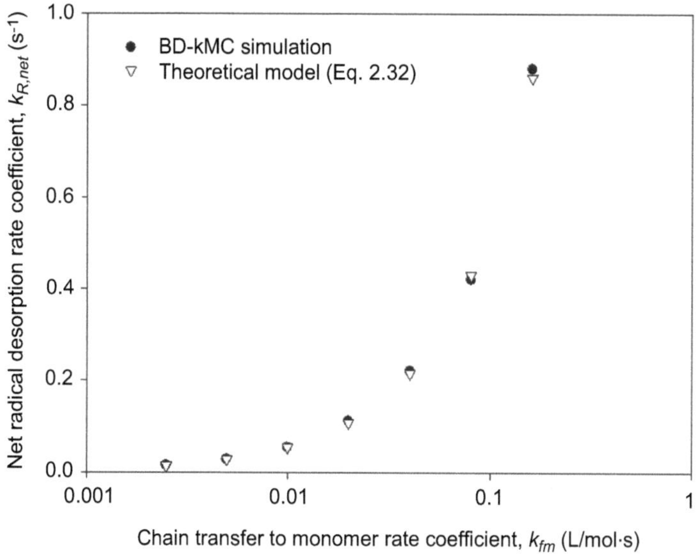

Figure 2.8 Net desorption rate coefficient (K_0) for different values of the chain transfer to monomer rate coefficient. $D_{\mathrm{p}} = 10^{-11}$ m^2/s, $d_{\mathrm{p}} = 100$ nm, $[M]_{\mathrm{p}} = 5.5$ mol/L, $k_{\mathrm{p}} = 240$ L/mol·s.

The net desorption rate coefficient in this case is given by

$$k_{\mathrm{R,net}} = \Psi_{\mathrm{s}} \frac{\Psi_{\mathrm{f}} N_{\mathrm{A}}}{\Psi_{\mathrm{p}} m} \tag{2.40}$$

Thickett and Gilbert [44] also presented a modified Smoluchowski equation for the diffusion of radicals through the surfactant layer. Based on this equation and Nomura's and Hansen and Ugelstad's models for desorption, they obtained the following expression for the net desorption rate coefficient (without reabsorption):

$$k_{\mathrm{R,net}} = k_{\mathrm{fm}}[M]_{\mathrm{p}} \frac{k_{\mathrm{R,eq}}}{k_{\mathrm{R,eq}} + k_{\mathrm{p}}[M]_{\mathrm{p}}} \tag{2.41}$$

$$k_{\mathrm{R,eq}} = \frac{12 D_{\mathrm{R,s}} D_{\mathrm{R,c}}}{d_{\mathrm{p}}^2} \frac{[M]_{\mathrm{c}}}{[M]_{\mathrm{p}}} \left(\frac{d_{\mathrm{p}} + 2s_{\mathrm{s}}}{d_{\mathrm{p}} D_{\mathrm{R,s}} + 2s_{\mathrm{s}} D_{\mathrm{R,c}}} \right) \tag{2.42}$$

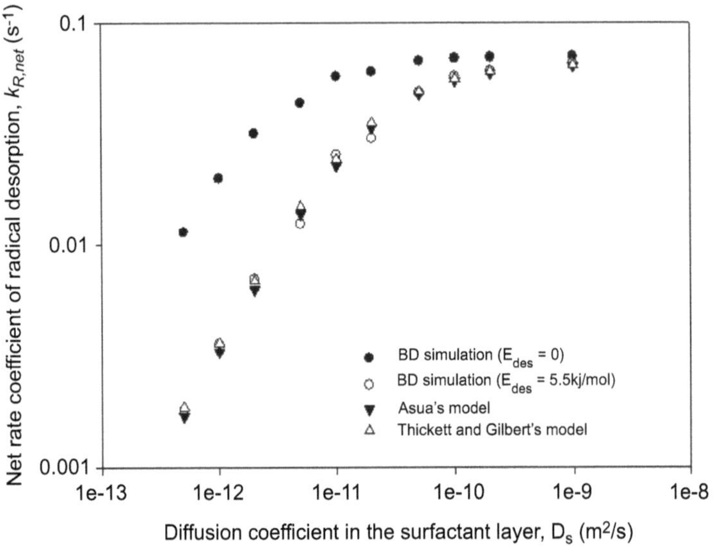

Figure 2.9 Effect of the diffusion coefficient of the surfactant layer on the net rate coefficient of radical desorption. Comparison between Asua's (Eq. 2.40) and Thickett and Gilbert's (Eq. 2.41) models, and BD simulation results.

In order to compare these two models (Eqs. 2.40 and 2.41) with the BD simulation method, the conditions presented by Asua [39] were used to determine the net desorption rate coefficient. The results are presented in Fig. 2.9. Initially, since the energy barrier for desorption in this example is unknown, the BD simulation was performed in the absence of energy barriers to obtain the rate coefficient of simple desorption. Clearly, if no energy barrier is present, the rates of radical desorption are much higher than under equilibrium conditions. However, by setting the energy barrier to 5.5 kJ/mol, the agreement between both models and the simulation results is very good. It can also be concluded that there are no significant differences in the results obtained with the models of Asua as well as Thickett and Gilbert when no reabsorption is considered. This example shows that the idea of an energy barrier for desorption can be safely used to describe the kinetics of molecular desorption in heterophase polymerization instead of macroscopic mass transfer coefficients through stationary films between the particles and the continuous phase.

2.2.3 Equilibrium

In the previous section, the concept of equilibrium desorption was introduced, taking into account the differences in solubility of a certain molecular species between the polymer particles and the continuous phase. This definition involves reaching a steady state in the molecular concentrations in both the particle and the continuous phase, and the establishment of concentration profiles on both sides of the interface (boundary layers) as depicted in Fig. 2.10.

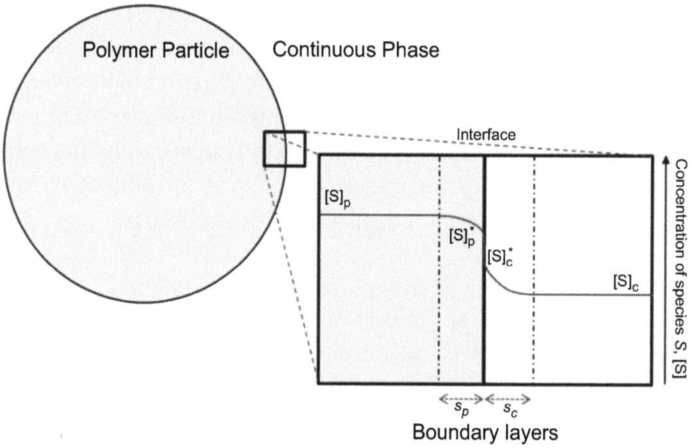

Figure 2.10 Equilibrium distribution of molecules of species *S* at the polymer particle–continuous phase interface.

Such steady state is denoted as ***chemical equilibrium***, or just equilibrium for simplicity (since no other types of equilibrium are considered in this discussion). At equilibrium, concentration profiles remain constant, from a macroscopic point of view. This can be seen as the result of the equilibration of chemical potentials on either side of the interface. Figure 2.10 sketches just one case for a species present in the particles at higher concentration than in the continuous phase; the opposite situation is possible as well (for instance radicals generated in the aqueous phase). At the molecular scale, molecules are in permanent transfer across the interface, being either absorbed or desorbed by the polymer particles. However, at equilibrium, the absorption and desorption rates are exactly the same. Since the environment provided by both the polymer particle and the continuous phase is different, the corresponding solubility of any molecule is expected to be also different.

Chemical equilibrium (for all components): A condition of a multiphase system in which the concentration of the molecular species present in each phase has no further tendency to change with time. In this situation, the chemical potential of each species is the same for all phases.

Particularly at the interface, the concentration of a molecular species is also expected to be different at both sides of it. Given that the molecular environment around the interface is fuzzy and chaotic, a certain transition is observed between the interface and the bulk of each phase. Such transition region is known as the ***boundary layer***. There is a boundary layer at each side of the interface.

Boundary layer: Region in the proximity of a wall or a boundary, where the behavior of a system is significantly different from the bulk behavior on either side. The boundary layers are heterogeneous, presenting concentration gradients in the direction normal to the interface.

The model of equilibrium desorption developed by Nomura et al. [37, 45–47] has been widely used to describe desorption in heterophase polymerization. However, some underlying assumptions involved in the derivation of this model may not be fulfilled by scarce molecular species, such as active molecules in chain-growth polymerization reactions.

The molecular flux of species S from the particles to the interface ($J_{S,p}$) is

$$J_{S,p} = K_{mp}\left([S]_p - [S]_p^*\right) = \frac{D_{S,p}}{s_p}\left([S]_p - [S]_p^*\right) \tag{2.43}$$

where K_{mp} is the mass transfer coefficient in the boundary layer of the polymer phase (of thickness s_p), $[S]_p$ is the concentration of molecules of species S in the bulk of the polymer particle, $[S]_p^*$ is the concentration of molecules of species S at the polymer particle side of the interface, and $D_{S,p}$ is the diffusion coefficient of species S inside the polymer particle.

Similarly, the molecular flux of species S from the interface to the bulk of the continuous phase ($J_{S,c}$) is given by

$$J_{S,c} = K_{mc}\left([S]_c^* - [S]_c\right) = \frac{D_{S,c}}{s_c}\left([S]_c^* - [S]_c\right) \tag{2.44}$$

where K_{mc} is the mass transfer coefficient in the boundary layer of the continuous phase (of thickness s_c), $[S]_c$ is the concentration of molecular species S in the bulk of the continuous phase, $[S]_c^*$ is the concentration of species S at the continuous-phase side of the interface, and $D_{S,c}$ is the diffusion coefficient of species S in the continuous phase.

It is possible estimating the thickness of the boundary layer in the continuous phase for the specific geometry and flow regime of the system from semi-empirical relations for mass transfer in spheres [48]:

$$\frac{s_c}{s_p} = \frac{1}{Sh} = \frac{1}{2 + 0.6Re^{1/2}Sc^{1/3}} \tag{2.45}$$

where Sh, Re, and Sc are the dimensionless numbers of Sherwood, Reynolds, and Schmidt, respectively. Assuming a stagnant boundary layer ($Re = 0$),

$$s_c = \frac{s_p}{2} \tag{2.46}$$

Thus, at equilibrium where $J_{S,p} = J_{S,c}$ and $[S]_p^* = m_S[S]_c^*$,

$$[S]_p = m_S[S]_c^* + 2\frac{D_{S,c}}{D_{S,p}}\left([S]_c - [S]_c^*\right) \tag{2.47}$$

where m_S is a partition coefficient of species S between the particle phase and the continuous phase. If $m_S \rightarrow 0$, then the molecules of S prefer staying in the continuous phase, resulting in $[S]_p^* \rightarrow 0$, $[S]_c^* \rightarrow [S]_c$, and

$$[S]_p \rightarrow 2\frac{D_{S,c}}{D_{S,p}}\left([S]_c - [S]_c^*\right) \rightarrow 0 \tag{2.48}$$

On the opposite side, if $m_S \rightarrow \infty$, $[S]_c^* \rightarrow 0$, and $[S]_p^* \rightarrow [S]_p$.

The time required by a system to reach equilibrium versus the characteristic time at which the conditions in the system are changing, with respect to concentrations, is crucial. During heterophase polymerization, the latter one is usually shorter, i.e., the conditions are changing before a new equilibrium can be established. Thus, in a polymerizing system (or reacting system in general), equilibrium conditions are virtually impossible to achieve, unless all reactions stop. However, the system will be permanently trying to reach the equilibrium state.

The theory of mass transfer by Fickian diffusion through stationary films presented above requires the presence of a large number of molecules of species S in the system in order to be valid. Thus, it may not be suitable for describing the behavior of scarce species, such as radicals, catalysts, and other active molecules. In the case of transfer of molecules at very low concentration, it is not possible to define a steady concentration profile around a single particle. The molecular motion of individual species results in large changes in molecular concentration, and thus chemical equilibrium is not possible. Nomura and Harada [37] were aware of this situation when they proposed their model of equilibrium radical desorption for emulsion polymerization. Although for a large number of identical particles, a probability distribution profile inside and outside the particles can be obtained, which may be interpreted as the average concentration profile for a representative particle, this probability distribution does not lead to the formation of a stationary film around the particles, as depicted in Fig. 2.10, because it would require the presence of a concentration gradient around each individual particle.

Another evidence of the inadequacy of the boundary layer model for polymer particles was presented by Grady [49]. He calculated the thickness of the stationary layer around polymer particles and found that when the volume fraction of particles increases, the boundary layers of neighboring particles begin to overlap, and around a volume fraction of 15%, the mass transfer boundary layer of the particles occupies the whole continuous phase volume. This implies that above this critical volume fraction, there is no "bulk" continuous phase. Overlapping of boundary layers for different particles also results in the coexistence of different concentration values in the overlapping regions.

It is, therefore, important to understand that for scarce species in heterophase polymerization, chemical equilibrium or steady-state concentration profiles around the particles cannot be reached. However, when considering the probability distribution of scarce molecules around all identical polymer particles in the system, a steady-state probability distribution profile is obtained, which behaves equivalently to the concentration profile of abundant species.

Another particular case occurs when abundant species are considered, and they are good solvents for the polymer particle. That is, when $m_S > 1$. The absorption of abundant solvent molecules by polymer particles is usually denoted as ***swelling***. On the other hand, desorption of solvent molecules from the particles is *deswelling*. The equilibrium between these two processes is known as *equilibrium swelling*. In most heterophase polymerizations, the monomeric species are good solvents for the polymer, and then the term *monomer swelling* is used.

Swelling: Uptake of a solvent by polymer, either bulk or in the form of particles, leading to an increase in the size of the polymer phase.

Swelling takes place as a result of motion through the continuous phase of individual molecules or molecular aggregates (of any size) to the surface of polymer particles and their subsequent absorption when the interfacial energy barrier is surpassed. Particles in close contact with swelling agent drops can also enter the drops driven by the equilibration of the chemical potential [11], also resulting in a swollen polymer particle. Under special conditions of heterophase polymerization, polymer particles can grow by swelling to sizes much larger than their original unswollen sizes [50]. However, as particles grow by swelling, the surface area increases as well, and thus swelling reduces the surfactant density at the interface and consequently the particles will lose stability.

From a thermodynamic point of view, the driving force for the swelling of a polymer by a solvent is the free energy of polymer–solvent mixing, which has both entropic and enthalpic components. Even if the monomer and the polymer are miscible in all proportions in bulk, only a limited amount of monomer can enter polymer particles from the continuous phase. Each particle can swell only to the extent where the free energy of mixing and the surface energy change on swelling exactly compensate each other and there is a well-defined swelling equilibrium. When emulsifiers are used, the decreased interfacial tension allows a substantial increase in the degree of swelling. For crosslinked particles, the swelling capacity is strongly reduced and depends inversely on the degree of crosslinking [51].

At equilibrium, the change in free energy is zero, and therefore

$$\Delta G_{\text{mix}} = \Delta H_{\text{mix}} - T\Delta S_{\text{mix}} = 0 \tag{2.49}$$

where ΔG_{mix} represents Gibbs free energy change during swelling, ΔH_{mix} is the enthalpy change, and ΔS_{mix} is the entropy change at temperature T. Using the **Flory–Huggins** approach [52, 53] to describe the enthalpy and entropy of mixing and including an additional term for the change in surface free energy of the particle, the **Morton–Kaizermann–Altier (MKA) equation** is obtained [54]:

$$\ln\varphi_{\text{m}} = \left(\frac{1}{j} - 1\right)\varphi_{\text{p}} - \chi\varphi_{\text{p}}^2 - b\frac{\upsilon_{\text{m}}\rho_{\text{p}}}{M_{\text{C}}}\left(\varphi_{\text{p}}^{1/2} - \frac{\varphi_{\text{p}}}{2}\right) - \frac{2\upsilon_{\text{m}}\sigma}{RTr_{\text{p}}} \tag{2.50}$$

where φ_{p} and φ_{m} are volume fractions of polymer and solvent (or monomer) inside the particle, χ is the Flory–Huggins interaction parameter between polymer and solvent and is a function of concentration, υ_{m} is the partial molar volume of the solvent, σ is the interfacial tension between the particle and the surrounding medium, r_0 is the unswollen particle radius, r_{p} is the swollen particle radius, b is a crosslinking parameter (b = 0 uncrosslinked, b = 1 crosslinked), ρ_{p} is the density of the polymer, and M_{C} is the average molecular mass between crosslinks.

The last term in Eq. 2.50 represents the resistance to the creation of new surface area upon swelling due to increasing the interfacial free energy. There is a pressure difference between the interior of the particle and the continuous phase (***Laplace pressure***), which is inversely proportional to the radius of curvature while proportional to the interfacial tension of the particle. For this reason, larger particles swell to a larger extent and also swell initially faster [55]. Please notice that the particle interfacial tension may increase as the particle size grows (assuming a constant amount of surfactant in the system), complicating the overall effect of particle size on the Laplace pressure.

Laplace pressure: It is the pressure difference across a curved interface caused by the capillary effect of interfacial tension.

The Laplace term also indicates that larger objects in coexistence with smaller ones will grow in size at the expense of the smaller objects, which have a tendency to dissolve. This effect is known in colloid science as ***Ostwald ripening***.

Ostwald ripening: Experimental effect observed in colloidal systems with a non-monodisperse size distribution describing the growth of larger objects on the expense of smaller ones, caused by the Laplace pressure.

When the MKA approach is used, it is possible to predict the equilibrium concentration of a solvent inside a polymer particle as a function of the average degree of polymerization, the partial molar volume of the solvent, the average radius of the polymer particles, the polymer-continuous phase interfacial tension, the solvent–polymer interaction parameter, and the temperature of the system. There are, however, two important drawbacks in this approach. The first drawback is that according to the MKA equation, the equilibrium concentration of solvent in the particles does not depend on the concentration of solvent in the continuous phase. In order to overcome this difficulty, the MKA equation is used only if the solvent is present above its saturation concentration in the continuous phase (this means that tiny drops of solvents exist in the continuous phase), whereas for lower concentrations, the following partition coefficient is defined:

$$K_{\mathrm{S,pc}} = \frac{[S]_{\mathrm{p}}^{*\mathrm{sat}}}{[S]_{\mathrm{c}}^{\mathrm{sat}}} \tag{2.51}$$

where $[S]_p^{*sat}$ is the equilibrium concentration of species S (solvent in this case) inside the particles obtained from the MKA equation, and $[S]_c^{sat}$ is the saturation concentration of the solvent in the continuous phase. Therefore, the equilibrium concentration of solvent inside the particles can be obtained from the concentration of solvent in the continuous phase below saturation:

$$[S]_p^* = K_{S,pc}[S]_c \tag{2.52}$$

As an alternative, Vanzo et al. [56] compared the free energy of monomer–polymer mixing inside the particles (MKA equation) to the free energy of the monomer in the continuous phase and found the following relationship, usually known as the **Vanzo equation**:

$$\ln\left(\frac{[S]_c}{[S]_c^{sat}}\right) = \ln(1-\varphi_p) + \varphi_p + \chi\varphi_p^2 + \frac{2\upsilon_m\sigma}{RTr_p} \tag{2.53}$$

The second drawback, and perhaps the most important, is that the MKA equation, as well as the Vanzo equation, cannot always be safely used to predict equilibrium concentrations inside polymer particles because (i) some of the parameters included in the model are not constant (such as the interaction parameter and the interfacial tension), and (ii) not all thermodynamic effects have been considered (for example the swelling pressure effect [15, 57]).

The MKA equation is actually based on the condition for polymer solubility in a pure solvent (no third phase present); the correction for a three-phase system with a different continuous environment can be made by introducing an interfacial tension (assumed to stay unchanged). Moreover, it holds only for equilibrium conditions. This is why the MKA equation is not really applicable for heterophase polymerization. There are even experimental results proving that diffusion of dissolved monomer molecules is not enough to reach equilibrium swelling [11]. It actually takes quite a long period of time to reach swelling equilibrium even under favored conditions.

The experimental values of swelling are much lower than described by the classical MKA equation. Antonietti et al. [58] observed a pronounced dependence of the swelling ratio on particle size. In order to explain this phenomenon, the authors presented a modified description that considered size-relevant effects using an additional swelling pressure term $\Delta\Pi$ (in analogy to swelling of macroscopic gels), which increases with the curvature of the particle size and counteracts swelling. Thus, the MKA equation cannot be trusted to accurately predict the concentration of monomer inside a particle as a function of the particle radius. The equation for swelling equilibrium including size-dependent corrections, and considering that χ also depends on the polymer volume fraction, presented by Kaspar [59], is the following:

$$\ln \varphi_{\mathrm{m}} = \left(\frac{1}{j} - 1\right)\varphi_{\mathrm{p}} - \chi\varphi_{\mathrm{p}}^{2} - b\frac{\upsilon_{\mathrm{m}}\rho_{\mathrm{p}}}{M_{\mathrm{C}}}\left(\varphi_{\mathrm{p}}^{1/2} - \frac{\varphi_{\mathrm{p}}}{2}\right) - \frac{\upsilon_{\mathrm{m}}}{RT}\left(\frac{2\sigma}{r_{\mathrm{p}}} + \Delta\Pi\right) \tag{2.54}$$

Alternatively, swelling has also been described using empirical equations relating the solvent concentration in both phases. One of the most important examples of the use of empirical expressions in heterophase polymerization has been presented by Ballard et al. [60]. The general empirical expression proposed is the following:

$$\frac{[S]_{\mathrm{c}}}{[S]_{\mathrm{c}}^{\mathrm{sat}}} = \left(\frac{[S]_{\mathrm{p}}}{[S]_{\mathrm{p}}^{\mathrm{sat}}}\right)^{y} \tag{2.55}$$

where the exponent y is an empirical parameter determined for each solvent–polymer system. For example, for the methyl methacrylate/poly(methyl methacrylate) system, a value of $y = 0.6$ was obtained from experimental data. A similar value has also been obtained for the vinyl acetate/poly(vinyl acetate) and the styrene/polystyrene systems [8].

A better alternative is the use of molecular simulation methods for describing swelling. From a molecular point of view, monomer or solvent molecules (and aggregates) are continuously absorbed by or desorbed from the polymer particles. If the rate of molecular absorption per particle is larger than the rate of desorption, swelling takes place; otherwise, deswelling occurs. Thus, swelling (or deswelling) is simply the result of the balance between the absorption and desorption of molecules (either individually or in the form of molecular aggregates). At equilibrium, the rates of desorption and absorption are exactly the same, and thus there is no net change in the number of molecules inside the particles. Since the rates of desorption and absorption depend on the diffusion coefficients inside the particles and in the continuous phase, their

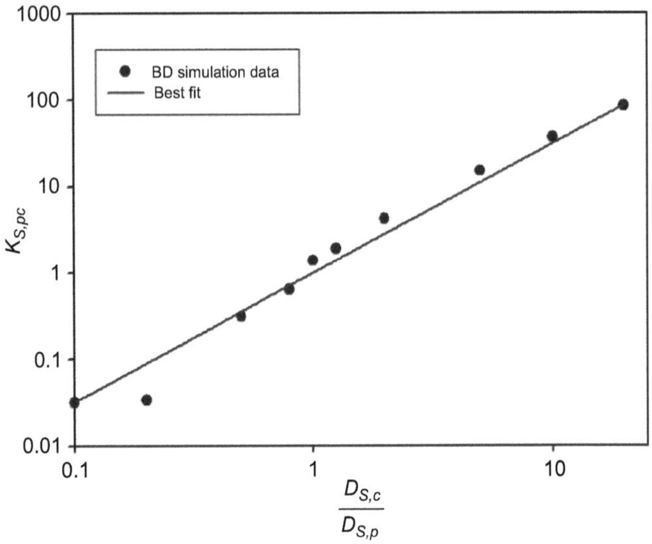

Figure 2.11 Equilibrium partition coefficients as a function of the ratio between diffusion coefficients, in the absence of energy barriers for absorption or desorption and at a constant volume fraction of spherical polymer particles of 10%. Solid line: Best fit using Eq. 2.56.

values are crucial for the equilibrium distribution of the molecules in the system. The effect of the diffusion coefficients on the partition coefficient (at equilibrium) in dispersions of polymer particles has been investigated using BD simulation [29]. It was observed that the molecules tend to accumulate in the phase where they present the lowest diffusion coefficient. This is reasonable because the fastest diffusing molecules may cross the interface more frequently causing an accumulation in the phase of lowest mobility. The effect of the ratio of diffusion coefficients on the equilibrium partition coefficient of the system can be expressed approximately using the following empirical expression:

$$K_{S,pc} \approx \left(\frac{D_{S,c}}{D_{S,p}}\right)^{\beta} \exp\left(-\frac{E_a^{abs} - E_a^{des}}{3RT}\right) \tag{2.56}$$

where β is positive and its value depends on the particular conditions such as the volume fraction of the particles in the dispersion, $E_a^{abs} \geq 0$ is the activation energy of the absorption process, $E_a^{des} \geq 0$ is the activation energy for the desorption process. If there is a net energy barrier for absorption, the molecules accumulate in the continuous phase. Similarly, if there is a net energy barrier for desorption, the molecules will concentrate inside the particles. If both activation energies are identical, the partition coefficient will only depend on the relative diffusivities of the molecule at each phase. Figure 2.11 illustrates the behavior of equilibrium partition coefficients determined by BD simulation in the absence of energy barriers, as a function of the ratio of diffusion coefficients, for constant polymer volume fraction.

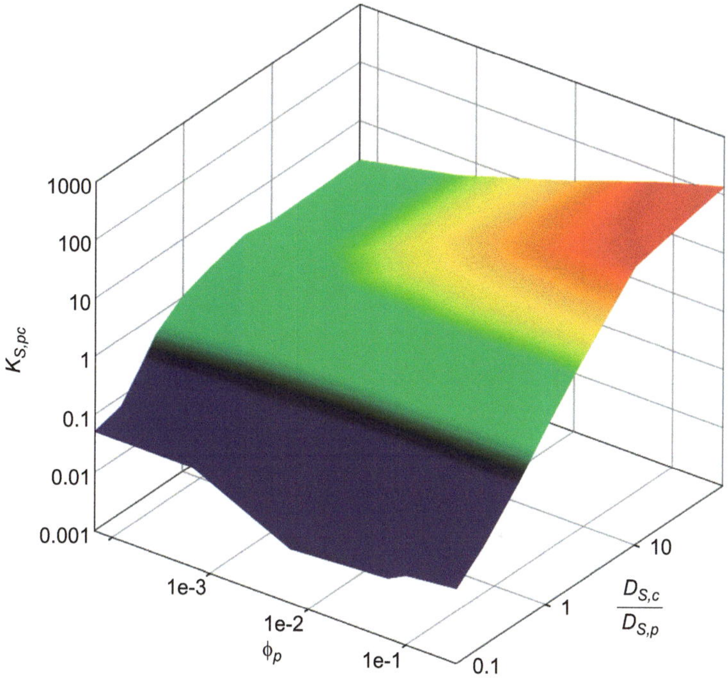

Figure 2.12 Equilibrium partition coefficients as a function of the ratio between diffusion coefficients and the volume fraction for monodisperse spherical particles in the absence of energy barriers for absorption or desorption.

In the case of irreversible absorption and desorption of molecules (no re-desorption or reabsorption is possible), a value of $\beta = 1$ would be expected. However, for the conditions considered in Fig. 2.11, it was found that $\beta \approx 1.6$. This deviation is caused by the fact that a single molecule at the interface can be absorbed and desorbed several times (by the same particle or by neighboring particles) before diffusing to the center of the particle or to the bulk of the continuous phase, which is not considered by the models of irreversible phase transfer, giving rise to different effective diffusion paths for phase transfer.

The simultaneous effect of volume fraction and diffusion coefficients on the equilibrium partition coefficient for monodisperse spherical particles is presented in Fig. 2.12. It can be noticed that the value of β (steepness of the slope in the y–z plane) decreases as the volume fraction of particles decreases. In the limit of infinitely diluted dispersions, β tends to 1. The functional dependence of β on the volume fraction of particles in the dispersion is highly nonlinear.

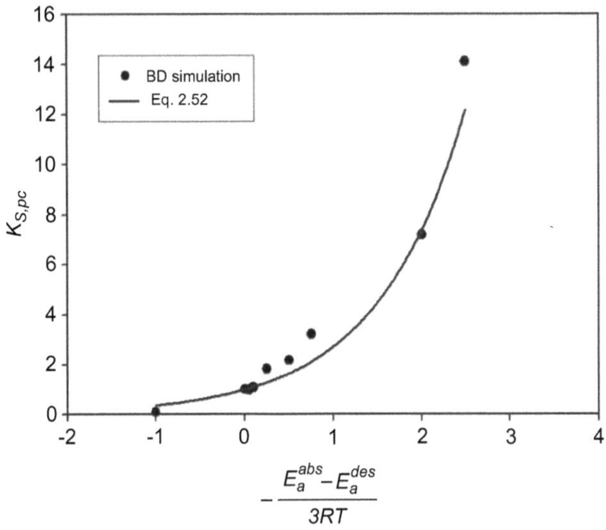

Figure 2.13 Effect of activation energies for phase transfer on the equilibrium distribution coefficient.

When energy barriers for absorption and desorption are considered, the energy difference will influence the equilibrium partition coefficient. Figure 2.13 shows the results of BD simulation for the simultaneous absorption and desorption when different activation energies are present, assuming that $D_{S,c} = D_{S,p}$. These results can be well described using Eq. 2.56.

Molecular simulation techniques can become very good alternatives for the prediction of non-equilibrium monomer concentration inside polymer particles in heterophase polymerization and could be very useful in systems where the swelling equilibrium assumption may not be valid, for example, in the widely used monomer-starved semi-batch processes in the industry. This approach could be further used to investigate the non-equilibrium uptake of any other type of molecule, such as primary radicals, oligomers, solvents, or any other compound that might be useful to modify colloidal particles.

2.3 Kinetics of Polymerization

In general, the mechanisms of reaction involved in heterophase polymerization are identical to those of a single-phase homogeneous polymerization. The difference in the overall kinetics of polymerization between homogeneous and heterogeneous polymerization systems is the result of the particular partition or distribution of the reacting species between the different phases coexisting in the system.

Thus, for heterophase polymerization systems, the kinetics of polymerization is determined not only by the rate of the specific reactions involved, but also by the transfer and balance of chemical constituents between the phases. Since the latter was already discussed in the previous section, only some additional remarks regarding the rate of polymerization will be presented in this section. Particularly, we will generally refer in this section to homopolymerization, since copolymerization involves an additional degree of complexity regarding the differences in distribution equilibria for the different monomers. In order to minimize these effects during copolymerization, semi-batch procedures with monomer feed under starved conditions are usually applied instead of batch-flooded polymerization. Such monomer-starved conditions allow instantaneous consumption of the monomers as they are fed into the reactor, preventing their accumulation.

Determining the kinetics of polymerization in any system requires carefully defining the *rate of polymerization* (r_p). While this might seem, at this point, quite straightforward, the fuzzy definition of a polymer and the random nature of polymer chains complicate such definition. For example, individual monomer units can usually be treated as polymer chains of chain length 1. Under such assumption, pure monomers would already be considered polymers. The same occurs with oligomers. A dimer can be considered a polymer chain, but the physical properties of a dimer are certainly different to both the monomer and the polymer. So where is the dividing line for considering that a certain chemical reaction is a polymerization reaction? Can a fixed minimum chain length limit be used for defining polymerization? Since these are still unresolved issues, we will just arbitrarily consider as a ***polymerization reaction*** any reaction that increases the chain length of a polymer, independently of the final chain length obtained. Please notice that in this sense, increasing the chain length of a polymer always involves the decrease in the number of polymer chains (including monomers).

Polymerization reaction: Any reaction resulting in an increase in the chain length of a polymer.

Depolymerization reaction: Any reaction resulting in a decrease in the chain length of a polymer.

Following the same line of thought, a ***depolymerization reaction*** is any reaction that results in a decrease in the chain length of a polymer, resulting in the appearance of smaller polymer chains in the system (including monomers). Depolymerization typically happens for chain polymerizations at elevated temperature, above the so-called *ceiling temperature,* which is determined when the Gibbs free energy change during polymerization (or depolymerization) becomes zero ($\Delta G = 0$).

In some cases, such as in chain-growth polymerization for example, different parallel reactions take place, which do not affect the chain length of the polymers involved (for example, the decomposition and recombination of initiator molecules, as well as chain transfer to monomer). Thus, these reactions are not polymerization (or depolymerization) reactions. However, since they have an effect on the availability of the reacting species that lead to polymer growth, they play a key role in the kinetics of polymerization.

The next conceptual challenge is quantifying the amount of polymer formed in order to determine a rate of polymerization. As it was previously mentioned, polymerization involves the combination of polymer chains, and thus the number of polymer chains in the system decreases as polymerization proceeds. In fact, the theoretical limit of polymerization is reaching a single polymer chain containing all monomer units initially available in the system. Thus, the rate of polymerization cannot be measured in terms of the increase in the number or concentration of polymer chains present in the system, as it is customary in conventional chemical reactions. Therefore, a different approach is needed.

The ***rate of polymerization*** is then defined in terms of the ***net*** number of individual polymerization events that are taking place per unit time per unit volume. Such *net* number of polymerization events will be determined as the total number of polymerization events taking place per unit time per unit volume, minus the total number of depolymerization events taking place in the same period and same volume. Thus, the rate of polymerization will measure the frequency of chain growth in a system of unit volume.

Polymerization rate: Net number of polymerization events taking place per unit time per unit volume (frequency of chain growth).

The relative advance of the polymerization reaction is determined by the ***conversion*** (X_P). The conversion indicates the progress of the reaction. A conversion value of 0 indicates that no polymerization events have taken place in the system. A conversion value of 1 indicates that all monomer units available became part of the polymer chains (molecules with chain length greater than or equal to 2). Thus, the rate of polymerization indicates how fast the conversion is increasing during the process.

Conversion: Ratio of the number of monomer units contributing to an increase in the polymer chain length to the initial number of monomers available.

The conversion of the polymerization reaction is also closely related (but not equivalent) to the ***degree of polymerization*** (DP). The degree of polymerization indicates the average number of monomer units incorporated to each polymer chain in the system (total number of polymerized monomer units divided by the total number of polymer chains). Note that for a single chain, the degree of polymerization corresponds to its chain length. As the conversion of the polymerization reaction increases, the degree of polymerization is also expected to increase. However, the relation between conversion and the degree of polymerization is not linear because the total number of chains (including monomer units) decreases during polymerization. Please notice that considering monomer units as polymer chains of chain length 1 has a significant effect on the determination of the degree of polymerization. Therefore, it is important to specify whether or not the monomer units are considered polymer chains in the determination of the degree of polymerization, and other properties of the polymer chain distribution.

Degree of polymerization: Average number of monomer units per polymer chain in the system considered.

Since each polymer chain can grow at a different rate, at any instant the chain length of the polymers can be different. Furthermore, the molecular mass of each polymer chain can be different (even when the chain length is the same, if different monomer units are present in the system as in the case of co-, ter-, and multi-polymerization). Since it is not possible to determine the individual properties of all polymer chains in a system, the polymer chain length and molecular mass are treated as random variables (using statistical methods). This way, both the polymer chain length and the polymer molecular mass will be described by a particular statistical distribution. In principle, those distributions are discrete, because the sets of possible values taken by the chain length or the molecular mass are finite. However, as the polymer chains grow larger, the set of possible values increases significantly allowing treating the distributions as continuous random variables. In order not to lose generality, a discrete formulation of the probability distributions for the chain length and molecular mass of the polymers will be considered.

The relative frequency of chains with a certain chain length l in the system will be denoted as the probability $p_{\mathrm{L}}(l)$. Similarly, the relative frequency of polymers with a molecular mass m is denoted as $p_{\mathrm{M}}(m)$. The set of values of the probability for each chain length or molecular mass represents the *polymer chain probability distribution*. These probability sets can then be used for estimating some representative properties of the polymer system. Those properties are commonly denoted as *moments of the distribution* and are defined as follows:

$$M_n(x) = E(x^n) = \sum_{i=1}^{N} x_i^n p_x(x_i) \tag{2.57}$$

where M_n represents the n-th moment of the distribution, x is the random variable, either the chain length (l) or the molecular mass (m), E represents the *expectancy* or expected value operator, i is a chain identification number, N is the total number of polymer chains in the system, and x_i is the corresponding value of the random variable for the i-th polymer chain (l_i for chain lengths, m_i for molecular masses).

The average value of a distribution corresponds to its first moment. For example, the average chain length (corresponding to the degree of polymerization) will be

$$\mathrm{DP} = M_1(l) = E(l) = \sum_{i=1}^{N} l_i p_{\mathrm{L}}(l_i) \tag{2.58}$$

On the other hand, the average molecular mass of the polymer (also denoted as number-average molecular mass) will be

$$\bar{M}_n = M_1(m) = E(m) = \sum_{i=1}^{N} m_i p_{\mathrm{M}}(m_i) \tag{2.59}$$

Other commonly used properties of the polymer chain distribution include the mass-average molecular mass (M_{w}):

$$\bar{M}_{\mathrm{w}} = \frac{M_2(m)}{M_1(m)} = \frac{E(m^2)}{E(m)} = \frac{\sum_{i=1}^{N} m_i^2 p_{\mathrm{M}}(m_i)}{\sum_{i=1}^{N} m_i p_{\mathrm{M}}(m_i)} \tag{2.60}$$

the Z-average molecular mass (M_{Z}):

$$\bar{M}_{\mathrm{Z}} = \frac{M_3(m)}{M_2(m)} = \frac{E(m^3)}{E(m^2)} = \frac{\sum_{i=1}^{N} m_i^3 p_{\mathrm{M}}(m_i)}{\sum_{i=1}^{N} m_i^2 p_{\mathrm{M}}(m_i)} \tag{2.61}$$

and the molecular mass *dispersity* (Δ_{M}):

$$\Delta_{\mathrm{M}} = \frac{\bar{M}_{\mathrm{w}}}{\bar{M}_n} = \frac{M_2(m)}{M_1^2(m)} = \frac{E(m^2)}{E^2(m)} = \frac{\sum_{i=1}^{N} m_i^2 p_{\mathrm{M}}(m_i)}{\left(\sum_{i=1}^{N} m_i p_{\mathrm{M}}(m_i)\right)^2} \quad (2.62)$$

The dispersity is a property related to the broadness of the MMD. It can be shown that the dispersity can equivalently be expressed as

$$\Delta_{\mathrm{M}} = 1 + \left(\frac{\sigma_{\mathrm{M}}}{\bar{M}_n}\right)^2 \quad (2.63)$$

where σ_{M} is the standard deviation in the MMD of the polymer chains. Clearly, Δ_{M} has a minimum value of 1 when all the chains have the same molecular mass. An equivalent definition for the dispersity of chain lengths also applies.

For all the previous calculations, it is important to clearly specify what is the minimum chain length required to be considered part of the N polymer chains. And the answer will be closely related to the experimental methods used to determine those properties. If, for example, the properties are measured for a solid polymer after filtration and drying, then monomer and oligomers in the liquid state should not be considered. This clarity is important when comparing polymerization models with experimental results.

Now, after introducing these basic polymerization concepts, we are ready to consider specific polymerization systems. Following Flory's classification of polymerization processes [61], step-growth and chain-growth polymerization kinetics will be discussed.

2.3.1 Step-Growth Polymerization Kinetics

In step-growth polymerization, the polymer chain grows with each reaction between the reactive functional groups at the end of the chains. A very general representation of the polymerization reaction in step-growth polymerization systems is the following:

$$P_i + P_j \rightarrow P_{i+j} + (R) \quad (2.64)$$

where two polymer chains (P_i and P_j) react to form a larger polymer chain (P_{i+j}) and optionally a residual molecule (R). A more detailed representation of the molecular structure of the chains would be

$$M_iE_a + M_jE_b \rightarrow M_{i+j}E_{a+b-2} + (R) \qquad (2.65)$$

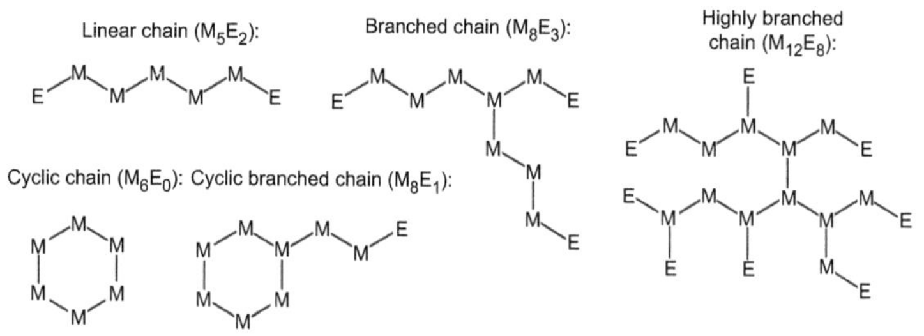

Figure 2.14 Selected examples of macromolecular chains containing different numbers of monomeric building blocks (*M*) and chain end fragments (*E*).

The first reacting polymer chain (P_i) is represented by i monomeric units (M) and a different chain ends (E). A linear chain has only two chain ends. A branched chain presents more than two different chain ends. Cyclic or closed chains will present less than two chain ends. This is graphically illustrated in Fig. 2.14. This notation, however, does not take into account the particular conformation of the chain. Furthermore, chain ends (E) can be either complementary functional groups (F) capable of reacting between them or inert groups (I). The second reacting polymer chain has j monomeric units and b chain ends. The polymer chain obtained after the reaction 2.65 has now a chain length of i+j and a+b-2 chain ends (either functional or inert). This means that all the monomeric building blocks of the chains are added in the resulting chain, but the number of chain ends is reduced by 2 (one functional chain end for each reacting chain). Reaction 2.65 can be more rigorously expressed as follows:

$$M_iE_{a-1}F^A + M_jE_{b-1}F^B \rightarrow M_{i+j}E_{a+b-2} + (R) \qquad (2.66)$$

In this case, functional groups F^A and F^B react forming a covalent bond connecting the chains and yielding an optional residue R. Since the presence of the residue is optional, it cannot always be used to measure the polymerization rate. On the other hand, the number of chain ends present in the system always decreases after each polymerization step. Thus, the most general expression for the instantaneous step-growth net polymerization rate (r_p) is

$$r_{\mathrm{p}} = \frac{1}{2}\left[\sum \frac{q_{\mathrm{in}}}{V}\left([E]_{\mathrm{in}} - [E]\right) - \sum \frac{q_{\mathrm{out}}}{V}\left([E]_{\mathrm{out}} - [E]\right) - \frac{d[E]}{dt}\right] \quad (2.67)$$

where $[E]$ represents the instantaneous molar concentration of total chain end groups in the reaction zone of volume V, which is assumed to be perfectly mixed, q_{in} represents the input volumetric flows to the reaction zone, with the respective total chain end groups concentration $[E]_{\mathrm{in}}$, and q_{out} represents the output volumetric flows with the respective total end groups concentration $[E]_{\mathrm{out}}$. The reaction zone can be the overall reactor volume, as long as it is homogeneous and perfectly mixed. Otherwise, the volume of reaction will represent only the volume of a dispersed phase or the volume of a local, perfectly mixed zone within the continuous phase. The volumetric input and output flows may represent overall inputs or outputs to the reactor, but also may indicate mass transfer terms between the different reaction zones inside the reactor. The term ½ emerges because two functional groups are consumed per polymerization reaction. Equation 2.67 assumes that chain end groups are only consumed by the polymerization reaction. Since only functional groups are consumed, assuming that the chains contain non-functional end groups, the decrease in total chain end groups is caused by the decrease in functional groups. Thus,

$$r_{\mathrm{p}} = \frac{1}{2}\left[\sum \frac{q_{\mathrm{in}}}{V}\left([F]_{\mathrm{in}} - [F]\right) - \sum \frac{q_{\mathrm{out}}}{V}\left([F]_{\mathrm{out}} - [F]\right) - \frac{d[F]}{dt}\right] \quad (2.68)$$

where $[F]$ is the molar concentration of functional end groups.

If the residue molecule is always generated at each polymerization step, then the rate of polymerization can alternatively be determined as

$$r_{\mathrm{p}} = \frac{d[R]}{dt} - \sum \frac{q_{\mathrm{in}}}{V}\left([R]_{\mathrm{in}} - [R]\right) + \sum \frac{q_{\mathrm{out}}}{V}\left([R]_{\mathrm{out}} - [R]\right) \quad (2.69)$$

where $[R]$ is the molar concentration of the residue molecule in the reaction zone considered. In continuous reactors, where $q = q_{\mathrm{in}} = q_{\mathrm{out}}$, the term q/V corresponds to the reciprocal of the *residence time*. Furthermore, q_{in}/V can also be interpreted as a *dilution rate*. Equation 2.69 assumes that R is not involved in any additional reaction (other than polymerization or depolymerization). Otherwise, additional reaction terms (consumption or generation) should be included.

From a mechanistic point of view, the kinetics of polymerization reactions taking place between functional groups F^A and F^B can be expressed as

$$r_{p(A,B)} = k_{p(A,B)}[F^A][F^B] \tag{2.70}$$

where $k_{p(A,B)}$ represents the reaction rate coefficient between functional groups F^A and F^B. Then, the total polymerization rate will be determined as the sum of the contributions from all possible reactions between pairs of functional groups taking place in the reaction volume:

$$r_p = \sum_{X,Y} r_{p(X,Y)} = \sum_{X,Y} k_{p(X,Y)}[F^X][F^Y] \tag{2.71}$$

Depolymerization, in this case, is the reverse reaction of Eq. 2.64. If residual molecules are involved, the rate of depolymerization (r_{dp}) will be given by

$$r_{dp} = k_{dp}[R]\sum_i (i-1)[P_i] \tag{2.72}$$

where $[P_i]$ represents the concentration of polymer chains of chain length i, and $i - 1$ indicates the number of susceptible bonds between monomeric units inside the chains. Of course, Eq. 2.72 neglects reactivity differences for the susceptible bonds in the polymer chains.

If no residual molecule is formed during polymerization, then the kinetics of depolymerization simplifies into

$$r_{dp} = k_{dp}\sum_i (i-1)[P_i] \tag{2.73}$$

where all internal bonds between monomeric units are susceptible to depolymerization, by thermal effects for example.

The local conversion (within the reaction zone considered) for step-growth polymerization reactions can be determined as follows:

$$X_p = 1 - \frac{[F_L]V}{[F_L]_0 V_0 + \sum q_{in}[F_L]_{in} - \sum q_{out}[F_L]_{out}} \tag{2.74}$$

where $[F_L]$ represents the present concentration of the limit functional groups, $[F_L]_0$ is their concentration at the beginning of the process, and V_0 is the initial volume of the reaction zone. The limit functional group represents the first reactant unit causing the polymerization to stop when it is completely consumed.

2.3.2 Chain-Growth Polymerization Kinetics

In the case of chain-growth polymerization, an active site is required for promoting bond breakage of a reaction partner (containing a susceptible bond), and subsequent bond formation between the active molecule and the reaction partner, resulting in chain growth. The susceptible bonds mainly include unsaturated bonds and rings. Even though other types of bonds may also be susceptible to reaction with the active site, they do not lead to a polymerization reaction, only to a molecular rearrangement without disappearance of molecules. The active sites may include radicals, ions, and transition metals. Polymer chains involved in chain-growth polymerization can be classified into: living polymer chains (containing at least one active site A), *macromers* (or macromonomers, containing at least one susceptible bond B), living macromers (containing at least one active site and one susceptible bond), and dead polymer chains (without active sites or susceptible bonds). The notation summarized in Table 2.1 is suggested for representing each type of polymer chains.

The simplest case of chain-growth polymerization involves a living polymer chain of chain length i (L_i) with one active site and a monomer molecule (M), as follows:

$$L_i + M \rightarrow L_{i+1} \qquad (2.75)$$

A more general version of the chain-growth polymerization reaction would involve a living polymer chain and a macromer:

$$P_iA_a + P_jB_b \rightarrow P_{i+j}A_aB_{b-1} \qquad a,b \geq 1 \qquad (2.76)$$

However, the presence of susceptible bonds in the living polymer chain, as well as the presence of active sites in the macromer, is also possible resulting in

$$P_iB_cA_a + P_jB_bA_d \rightarrow P_{i+j}B_{c+b-1}A_{a+d}, \qquad a,b \geq 1; \quad c,d \geq 0 \quad (2.77)$$

Thus, an active site of one chain reacts with a susceptible bond of the other chain resulting in chain growth and the disappearance of the susceptible bond.

Please notice that the main difference between step-growth and chain-growth polymerization reactions is that for the latter, the structural units responsible for the polymerization (active sites) do not necessarily disappear, whereas in the former both reactive functional groups are necessarily consumed. However, this does not mean that the active sites in chain-growth polymerization cannot disappear; just that it is not a mandatory condition of the polymerization reaction. They may disappear in other side reactions or in certain particular polymerization reactions (for example, in termination and deactivation reactions).

Table 2.1 Notation used for the different types of polymer chains present in chain-growth polymerization

Polymer type	Simplified notation*	Detailed notation*	Description
Dead polymer chain	D_i	P_i	Polymer chain containing *i* monomeric building blocks, without any active site or susceptible bond
Living polymer chain	$L_{i,a}$	P_iA_a	Polymer chain containing *i* monomeric building blocks and *a* active sites, but no susceptible bonds
Macromer chain	$M_{i,b}$	P_iB_b	Polymer chain containing *i* monomeric building blocks and *b* susceptible bonds, but no active sites
Living macromer chain	$LM_{i,b,m}$	$P_iB_bA_m$	Polymer chain containing *i* monomeric building blocks, *b* susceptible bonds, and *a* active sites

*Subscripts represent positive integer values. Subscripts may be omitted whenever they present a value of 1.

When the active sites of the polymerization are *radicals*, also the following reaction may take place and lead to chain growth:

$$P_iB_cA + P_jB_bA \rightarrow P_{i+j}B_{c+b} \qquad (2.78)$$

This reaction is commonly known as *termination by recombination*. However, it should be considered a polymerization reaction because it results in an increase in the chain length of one polymer and the disappearance of the other. Alternatively, the reactants in 2.78 may result in different products. For example:

$$P_iB_cA + P_jB_bA \rightarrow \begin{cases} P_iB_{c+1} + P_jB_b \\ P_iB_c + P_jB_{b+1} \end{cases} \tag{2.79}$$

This reaction is known as *termination by disproportionation*, and it is not a polymerization reaction.

The polymerization reaction (Eq. 2.76), also known as *propagation*, is characterized by the disappearance of one susceptible bond per reaction. On the other hand, recombination (Eq. 2.78) is characterized by the disappearance of two active sites. However, such effect is not exclusive of Eq. 2.78 since primary radicals (radicals without any monomer unit or susceptible bond) may also recombine without polymerization. Even though recombination has an important effect on the polymer chain distribution, it is usually not considered in the kinetics of free-radical chain-growth polymerization because it cannot easily be traced during the reaction.

Thus, in general, the rate of chain-growth polymerization will be given by

$$r_\mathrm{p} = \sum \frac{q_\mathrm{in}}{V}\left([B]_\mathrm{in} - [B]\right) - \sum \frac{q_\mathrm{out}}{V}\left([B]_\mathrm{out} - [B]\right) - \frac{d[B]}{dt} \tag{2.80}$$

where $[B]$ represents the concentration of susceptible bonds in the reaction zone of volume V.

Neglecting the termination by recombination reaction, the rate of polymerization (here the rate of chain growth) can be determined from the rate of propagation as

$$r_\mathrm{p} = k_\mathrm{p} \cdot [B] \cdot [A] \tag{2.81}$$

where $[A]$ represents the concentration of active sites present in the reaction zone. When only the monomer units contain susceptible bonds, and there is only one susceptible bond per monomer M, then $[B] = [M]$. When some monomers contain two or more susceptible bonds in their structure, the polymer chains may also contain susceptible bonds. When monomer mixtures are used, if the proportion of monomers with two or more susceptible bonds is small, $[B] \approx [M]$ will remain a good approximation.

The local conversion (within the reaction zone considered) for chain-growth polymerization processes can be expressed in terms of the consumption of susceptible bonds as follows:

$$X_{\mathrm{P}} = 1 - \frac{[B]V}{[B]_0 V_0 + \sum q_{\mathrm{in}} [B]_{\mathrm{in}} - \sum q_{\mathrm{out}} [B]_{\mathrm{out}}} \tag{2.82}$$

where $[B]$ represents the actual concentration of susceptible bonds in the system, and $[B]_0$ is their initial concentration when the volume of the reaction zone was V_0.

While the conversion depends only on the rate of polymerization r_{p}, this rate depends on the concentration of active growing chains. Now, the number of active sites will depend on the rate of initiation r_{i} (which depends on the rate of generation of active molecules), and on the rate of termination r_{t} of active chains (in the case of radicals). This is mathematical expressed as follows:

$$\frac{d[A]}{dt} = r_{\mathrm{i}} - r_{\mathrm{t}} + \sum \frac{q_{\mathrm{in}}}{V}\left([A]_{\mathrm{in}} - [A]\right) - \sum \frac{q_{\mathrm{out}}}{V}\left([A]_{\mathrm{out}} - [A]\right) \tag{2.83}$$

The rate of initiation can be expressed as

$$r_{\mathrm{i}} = k_{\mathrm{i}} \cdot [B] \cdot [L_0] \tag{2.84}$$

where k_{i} is a kinetic rate coefficient of initiation (sometimes considered identical to k_{p}), and $[L_0]$ is the concentration of primary active molecules (with chain length zero). This concentration of primary active molecules is given by

$$\frac{d[L_0]}{dt} = r_{\mathrm{g}} - r_{\mathrm{i}} + \sum \frac{q_{\mathrm{in}}}{V}\left([L_0]_{\mathrm{in}} - [L_0]\right) - \sum \frac{q_{\mathrm{out}}}{V}\left([L_0]_{\mathrm{out}} - [L_0]\right) \tag{2.85}$$

where r_{g} is the rate of generation of primary active molecules. If the primary active molecules are generated from a single initiating molecule (I), then such rate can be expressed as

$$r_{\mathrm{g}} = f k_{\mathrm{g}} [I] \tag{2.86}$$

where f is an efficiency factor for the generation of primary active molecules (with values between 0 and 1), k_{g} is a rate coefficient, and $[I]$ is the concentration of initiating molecules. The rate coefficient k_{g} already takes into account the stoichiometric coefficient of the number of active molecules generated per reaction event. If the primary active molecules are generated by the reaction of two initiating molecules (I_1, I_2), such as in redox initiation systems, then the rate of generation is

$$r_{\mathrm{g}} = f k_{\mathrm{g}} [I_1][I_2] \tag{2.87}$$

It is also possible to incorporate two or more initiators with different decomposition rates (frequently used in suspension polymerization), each one of which will be represented by its own decomposition equation (Eq. 2.86).

Returning to Eq. 2.83, the rate of termination (either by recombination or disproportionation) of active growing chains is given by

$$r_t = k_t \cdot [A]^2 \quad (2.88)$$

where k_t is the kinetic rate coefficient of termination, which already includes the value of the stoichiometric coefficient indicating the number of active chains suppressed per termination reaction (with a value of 2). Please recall that this termination reaction may not be present in all types of chain-growth polymerization processes.

Applying the quasi-steady-state assumption (Bodenstein's principle [62]), the concentration of active sites $[A]$ can be approximately expressed as (neglecting the transfer of active sites between the reaction zones):

$$[A] \approx \sqrt{\frac{r_g}{k_t}} \quad (2.89)$$

where r_g is the total rate of generation of primary active molecules, independent of the mechanism of generation.

Using Eq. 2.89 in Eq. 2.81 greatly simplifies the determination of the rate of polymerization (when bimolecular termination of active chains occurs).

Another important concept in chain-growth polymerization reactions is the kinetic chain length (υ), defined as the average number of monomer units added to the growing chain until its activity is lost. When the activity of the growing chain is lost only by termination, then

$$\nu = \frac{k_p[B]}{k_t[A]} \approx \frac{k_p[B]}{\sqrt{k_t r_g}} \quad (2.90)$$

Then, the degree of polymerization DP can be related to the kinetic chain length by

$$\mathrm{DP} = f_t \nu \quad (2.91)$$

where f_t is a termination factor corresponding to 1 for termination by disproportionation only, 2 for termination by recombination only, and intermediate values when both types of termination are present.

The simple relation between the degree of polymerization and the kinetic chain length, as given by Eq. 2.91, is not fulfilled if chain transfer is an important kinetic event during the polymerization as for the first time discussed comprehensively by Flory [63].

If transfer reactions are important, the average degree of polymerization will be

$$\mathrm{DP} \approx f_t \frac{k_p[B]}{k_{tr}[T]+\sqrt{k_t r_g}} \tag{2.92}$$

where k_{tr} is the rate constant of transfer of the active site to another compound T. This transfer reaction can be expressed, in general, as follows:

$$P_iB_bA_a + T \rightarrow P_iB_bA_{a-1} + TA \tag{2.93}$$

where TA is a new active compound, capable of initiating the growth of a new chain comparable to a primary active molecule. Furthermore, the transfer agent can also be a growing polymer chain [64]. In addition, the rate of transfer of the active site is

$$r_{tr} = k_{tr}\,[A][T] \tag{2.94}$$

Additional discussions on the kinetics of chain-growth polymerization reactions can be found in the scientific literature, particularly for free-radical polymerization [65–69].

2.3.3 Diffusion-Controlled Polymerization Kinetics

As it was shown in the previous sections, most reactions involved in polymerization are bimolecular. A bimolecular reaction can take place only after both molecules find each other in the local reaction zone. **Please note that reactions are local events** occurring at the molecular level, only when all the reactants involved are in the right place, in the right position, and in the right time. Such local reaction zone is assumed to be perfectly mixed, and therefore bimolecular reactions will depend only on the spatial probability of collision. However, when the reactants are in different reaction zones, molecular transfer between these zones must occur first. Such transfer will be determined by molecular diffusion, as it was previously mentioned in Chapter 1. Now, if the rate of reaction is slow compared to the rate of mass transfer by molecular diffusion, the spatial composition of the system will behave as homogeneous and perfectly mixed. On the other hand, if the rate of mass transfer (by diffusion) is slower than the rate of reaction, the reacting system becomes diffusion controlled and the local spatial composition will not be homogeneous (imperfect mixing) [70]. Then, different reaction zones will be present in the reactor. Those reaction zones will behave equivalently as dispersed phases in a heterophase polymerization system. The difference is that not all segregated reaction zones can be easily distinguished, and they do not necessarily result in the formation of polymer particles.

Due to the high viscosity of polymer media (low free volume), and to the lower mobility of oligomeric and polymeric chains, low molecular diffusivities of the reactants can easily result in diffusion control or imperfect mixing, which seriously affects the final conversion and MMD of the polymer formed as a result of the competitive nature of the polymerization reactions involved. Since the mobility of the chains and their neighboring molecules is significantly reduced as they grow, it is expected that the kinetics of polymerization changes as conversion increases. This is reflected in the decrease in the volume of perfectly mixed reaction zones (increasing their number), and a reduced mass transfer between reaction zones.

At some point, all reactants are eventually isolated in small reacting zones within the polymer phase formed, and polymerization completely stops before reaching full conversion (glass effect) [71].

In the case of radical chain-growth polymerization, it is possible to observe an increase in the recombination of primary active molecules due to some confinement or "cage" effect in the reaction zones, decreasing the initiation efficiency, reducing the overall polymerization rate and resulting in lower final monomer conversions as well as lower molecular weight polymers. On the other hand, the increase in the viscosity of the medium may also cause isolation of the growing chains, suppressing all termination reactions and giving rise to the *Trommsdorff* or gel effect [72], which consists in a sudden acceleration of the polymerization rate, just before polymerization stops by the glass effect. Another important effect of diffusional limitations on the kinetics of polymerization is the dependence of the reaction rate coefficients involving macromolecules on the chain length [73]. Thus, smaller chains tend to react more readily than larger chains.

2.3.4 Heterophase Polymerization Kinetics

So far we have discussed the kinetics of polymerization in a certain perfectly mixed reaction zone. However, the main interest of kinetics is predicting the overall (macroscopic) rate of a reaction. The same applies to heterophase polymerization. The overall rate of polymerization $\left(\bar{r}_{\mathrm{p}}\right)$ in a heterophase polymerization system will be given by

$$\bar{r}_{\mathrm{p}} = \frac{\sum_{i=1}^{Z} r_{\mathrm{p},i} V_i}{V_{\mathrm{T}}} = \sum_{i=1}^{Z} r_{\mathrm{p},i} \varphi_i \tag{2.95}$$

where Z represents the total number of reacting zones within the reactor of total volume V_{T}, $r_{\mathrm{p},i}$ is the polymerization rate in the i-th reacting zone of volume V_i, and φ_{i} is the volume fraction occupied by the i-th reacting zone with respect to the total reactor volume. In other words, the overall polymerization rate is the volume-average polymerization rate of the individual reacting zones.

The main difficulty in determining the overall polymerization rate is that the polymerization rates of all individual reacting zones are needed. And the individual polymerization rates will depend on the local reactants concentration, which also depends on molecular diffusion and mass transfer phenomena.

It is possible, however, to make certain assumptions depending on the particular type of heterophase polymerization considered. For example, if the monomeric species have a low solubility in the continuous phase and polymerization therein can be neglected. If the dispersed phases are small and spatially separated over certain distance (needed to consider the particles as isolated), each can be considered an individual perfectly mixed reacting zone. Additionally, if the size and composition of the dispersed phases are similar, the overall polymerization rate simply becomes

$$\bar{r}_{\mathrm{p}} = N_{\mathrm{p}} r_{\mathrm{p}} V_{\mathrm{p}} \tag{2.96}$$

where N_{p} is the number concentration of dispersed phases in the system, V_p is the volume of each dispersed phase, and r_p is the individual polymerization rate.

For step-growth polymerization reactions (assuming only two functional groups), we get

$$\bar{r}_p = N_p k_p [F^A][F^B] V_p \tag{2.97}$$

whereas for chain-growth polymerization reactions, we have

$$\bar{r}_p = N_p k_p [A][B] V_p \tag{2.98}$$

The similitude between the two mechanisms of polymerization is evident, if the susceptible bonds and the active sites are considered reactive functional groups. However, let us recall that the active site may survive the polymerization reaction, whereas the functional groups are consumed. The concentrations used in Eqs. 2.97 and 2.98 correspond to average concentrations inside the dispersed phases.

The overall reaction rate of chain-growth polymerization can also be expressed as

$$\bar{r}_p = \bar{n}_A N_p k_p [B] \tag{2.99}$$

where

$$\bar{n}_A = [A] V_p \tag{2.100}$$

represents the average number of active sites inside each dispersed phase.

2.4 Particle Dynamics

Heterophase polymerization processes are characterized by the emergence of polymer particles dispersed in a continuous phase (which is not a good solvent for the polymer). One of the most important properties of such polymer dispersion is the probability distribution of particle sizes. The particle size distribution (PSD) is a statistical distribution analogous to the chain length distribution (CLD) or MMD of the polymer chains discussed in the previous section. The nature of the border between the particles and the continuous phase can be quite sharp, i.e., only negligible parts of the polymer molecules close to the interface stretch into the continuous phase. However, the interfacial layer can also be quite thick, particularly if polymeric stabilizers are employed or if the particles contain copolymers with hydrophilic components. In the latter case, it is important which sizing technique is applied, e.g., dynamic light scattering (giving the hydrodynamic size) or electron microscopy (giving almost the size of the polymer core because the hydrophilic parts collapse in the dried state and form only a thin layer around the core). The difference in the determined diameter can be quite large up to some 50 nm and corresponds to the hydrodynamic layer thickness (cf. stabilization of particles discussed below). On the other hand, particle size is not a discrete variable. Thus, the PSD can be described by a probability density function ρ_S. It can also be partially described by its representative moments and functions of its moments, including average particle sizes, standard deviation, dispersity, etc. In general, the n-th moment of a particle size distribution can be expressed as

$$M_n(s) = E\left(s^n\right) = \int_0^\infty s^n \rho_S(s)ds \approx \sum_{i=1}^{m} \bar{s}_i^n p\left(s_{i-1} < s \le s_i\right) \quad (2.101)$$

where s represents a measure of particle size. For spherical particles, s can represent either the radius or the diameter of the particle. For non-spherical particles, the particle size can be expressed as the radius or diameter of an equivalent spherical particle of the same volume, of the same surface area, or of the same mass; as the radius or the diameter of gyration of the particle; as the largest distance between two points at the surface of the particle; etc. Equation 2.101 also shows that the continuous size distribution can be approximated using a discretization method [74], involving division of the range of particle sizes in m discrete particle size intervals, where the class mark of the i-th interval is given by $\bar{s}_i$, and $p(s_{i-1} < s \le s_i)$ is the relative frequency of particles found in such interval.

It is also possible to define different average values of the PSD. For example, the number-average particle size D_1 (corresponding to the first moment of the particle size distribution) will be

$$D_1 = M_1(s) = E(s) = \int_0^\infty s\rho_S(s)ds \approx \sum_{i=1}^{m} \bar{s}_i p\left(s_{i-1} < s \le s_i\right) \quad (2.102)$$

In general, it is possible to define different average sizes of the particle using the following expression:

$$D_j = \frac{M_j(s)}{M_{j-1}(s)} = \frac{E\left(s^j\right)}{E\left(s^{j-1}\right)} = \frac{\int_0^\infty s^j \rho_S(s)ds}{\int_0^\infty s^{j-1} \rho_S(s)ds} \approx \frac{\sum_{i=1}^{m} \bar{s}_i^j p\left(s_{i-1} < s \le s_i\right)}{\sum_{i=1}^{m} \bar{s}_i^{j-1} p\left(s_{i-1} < s \le s_i\right)} \quad (2.103)$$

Particularly, D_3 is denoted as the surface-average particle size (or surface mean size), and D_4 is the volume-average particle size (or volume mean size). Similarly, D_2 would correspond to a perimeter-average particle size. It is also common to find a Z-average particle size approximately corresponding to D_6.

The standard deviation in particle size σ_S, describing the broadness of the distribution, can be determined as

$$\sigma_S = \sqrt{M_2(s) - M_1^2(s)} = \sqrt{D_2 D_1 - D_1^2} \tag{2.104}$$

Another concept describing broadness is the particle size dispersity (Δ_S). The term dispersity was previously used to describe CLD and MMD of polymers. However, for particle size distributions, a preferred definition of dispersity is

$$\Delta_{S,j} = 1 + \left(\frac{\sigma_S}{D_j}\right)^2 \tag{2.105}$$

where $\Delta_{S,j}$ represents the dispersity of the particle size distribution with respect to the j-th average particle size (D_6, for example [75]).

The size distribution of the particles in heterophase polymerization is influenced by particle formation (Section 2.1), molecular transfer (Section 2.2), and polymerization (Section 2.3), but also by particle aggregation and breakage. Chung-li et al. [76] found that the rate of swelling of polymer particles was too slow to account for the rate of particle growth, and that aggregation played an important role in particle growth. There are two types of aggregation of polymer particles: coagulation and coalescence. During coagulation, the total surface of the particles remains almost constant, whereas during coalescence, the total particle surface significantly decreases as a result of the migration of the chains trying to minimize the free energy of the system (minimizing repulsive forces at the polymer/continuous phase interface). However, coagulation can be considered the first step in the coalescence process because it brings the particle close enough for further action, and for that reason, it will be the main focus of this section.

Despite this restriction, important differences between coagulation and coalescence should be briefly discussed. First, coagulation is mainly an issue connected with the colloidal instability of the particles, but coalescence is not. Second, coagulation leads finally to non-spherical objects, but coalescence again to spherical objects. Coalescence is observed in emulsions and during emulsion polymerization of monomers with high tendency to chain transfer to monomer. It was experimentally shown that the particle coalescence during the emulsion polymerization of vinyl chloride is connected with the polymerization reaction. Terminating the polymerization by the addition of radical scavengers stops the decrease in the particle number as long as the scavenger is used up [77, 78]. A model for particle coalescence was developed assuming that radical desorption is the crucial step (reaction-induced coalescence) and successfully applied for modeling batch and continuous emulsion polymerization [79, 80].

Coagulation depends on the forces acting between different particles. When the repulsive potential between particles is stronger than the kinetic energy of their collision, the particles are stable and do not coagulate; if the repulsive forces are weak, then particle aggregation (coagulation or flocculation) may take place. Flocculation can be regarded as a reversible (or weak) aggregation process, whereas coagulation is irreversible (strong aggregation).

Unless very strong mechanical forces or very weak internal cohesive forces (e.g., flocculation) are present in the heterophase polymerization system, the breakage of polymer particles can usually be neglected.

The nature of the interparticle forces responsible for the stabilization, and also destabilization, of the colloidal particles can be classified into electrostatic, steric, and electrosteric.

Deryaguin–Landau–Verwey–Overbeek theory: Mathematical model describing the interaction forces between two particles based on the consideration of

- Their van der Waals attraction,
- The electrostatic interaction of their charged layers.

The theory of electrostatic stabilization of particles is called the ***Deryaguin–Landau–Verwey–Overbeek (DLVO) theory*** [81] after the scientists who developed it. The DLVO theory relies on the consideration of both electrostatic forces and van der Waals forces. The principle of electrostatic stabilization is the presence of a net electric charge at the surface of the particles dispersed in a polar medium (Fig. 2.15). Around these charges, a well-defined layer (*Stern layer*) of ions of opposite sign to that of the surface ions (counter ions) is formed. In addition, as a result of electrostatic interactions and thermal motion of the molecules, a non-uniform diffuse second layer develops around the particles, which is composed mainly of counter ions, but may contain also ions of the same sign as the surface (co-ions). This layer, called the diffuse electrical layer, can be described mathematically by the *Poisson–Boltzmann equation* [82].

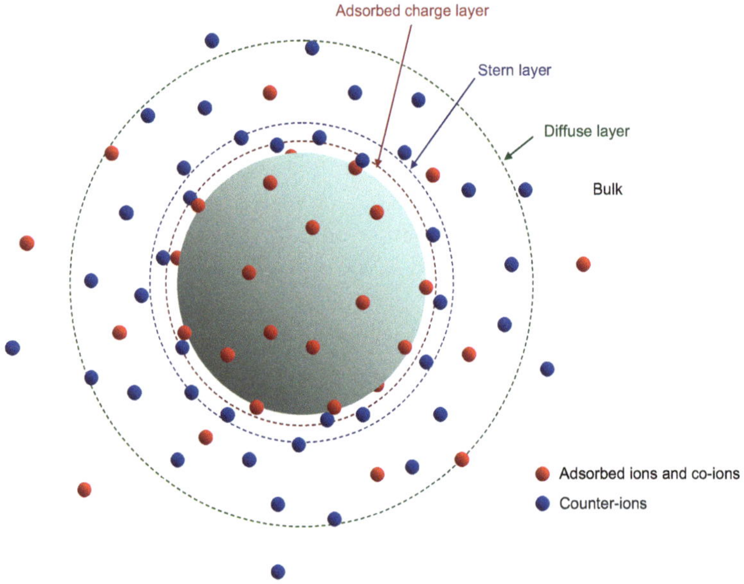

Figure 2.15 Ionic layers around a polymer particle dispersed in a polar continuous phase.

The theory of electrical double layers at interfaces is also called the ***Gouy–Chapman theory*** [83]. When two different particles electrically charged at their surfaces with ions of the same sign approach each other, they will experience a net repulsion force as a result of the interaction between the ions located at their diffuse layers. If the net interaction potential between the particles is repulsive and larger than the kinetic energy of the collision, they will not approach each other close enough to coagulate.

Gouy–Chapman theory: This theory considers that the second layer around a surface is diffusive, with its electrical potential (given by Poisson's equation) decreasing exponentially (given by a Boltzmann-type distribution).

The electrostatic stability of a dispersion depends not only on the magnitude of the electrical charge surface density of the particles but also on the dielectric properties of the medium, on its ionic strength, on the valence of the ions in the double layer, on the size of the particles, and on the temperature of the system (only slightly). The total interaction potential between two spherical particles charged by a single type of ions at the surface can be determined using the DLVO equation:

$$\psi^{\mathrm{DLVO}}(r) = \frac{32\pi k_\mathrm{B} T d_\mathrm{p} \rho_\infty \gamma^2}{\kappa^2} \exp(-\kappa r) - \frac{A d_\mathrm{p}}{12r} \qquad (2.106)$$

where ρ_∞ is the number density of ions in the bulk continuous phase, d_p is the diameter of the spheres, A is the ***Hamaker constant***, r is the minimum distance between the surface of the particles, κ is the reciprocal of the ***Debye length***, given by

Hamaker constant: Represents a measure of the intensity of the van der Waals interaction between two bodies.

Debye length: Characteristic thickness of the diffuse layer, corresponding to the distance from the surface over which electric field screening, caused by mobile ions, takes place.

$$\kappa = \sqrt{\frac{\rho_\infty e^2 z^2}{\varepsilon \varepsilon_0 k_\mathrm{B} T}} \qquad (2.107)$$

and γ is obtained from the Gouy–Chapman theory as

$$\gamma = \tanh\left(\frac{z e \psi_0}{4 k_\mathrm{B} T}\right) \qquad (2.108)$$

where z is the valence of the ions at the surface, e is the charge of the electron, ε is the relative permittivity, ε_0 is the permittivity of vacuum, and ψ_0 is the electrostatic potential at the surface of the spheres.

The ionic strength has a strong influence on the Debye screening length, which determines at which distance a particle recognizes a second particle in its neighborhood. At low ionic strength, the Debye screening length is large and causes a crystal-like arrangement of electrostatically stabilized monodisperse colloids, including polymer particles (cf. Ref. [84] and references therein) and polyelectrolytes [85].

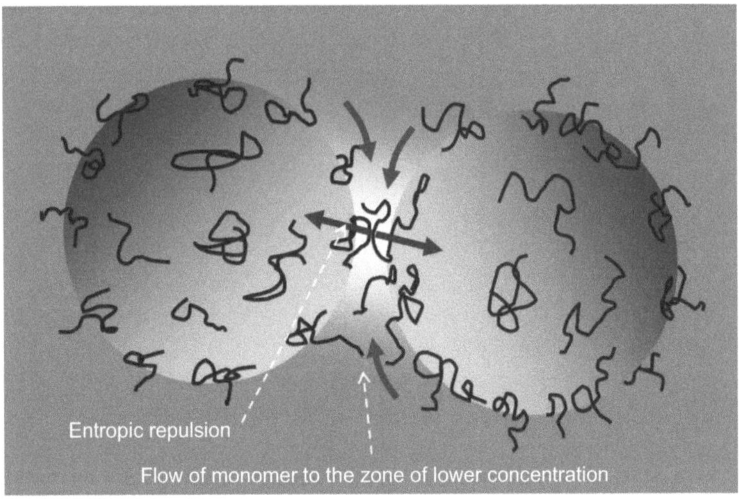

Figure 2.16 Principles of steric stabilization.

Electrically stabilized colloidal dispersions are very sensitive to the addition of electrolytes. If the concentration of ions in the solution increases, the thickness of the diffuse layer (*Debye length*) decreases as a result of both entropic and electrical screening effects, leading to a reduction in the repulsive potential. On the other hand, particles dispersed in organic media (low dielectric constant) cannot be electrically stabilized because the electrostatic forces become extremely short ranged. In these cases, steric stabilization is recommended. Steric stabilization is imparted by nonionic amphiphilic molecules (usually polymeric molecules). The lyophobic moiety of the amphiphiles will adsorb on the surface of polymer particles, while its lyophilic moiety will be extended in the continuous phase. When two sterically stabilized particles approach each other, the concentration of the lyophilic segments of the amphiphile in the continuous phase between the particles begins to increase and the concentration of solvent decreases. The difference in local solvent concentration around the particles results in an increase in the osmotic pressure and a net flow of solvent toward the interaction zone (Fig. 2.16) due to the tendency to equilibrate the chemical potential. At the same time, the increased concentration of the polymer segments leads to a reduction in the configurational entropy of the chains causing repulsion between the particles. Both effects (osmotic and entropic) restore the chemical potential equilibrium by separating the particles. The efficiency of steric stabilization relies on the solubility of the lyophilic moiety of the stabilizer in the continuous phase. Since the solubility is temperature dependent, sterically stabilized colloidal dispersions are very sensitive to the temperature of the system.

The third type of stabilization, electrosteric stabilization, is the combination of the two previous mechanisms. In this case, the colloidal particles can be made stable to the presence of electrolytes thanks to the steric repulsion and to changes in temperature thanks to the electrostatic interaction. Electrosteric stabilization can be achieved simply by using a mixture of stabilizers (one electrostatic and one steric), or by using only one single type of stabilizer, such as ***polyelectrolytes***. In particular, polyelectrolytes grafted to the polymer particles offer an extraordinary stability to colloidal objects in the presence of electrolytes [86]. The reason is that the polyelectrolyte chains are stretched away from the particle surface due to the osmotic force, which is balanced by the elastic force of the chain; if the polyelectrolyte chain is longer than the radius of particles, their thickness scales with the power 1/5 inversely with the added salt. The polyelectrolyte brush collapses only at high salt concentrations causing flocculation of the colloid. This prediction by Pincus [86] was experimentally proven [87].

Polyelectrolyte: Polymer containing an electrolyte group as a repeating unit of its chain. In aqueous solution, they are electrically conductive as a result of electrolyte dissociation.

If the repulsion potentials between the particles had been infinite, the particles would be stable forever. However, since the repulsion potentials are finite, there is always the probability of particle aggregation depending on the thermal fluctuations of the system. The rate of particle coagulation will be a function of the frequency at which the particles encounter each other and of the probability of coagulation at each collision. In the absence of a repulsive potential, the particles aggregate at the same rate at which they encounter by diffusion through the continuous phase. This rate is called the **Brownian collision rate**, or the **Smoluchowski fast coagulation rate** [88]:

$$r_{\mathrm{coag}}^{\mathrm{fast}} = 8\pi D_{\mathrm{p}} d_{\mathrm{p}} N_{\mathrm{p}} = \frac{8k_{\mathrm{B}}TN_{\mathrm{p}}}{3\eta} \tag{2.109}$$

where D_{p} is the diffusion coefficient of the particles, d_{p} is the diameter of the particles, N_{p} is the concentration of particles per unit volume of dispersion, and η is the viscosity of the continuous phase. In the absence of stabilizer, coagulation proceeds very rapidly even in fairly dilute dispersions.

In the presence of a net repulsion potential or energy barrier, **slow coagulation** takes place at a rate depending on the magnitude of the barrier. In general, the rate of slow coagulation can be expressed as:

$$r_{\mathrm{coag}}^{\mathrm{slow}} = \frac{r_{\mathrm{coag}}^{\mathrm{fast}}}{W} \approx r_{\mathrm{coag}}^{\mathrm{fast}} \exp\left(-\frac{\psi_{\mathrm{r}}}{k_{\mathrm{B}}T}\right) \tag{2.110}$$

where W is the *Fuchs stability ratio*, and ψ_{r} is the net repulsion energy barrier.

Given that in the absence of particle formation and breakage,

$$\frac{dN_{\mathrm{p}}}{dt} = -r_{\mathrm{coag}}^{\mathrm{slow}} N_{\mathrm{p}} \tag{2.111}$$

the temporal development of the number of particles after nucleation in the dispersion can be obtained from

$$N_{\mathrm{p}}(t) = \frac{N_0}{1 + \frac{8k_{\mathrm{B}}TN_0}{3\eta W} t} \tag{2.112}$$

The general expression for the dynamics of the number of particles in the system will be

$$\frac{dN_{\mathrm{p}}}{dt} = r_{\mathrm{form}} - k_{\mathrm{coag}} N_{\mathrm{p}}^2 + k_{\mathrm{break}} N_{\mathrm{p}} - \frac{N_{\mathrm{p}}}{V_{\mathrm{d}}} \frac{dV_{\mathrm{d}}}{dt} \tag{2.113}$$

where r_{form} is the rate of particle formation per unit volume of dispersion, k_{coag} is the rate coefficient of biparticle coagulation (which depends on temperature, diffusivity of polymer particles, particle size, and energy barrier), k_{break} is the rate coefficient of particle breakage (which depends on mechanical energy input and intraparticle cohesive forces, and also includes the average number of particle fragments released per breakage event), V_d is the total volume of the dispersion, and the last term in the right-hand side of Eq. 2.113 represents a dilution effect for non-batch polymerization systems. If multi-particle coagulation events are also considered, the second term in the right hand of Eq. 2.113 must be corrected accordingly, including higher-order effects of the particle concentration.

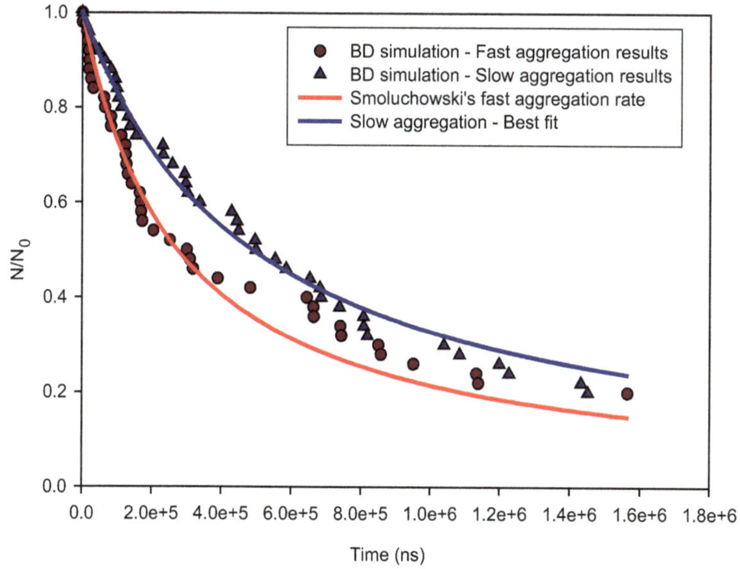

Figure 2.17 Kinetics of particle aggregation.

Smoluchowski's rate of coagulation assumes that particles are highly diluted in the system. As the number concentration increases, the rate of coagulation is expected to depend also on multi-particle collisions, increasing the influence of the particle concentration on the rate of coagulation.

The mechanism of particle aggregation is basically the same as that of molecular capture or molecular absorption: the collision between two different entities after they diffuse toward each other following random-walk trajectories. For that reason, BD can be used for the simulation of fast and slow particle aggregation. Rzepiela et al. [26] used BD to simulate the aggregation of colloid particles and found that for concentrated systems, the fast coagulation rate is even faster than predicted by Eq. 2.105, in agreement with the results obtained for molecular capture [24]. It is possible to include interaction forces, either by solving the Langevin equation for Brownian motion considering the interaction force term or by assuming activation energies for aggregation equal to the interaction potential energy between the particles. A very important consideration during the simulation of particle aggregation is that the particles form clusters as they aggregate. The new entity, the cluster of particles, will behave as a single, larger Brownian entity. Therefore, instead of simulating the Brownian motion of each particle independently, all the particles in the cluster will follow the same trajectory and will have the same velocity.

In the following example, BD is used to simulate the fast and slow aggregation of colloid particles in polymer dispersions (Fig. 2.17). The conditions considered are summarized in Table 2.2. Fast particle aggregation is obtained by neglecting both electrostatic and van der Waals interaction potentials.

Table 2.2 Simulation conditions for the simulation of aggregation of initially monodisperse particles

Parameter	Value
Temperature	80°C
Initial particle diameter	50 nm
Initial particle number concentration	10^{20} part/m^3
Particle density	1.06 g/cm^3
Water viscosity	3.55×10^{-4} Pa·s
Surface potential of the particles	10 mV
Average valence at the surface	3
Ionic concentration	0.01 M
Hamaker constant	0.58×10^{-21} J

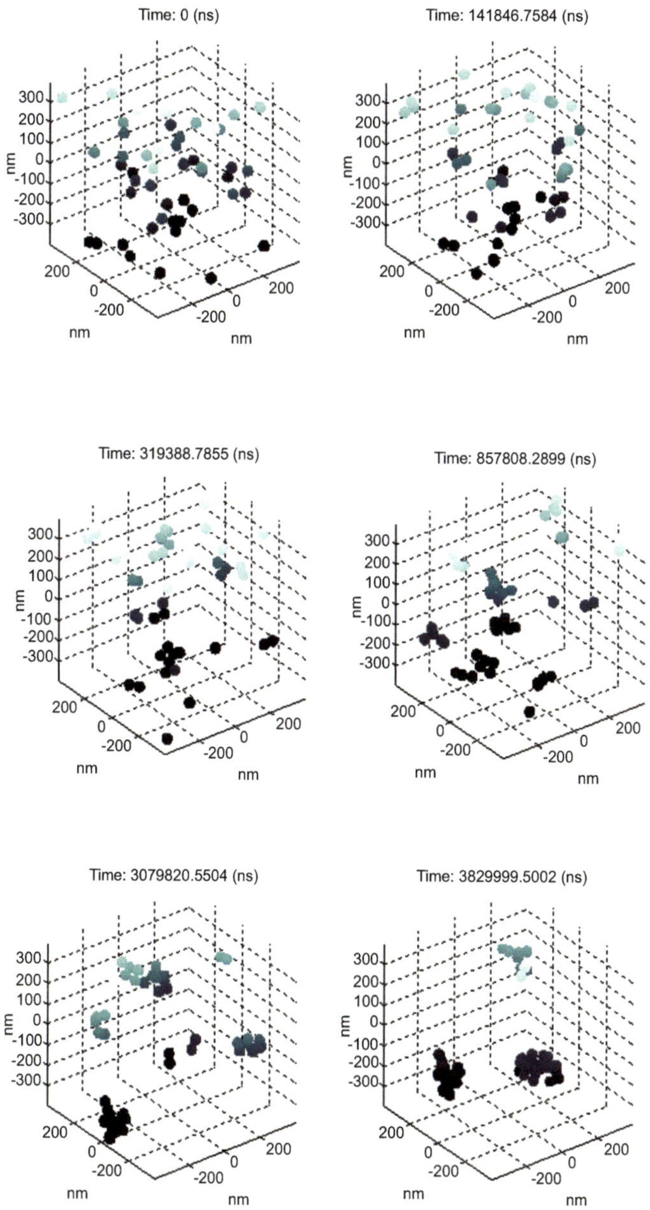

Figure 2.18 Evolution of particle aggregation. Color intensity indicates the relative z-position of the particles.

Fast particle aggregation results obtained using BD simulations are in good agreement with Smoluchowski fast coagulation equation (Eq. 2.109), especially at the beginning of the simulation. Since the example considered a diluted polymer particle dispersion (0.65% solids), significant deviations from Eq. 2.109 were not expected. On the other hand, slow aggregation was simulated considering both electrostatic and van der Waals interactions. The surface potential considered for the particles corresponds to a repulsive electrostatic energy of 4.8×10^{-21} J, while the van der Waals interaction (Hamaker constant) corresponds to 5.8×10^{-22} J. The results obtained by simulation were fitted to Eq. 2.110, and a value of $W = 1.784$ was obtained, corresponding to a net repulsion energy of $\psi_r = 2.82 \times 10^{-21}$ J. An example of the time evolution of the system during particle aggregation for a representative volume of 6.4×10^{-17} L is presented in Fig. 2.18. In this figure, it is possible to observe the formation and growth of clusters of particles as a result of the aggregation of particles.

References

1. Becker, R. and Döring, W. (1935). Kinetische Behandlung der Keimbildung in übersättigten Dämpfen, *Annalen der Physik*, **24**, pp. 719–752.

2. Laaksonen, A., Talanquer, V., and Oxtoby, D. W. (1995). Nucleation: Measurements, theory, and atmospheric applications, *Annual Review of Physical Chemistry*, **46**(1), pp. 489–524.

3. Barrett, K. E. J. and Thomas, H. R. (1975). Kinetics and mechanism of dispersion polymerization. In: Barrett, K. E. J., *Dispersion Polymerization in Organic Media* (John Wiley & Sons, UK), pp. 115–200.

4. Fitch, R. M. and Tsai, C. H. (1971). Particle formation in polymer colloids, III: Prediction of the number of particles by a homogeneous nucleation theory. In Fitch, R. M., *Polymer Colloids* (Springer, USA), pp. 73–102.

5. Ugelstad, J. and Hansen, F. K. (1976). Kinetics and mechanism of emulsion polymerization, *Rubber Chemistry and Technology*, **49**(3), pp. 536–609.

6. Tauer, K. and Kühn, I. (1995). Modeling particle formation in emulsion polymerization: An approach by means of the classical nucleation theory, *Macromolecules*, **28**, pp. 2236–2239.

7. Feeney, P. J., Napper, D. H., and Gilbert, R. G. (1984). Coagulative nucleation and particle size distributions in emulsion polymerization, *Macromolecules*, **17**(12), pp. 2520–2529.

8. Gilbert, R. G. (1995). *Emulsion Polymerization* (Academic Press, USA).

9. Hansen, F. K. and Ugesltad, J. (1979). Particle nucleation in emulsion polymerization IV. Nucleation in monomer droplets, *Journal of Polymer Science: Polymer Chemistry Edition*, **17**(10), pp. 3069–3082.

10. El-Aasser, M. S. and Sudol, E. D. (2004). Miniemulsions: Overview of research and applications, *JCT Research*, **1**(1), pp. 21–32.

11. Tauer, K., Hernandez, H., Kozempel, S., Lazareva, O., and Nazaran, P. (2008). Towards a consistent mechanism of emulsion polymerization: New experimental details, *Colloid & Polymer Science*, **286**, pp. 499–515.

12. Hernandez, H. (2017). Analysis of temperature fluctuations in ideal gases: From the macroscopic to the molecular scale, *ForsChem Research Reports*, **2**, 2017-3. doi: 10.13140/RG.2.2.19208.83203.

13. Fitch, R. M. and Tsai, C. H. (1971). Particle formation in polymer colloids IV: The role of soluble oligomeric radicals. In: Fitch, R. M., *Polymer Colloids* (Plenum Press, USA), pp. 103–116.

14. Ferguson, C. J., Russell, G. T., and Gilbert, R. G. (2002). Modelling secondary particle formation in emulsion polymerisation: Application to making core-shell morphologies, *Polymer*, **43**, pp. 4557–4570.

15. Krüger, K., Wei, C., Nuasaen, S., Höhne, P., Tangboriboonrat, P., and Tauer, K. (2015). Heterophase polymerization: Pressures, polymers, particles, *Colloid and Polymer Science*, **293**(3), pp. 761–776.

16. Harkins, W. D. (1947). A general theory of the mechanism of emulsion polymerization, *Journal of the American Chemical Society*, **69**, pp. 1428–1444.

17. Nazaran, P. (2008). *Nucleation in Emulsion Polymerization. Steps towards a Non-micellar Nucleation Theory*. Doctoral Dissertation (Universität Potsdam, Germany).

18. von Smoluchowski, M. (1917). Versuch einer mathematischen Theorie der Koagulationskinetik kolloider Lösungen, *Zeitschrift für Physikalische Chemie*, **92**, pp. 129–168.

19. Nomura, M., Tobita, H., and Suzuki, K. (2005). Emulsion polymerization: Kinetic and mechanistic aspects, *Advances in Polymer Science*, **175**, pp. 1–128.

20. Penboss, I. A., Gilbert, R. G., and Napper, D. H. (1986). Entry rate coefficients in emulsion polymerization systems, *Journal of the Chemical Society, Faraday Transactions I*, **82**, pp. 2247–2268.

21. Cheong, I.-W. and Kim, J.-H. (1998). Simulation of secondary particle formation in seeded emulsion polymerization: The effect of surface charge density, *Macromolecular Theory and Simulations*, **7**, pp. 49–57.
22. Maxwell, I. A., Morrison, B. R., Napper, D. H., and Gilbert, R. G. (1991). Entry of free radicals into latex particles in emulsion polymerization, *Macromolecules*, **24**, pp. 1629–1640.
23. Sajjadi, S. (2003). Particle formation under monomer-starved conditions in the semibatch emulsion polymerisation of styrene. Part II. Mathematical modelling, *Polymer*, **44**, pp. 223–237.
24. Tauer, K., Nozari, S., and Ali, A. M. I. (2005). Experimental reconsideration of radical entry into latex particles, *Macromolecules*, **38**, pp. 8611–8613.
25. Herrera-Ordóñez, J. and Olayo, R. (2000). On the kinetics of styrene emulsion polymerization above CMC. I. A Mathematical model, *Journal of Polymer Science Part A: Polymer Chemistry*, **38**, pp. 2201–2218.
26. Hernandez, H. F. and Tauer, K. (2007). Brownian dynamics simulation of the capture of primary radicals in dispersions of colloidal polymer particles, *Industrial & Engineering Chemistry Research*, **46**, pp. 4480–4485.
27. Hernandez, H. F. and Tauer, K. (2007). Brownian dynamics simulation studies on radical capture in emulsion polymerization, *Macromolecular Symposia*, **259**, pp. 274–283.
28. Rzepiela, A. A., van Opheusden, J. H. J., and van Vliet, T. (2001). Brownian dynamics simulation of aggregation kinetics of hard spheres with flexible bonds, *Journal of Colloid and Interface Science*, **244**(1), pp. 43–50.
29. Hernandez, H. F. (2008). *Multiscale Simulation of Heterophase Polymerization: Application to the Synthesis of Multicomponent Colloidal Polymer Particles*, Doctoral Dissertation (Universität Potsdam, Germany).
30. Yeliseyeva, V. I. (1982). Polymerization of polar monomers. In: Piirma, I., *Emulsion Polymerization* (Academic Press, USA), pp. 247–288.
31. Asua, J. M. and de la Cal, J. C. (1991). Entry and exit rate coefficients in emulsion polymerization of styrene, *Journal of Applied Polymer Science*, **42**, pp. 1869–1877.
32. López de Arbina, L., Barandiaran, M. J., Gugliotta, L. M., and Asua, J. M. (1996). Emulsion kinetics polymerization: Particle growth kinetics, *Polymer*, **37**(26), pp. 5907–5916.
33. Liotta, V., Georgakis, C., Sudol, E. D., and El-Aasser, M. S. (1997). Manipulation of competitive growth for particle size control in

emulsion polymerization, *Industrial & Engineering Chemistry Research*, **36**, pp. 3252–3263.

34. Herrera-Ordóñez, J. and Olayo, R. (2000). On the kinetics of styrene emulsion polymerization above CMC. II. Comparison with experimental results, *Journal of Polymer Science Part A: Polymer Chemistry*, **38**, pp. 2219–2231.
35. Sood, A. and Awasthi, S. K. (2004). Population balance model for miniemulsion polymerization 2, *Macromolecular Theory and Simulations*, **13**(7), pp. 615–628.
36. Ugelstad, J., Mork, P. C., Dahl, P., and Rangnes, P. (1969). A kinetic investigation of the emulsion polymerization of vinyl chloride, *Journal of Polymer Science Part C*, **27**, pp. 49–68.
37. Nomura, M. and Harada, M. (1981). Rate coefficient for radical desorption in emulsion polymerization, *Journal of Applied Polymer Science*, **26**, pp. 17–26.
38. Asua, J. M., Sudol, E. D., and El-Aasser, M. S. (1989). Radical desorption in emulsion polymerization, *Journal of Polymer Science Part A: Polymer Chemistry*, **27**, pp. 3903–3913.
39. Asua, J. M. (2003). A new model for radical desorption in emulsion polymerization, *Macromolecules*, **36**, pp. 6245–6251.
40. Hernandez, H. F. and Tauer, K. (2008). Radical desorption kinetics in emulsion polymerization. 1. Theory and simulation, *Industrial & Engineering Chemistry Research*, **47**(24), pp. 9795–9811.
41. Tauer, K. and Hernandez, H. (2010). Mechanism and modeling of emulsion polymerization: New ideas and concepts–2. Modeling strategies, *Macromolecular Symposia*, **288**(1), pp. 9–15.
42. Hernandez, H. F. and Tauer, K. (2010). Radical desorption kinetics in emulsion polymerization 2: Brownian dynamics simulation of radical desorption in non-homogeneous particles, *Macromolecular Theory and Simulations*, **19**(5), pp. 249–257.
43. Gillespie, D. T. (2007). Stochastic simulation of chemical kinetics, *Annual Review of Physical Chemistry*, **58**, pp. 35–55.
44. Thickett, S. C. and Gilbert, R. G. (2006). Rate-controlling events for radical exit in electrosterically stabilized emulsion polymerization systems, *Macromolecules*, **39**, pp. 2081–2091.
45. Harada, M., Nomura, M., Eguchi, W., and Nagata, S. (1971). Studies of the effect of polymer particles on emulsion polymerization, *Journal of Chemical Engineering of Japan*, **4**, pp. 54–60.

46. Nomura, M. (1982). Desorption and reabsorption of free radicals in emulsion polymerization. In: Piirma, I., *Emulsion Polymerization* (Academic Press, USA), pp. 191–219.

47. Nomura, M., Yamamoto, K., Horie, I., Fujita, K., and Harada, M. (1982). Kinetics of emulsion copolymerization. II. Effect of free radical desorption on the rate of emulsion copolymerization of styrene and methyl methacrylate, *Journal of Applied Polymer Science*, **27**(7), pp. 2483–2501.

48. Bird, R. B., Stewart, W. E., and Lightfoot, E. N. (2007). *Transport Phenomena*, Revised 2nd edition (John Wiley & Sons, USA).

49. Grady, M. C. (1996). *Preparation of Omega-Unsaturated Oligo(Methyl Methacrylate) Macromer and Its Application in Emulsion Polymerization: Key Learnings about Radical Desorption*. Doctoral Dissertation (ETH, Switzerland).

50. Ugelstad, J., Mfutakamba, H. R., Mørk, P. C., Ellingsen, T., Berge, A., Schmid, R., Holm, L., Jørgedal, A., Hansen, F., and Nustad, K. (1985). Preparation and application of monodisperse polymer particles, *Journal of Polymer Science: Polymer Symposia*, **72**(1), pp. 225–240.

51. Vanzo, E., Marchessault, R. H., and Stannett, V. (1965). The solubility and swelling of latex particles, *Journal of Colloid Science*, **20**, pp. 62–71.

52. Flory, P. J. (1942). Thermodynamics of high polymer solutions, *The Journal of Chemical Physics*, **10**(1), pp. 51–61.

53. Huggins, M. L. (1942). Theory of solutions of high polymers, *Journal of the American Chemical Society*, **64**(7), pp. 1712–1719.

54. Morton, M., Kaizerman, S., and Altier, M. W. (1954). Swelling of latex particles, *Journal of Colloid Science*, **9**, pp. 300–312.

55. Tauer, K., Nozari, S., Ali, A. I., and Kozempel, S. (2005). Sorption of hydrophobic organic compounds by aqueous latexes, *Macromolecular Rapid Communications*, **26**(15), pp. 1228–1232.

56. Vanzo, E., Marchessault, R. H., and Stannett, V. (1965). The solubility and swelling of latex particles, *Journal of Colloid Science*, **20**, pp. 62–71.

57. Tauer, K., Kaspar, H., and Antonietti, M. (2000). Equilibrium swelling of colloidal polymeric particles with water-insoluble organic solvents, *Colloid and Polymer Science*, **278**(9), pp. 814–820.

58. Antonietti, M., Kaspar, H., and Tauer, K. (1996). Swelling equilibrium of small polymer colloids: Influence of surface structure and a size-dependent depletion correction, *Langmuir*, **12**(26), pp. 6211–6217.

59. Kaspar, H. (1996). *Untersuchungen zur Gleichgewichtsquellung polymerer Nanopartikel*. Doctoral Dissertation (Universität Potsdam, Germany).

60. Ballard, M. J., Napper, D. H., and Gilbert, R. G. (1984). Kinetics of emulsion polymerization of methyl methacrylate, *Journal of Polymer Science: Polymer Chemistry Edition*, **22**(11), pp. 3225–3253.

61. Flory, P. J. (1953). *Principles of Polymer Chemistry* (Cornell University Press, USA).

62. Bodenstein, M. (1913). Eine theorie der photochemischen reaktionsgeschwindigkeiten, *Zeitschrift für Physikalische Chemie*, **85**(1), pp. 329–397.

63. Flory, P. J. (1937). The mechanism of vinyl polymerizations, *Journal of the American Chemical Society*, **59**, pp. 241–253.

64. Ahmad, N. M., Heatley, F., and Lovell, P. A. (1998). Chain transfer to polymer in free-radical solution polymerization of n-butyl acrylate studied by NMR spectroscopy, *Macromolecules*, **31**(9), pp. 2822–2827.

65. O'Driscoll, K. F. (1981). Free radical polymerization kinetics: Revisited, *Pure and Applied Chemistry*, **53**(3), pp. 617–626.

66. Stickler, M. (1987). Experimental techniques in free radical polymerization kinetics, *Macromolecular Symposia*, **10**(1), pp. 17–69.

67. Chern, C. S. (2006). Emulsion polymerization mechanisms and kinetics, *Progress in Polymer Science*, **31**(5), pp. 443–486.

68. Tauer, K., Mukhamedjanova, M., Holtze, C., Nazaran, P., and Lee, J. (2007). Unusual kinetics in aqueous heterophase polymerizations, *Macromolecular Symposia*, **248**(1), pp. 227–238.

69. Tauer, K. and Hernandez, H. F. (2010). Molecular aspects of radical polymerizations: The propagation frequency, *Macromolecular Rapid Communications*, **31**(5), pp. 419–442.

70. Hernandez, H. F. and Tauer, K. (2008). Stochastic simulation of imperfect mixing in free radical polymerization, *Macromolecular Symposia*, **271**(1), pp. 64–74.

71. Achilias, D. and Kiparissides, C. (1988). Modeling of diffusion-controlled free-radical polymerization reactions, *Journal of Applied Polymer Science*, **35**(5), pp. 1303–1323.

72. Tulig, T. J. and Tirrell, M. (1981). Molecular theory of the Trommsdorff effect, *Macromolecules*, **14**(5), pp. 1501–1511.

73. Olaj, O. F., Vana, P., Zoder, M., Kornherr, A., and Zifferer, G. (2000). Is the rate constant of chain propagation kp in radical polymerization really chain-length independent? *Macromolecular Rapid Communications*, **21**(13), pp. 913–920.

74. Hernandez, H. (2019). Discretization of probability distributions: Random, deterministic and randomistic sampling, *ForsChem Research Reports*, **4**, 2019-11. doi: 10.13140/RG.2.2.11389.92643.

75. Doncom, K. E., Blackman, L. D., Wright, D. B., Gibson, M. I., and O'Reilly, R. K. (2017). Dispersity effects in polymer self-assemblies: A matter of hierarchical control, *Chemical Society Reviews*, **46**(14), pp. 4119–4134.

76. Chung-li, Y., Goodwin, J. W., and Ottewill, R. H. (1976). Studies on the preparation and characterisation of monodisperse polystyrene latices 4: The preparation of latex particles with a size greater than 1 μm, *Progress in Colloid & Polymer Science*, **60**, pp. 163–175.

77. Goebel, K. H., Schneider, H. J., Jaeger, W., and Reinisch, G. (1982). Reaktionsabhängige Teilchenvereinigung bei der emulsionspolymerisation von vinylchlorid, *Acta Polymerica*, **33**(1), pp. 49–62.

78. Tauer, K., Paulke, B. R., Müller, I., Jaeger, W., and Reinisch, G. (1982). Zur zeitlichen Änderung der Teilchenzahlen bei der Emulsionspolymerisation von Vinylchlorid, *Acta Polymerica*, **33**(4), pp. 287–289.

79. Tauer, K. and Müller, I. (1984). Mathematische Modellierung der Emulsionspolymerisation von Vinylchlorid. I. Modellierung des Verlaufs der diskontinuierlichen Emulsionspolymerisation von Vinylchlorid, *Acta Polymerica*, **35**(6), pp. 478–480.

80. Tauer, K., Reinisch, G., Gajewski, H., and Müller, I. (1991). Modeling of emulsion polymerization of vinyl chloride, *Journal of Macromolecular Science: Part A Chemistry*, **28**(3–4), pp. 431–460.

81. Derjaguin, B. V., Churaev, N. V., and Muller, V. M. (1987). *Surface Forces* (Springer, USA), pp. 293–310.

82. Fixman, M. (1979). The Poisson–Boltzmann equation and its application to polyelectrolytes, *The Journal of Chemical Physics*, **70**(11), pp. 4995–5005.

83. Gouy, M. (1910). Sur la constitution de la charge électrique à la surface d'un electrolyte, *Journal of Physics: Theories and Applications*, **9**(1), pp. 457–468.

84. Okubo, T. (2003). Electro-optic properties of giant colloidal crystals, *Progress in Colloid and Polymer Science*, **124**, pp. 112–115.

85. Tadmor, R., Hernández-Zapata, E., Chen, N., Pincus, P., and Israelachvili, J. N. (2002). Debye length and double-layer forces in polyelectrolyte solutions, *Macromolecules*, **35**(6), pp. 2380–2388.

86. Pincus, P. (1991). Colloid stabilization with grafted polyelectrolytes, *Macromolecules*, **24**(10), pp. 2912–2919.

87. Tauer, K., Müller, H., Rosengarten, L., and Riedelsberger, K. (1999). The use of polymers in heterophase polymerizations, *Colloids and Surfaces A: Physicochemical and Engineering Aspects*, **153**(1–3), pp. 75–88.

88. Smoluchowski, M. (1916). Drei Vorträge über Diffusion, Brownsche Molekularbewegung und Koagulation von Kolloidteilchen, *Zeitschrift für Physik*, **17**, pp. 557–585.

Chapter 3

Multiscale Modeling of Heterophase Polymerization

Heterophase polymerization is a dynamic process in which many different simultaneous and sometimes competitive chemical and physical events occur at very different time and length scales. The events taking place in heterophase polymerization processes may occur at rates ranging between 10^0 and 10^{18} s^{-1} and involving entities of very different length scales, such as atoms, ions, and molecules (<1 nm); macromolecules (1–10 nm); and polymer particles and monomer droplets (10 nm to 10 mm). The multiscale nature of heterophase polymerization can be appreciated in Fig. 3.1, where at least seven relevant different length scales can be identified, which will be described in further detail in this chapter.

In heterophase polymerization, there is always a continuous phase that contains dissolved molecules and/or macromolecules and also contains all dispersed phases. The dispersed phases are present in a wide range of sizes, from nanometer-scale molecular aggregates (e.g., precipitated polymers, amphiphilic aggregates, etc.) to millimeter-sized monomer droplets and polymer particles. All these dispersed phases can be considered clusters or aggregates of molecules (and macromolecules), and all of them are correspondingly subjected to the processes of diffusive and convective motion, aggregation (including absorption), and dissociation (including desorption).

Heterophase Polymerization: Basic Concepts and Principles
Hugo Hernandez and Klaus Tauer

ISBN 978-981-4877-32-9 (Hardcover), 978-1-003-11929-6 (eBook)
www.jennystanford.com

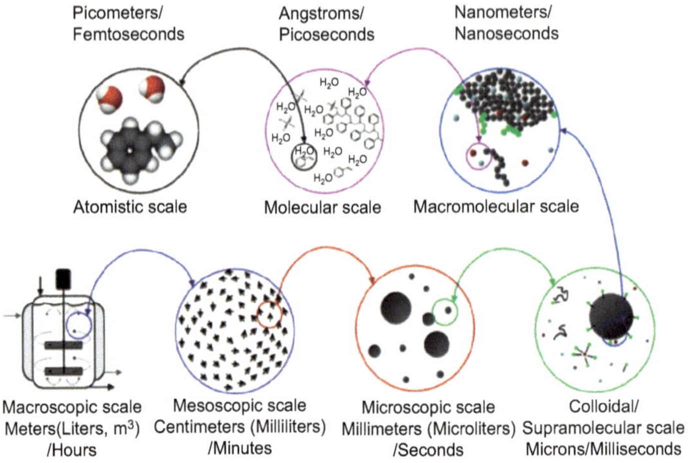

Figure 3.1 Multiscale representation of heterophase polymerization.

In previous chapters, it was shown how the main factors determining the motion, dissociation, and aggregation of clusters and individual molecules are the magnitude and direction of intermolecular forces acting between neighboring molecules. Because of this, it would be possible to get a complete picture of heterophase polymerization just by simulating the dynamics of the molecules in the system (i.e., by ***molecular dynamics simulation***); however, the computational requirement of performing a simulation for reasonable time and length scales is extremely high and at present, such a task is practically not viable. Much larger time and length scales can be reached using Brownian dynamics as it was exemplified in previous chapters for many different processes taking place in heterophase polymerization. However, these methods are not enough to simulate processes at an industrial scale. For these purposes, mesoscopic and/or macroscopic methods, such as *stochastic kinetic Monte Carlo* (kMC) or deterministic numerical integration of differential equations, can be used. Although **deterministic methods offer fast estimations of average values, stochastic methods can handle the variability of the process**.

Molecular dynamics simulation: Computational method used to simulate the motion of atoms and/or molecules in space, using Newton's equations of motion.

In general, polymer systems are difficult to simulate because of the wide spectrum of time and length scales characterizing their dynamics and structure. In mathematical terms, such systems are considered *stiff*. If the scale of the simulation is increased in order to observe slower dynamics, the fast dynamics at lower scales must be either neglected or approximated. Depending on the type of approximation and its accuracy, the results obtained in the simulation can or cannot represent the real process. In this sense, a suitable ***multiscale simulation*** algorithm capable of simultaneously considering all relevant dynamics would be desirable.

Multiscale simulation: Computational strategy used to simulate a physical system at different scales or levels of information.

Multiscale simulation can be defined as the enabling technology of science and engineering that links phenomena, models, and information between various scales of complex systems [1]. A very important aspect of multiscale modeling is the processing and exchange of information between the different scales. Linking widely different scales (such as those presented in Fig. 3.1) is done following different modeling strategies: the parallel or concurrent, the serial or sequential, and the adaptive resolution strategy [2, 3]. In the parallel approach, different-scale techniques are implemented simultaneously in the same *computational domain*, that is, calculations at each scale are performed concurrently during the same simulation. In the serial approach, lower-scale models require information about the state of the system (e.g., temperature, velocity, composition, etc.), which is determined after simulating a higher scale, while at the same time the upper-scale model requires parametric and structural information of the system (e.g., diffusion coefficients, molecular mass distributions, particle size distributions (PSD), etc.) obtained from simulations at the lower scale. Therefore, top-down and bottom-up information exchange procedures must be clearly defined [3–5]. In the top-down procedure, a suitable grid decomposition method based on the distribution of states of the corresponding system scale must be used, while in the bottom-up procedure, the integration of the lower-scale results must be performed. The challenge is the seamless coupling between the various models while meeting conservation laws, numerical convergence, and stability.

Some of the most important simulation methods relevant to heterophase polymerization include molecular dynamics (MD) simulation, Brownian dynamics (BD) simulation, dissipative particle dynamics (DPD), kMC simulation, quantum mechanics (QM), Monte Carlo (MC) simulation, Lattice–Boltzmann (LB), coarse-grained (CG) simulation, and finite element methods (FEMs). Properly executed computer simulations can provide the solution to any well-defined problem; thus, they can be used to test the validity and accuracy of analytic theories. Of course, the correctness of the simulation results depends on the use of the correct values of the simulation parameters, which can only be established by comparison with real experiments.

The modeling of each scale in heterophase polymerization will be considered in detail in the following sections, considering a bottom-up approach, i.e., starting with the atomistic scale and finishing with the macroscopic scale. A graphical representation of the different time and length scales (approximately) considered at each scale is presented in Fig. 3.2. Finally, an overview of multiscale integration will be discussed in Section 3.8.

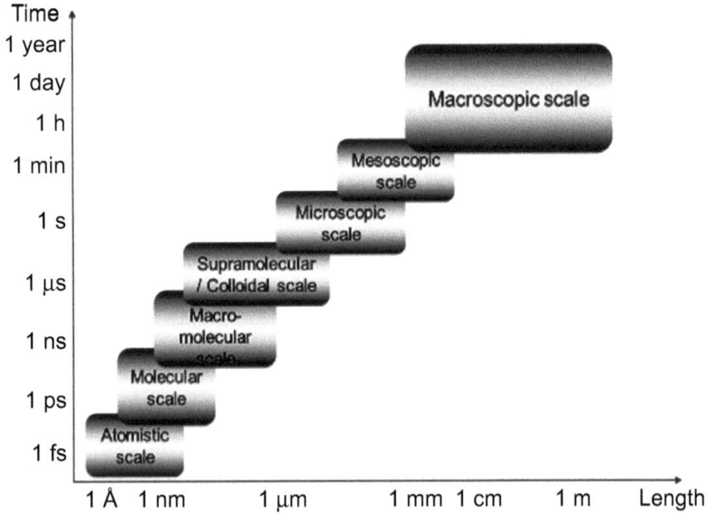

Figure 3.2 Superposition of relevant scales for heterophase polymerization. Characteristic time and length scales are included.

3.1 Atomistic Scale

Many macroscopic phenomena originate from elementary processes that take place in the atoms. The atomistic scale considers the internal dynamics of atoms, i.e., the behavior of matter at a subatomic scale. Table 3.1 shows the properties of the most relevant subatomic particles, considered in polymerization processes. The knowledge of the forces between subatomic particles should, in principle, be sufficient to determine the behavior of matter at all scales. The corresponding space and time scales considered are in the order of 10^{-15} m to 10^{-10} m (corresponding to the range of sizes from the nucleus to the electronic cloud around a single atom) and 10^{-18} s to 10^{-15} s (from the relaxation time of an electron to the vibration period of a molecular bond). At this scale, the governing equation describing the interaction between subatomic particles is the **Schrödinger equation** (Eq. 3.1). It is worth noticing that the Schrödinger equation is a fundamental equation, which does not involve any empirical parameter but only fundamental constants such as the mass and charge of the electron, and Planck constant. The more general time-dependent Schrödinger equation is expressed as follows:

$$i\hbar\frac{\partial}{\partial t}\Psi(t)=\widehat{H}\Psi(t) \tag{3.1}$$

where $\Psi(t)$ is the wavefunction of the system, ($|\Psi^2|$ describes the probability of finding a particle in space), $\widehat{H}$ is the Hamiltonian operator (representing the total energy of the system, as a function of its wavefunction), $\hbar$ is the reduced Planck constant (1.055×10^{-34} J·s), i is the unit imaginary number ($\sqrt{-1}$), t is time, and $\partial/\partial t$ is the partial derivative with respect to time.

Table 3.1 Relevant subatomic particles and their properties

Subatomic particle	Particle type	Mass (g)	Charge (C)
Electron	Lepton	9.107×10^{-28}	-1.602×10^{-19}
Proton	Baryon	1.6725×10^{-24}	$+1.602\times10^{-19}$
Neutron	Baryon	1.6725×10^{-24}	0

At stationary states of the system, the Schrödinger equation can be simplified into its time-independent form:

$$E\Psi = \widehat{H}\Psi \tag{3.2}$$

where E is the energy of the wavefunction at the stationary state.

Even though Eq. 3.1 and Eq. 3.2 are relatively simple expressions, due to the analytical complexity of the wavefunction (Ψ), exact analytical solutions to the Schrödinger equation are only possible for just a few cases, the most notable one being the Hydrogen atom [6]. On the other hand, the numerical solution to the Schrödinger equation for a particular system becomes difficult as the size of the system increases, because one equation is needed for each electron or nucleus (conformed by protons and neutrons) present in the system. In addition, for larger systems, the range of characteristic energies considered is wider, leading to larger numerical errors. Thus, the numerical solution to the Schrödinger equation is usually restricted by computational capability to just individual atoms or very small molecules. For these reasons, some strategies have been proposed for efficiently solving the Schrödinger equation in larger systems, including the *Born–Oppenheimer* (BO) [7] and the *Hartree–Fock* (HF) [8] approximations.

The BO approximation is based on the fact that the mass of nuclei is three orders of magnitude larger than the mass of electrons, and since their momenta is in the same order of magnitude, the velocities of the nuclei are negligible compared to the velocities of the electrons. Thus, the BO approximation assumes that all nuclei are stationary (classical point-like particles with zero kinetic energy) and calculates the energy of the system for this ***electronic ground-state*** configuration. The BO approximation is the basis of the concept of potential surface energy from which the geometry of chemical structures is extracted [9].

Electronic ground state: State of lowest energy of the electrons in the system.

The solution to the problem after the BO approximation is still difficult because the electrons are correlated as a result of inter-electronic interactions. In the HF approximation, the wavefunction of the system is assumed to be the sum of separable basis functions, representing each electron. Since the exact function representing each electron is not known, a set of basis functions is assumed, and the total energy of the system is determined. Then, by means of an iterative process, the set of basis functions is modified until a minimum value for the ground-state energy is found, which would be the best approximation to the exact ground-state energy of the system. The basis functions are usually based either on *Slater-type orbitals* (STO), which are radial functions containing a long-range exponential decay contribution term (Eq. 3.3), or on *Gaussian-type orbitals* (GTO), which contains a Gaussian distribution term (Eq. 3.4).

$$R(r) = Ar^{n-1}e^{-\alpha r} \tag{3.3}$$

$$R(r) = Ar^{n-1}e^{-\alpha r^2} \tag{3.4}$$

where $R(r)$ describes the radial orbital function, r is the distance of the electron to the nucleus, A is a normalization constant, n is a natural number, and α is a constant related to the effective charge of the nucleus.

Methods using the direct solution to the Schrödinger equation or any of the previous approximations (BO or HF) are usually denoted as *ab initio* methods, as they do not require any empirical parameter, but only the first principles of quantum mechanics and fundamental physical constants. Due to computational limitations, *ab initio* methods are suitable only for systems of up to thousands of atoms during up to just a few picoseconds (10^{-12} s) of simulation.

Some additional approximations allow the simulation of larger systems, but at the price of reduced accuracy in the calculations. These methods are usually known as *parameterized methods* since they incorporate experimental parameters, such as atom sizes, bond lengths, bond angles, and torsion angles. The parameterized methods can be either *semi-empirical* or *empirical.* The semi-empirical methods include the free-electron molecular orbital (MO) method, the Pariser–Parr–Pople (PPP) MO method, the Hückel MO method, and the extended Hückel method [10]. Empirical methods only consider the nuclei, completely neglecting the electrons in the atoms, and thus, they belong to the next scale: the molecular scale.

An additional method at the atomistic scale is the *density functional theory* (DFT) method, where the wavefunction of the system is not calculated directly, but instead the electronic probability density (ρ) for the system is determined and then the electronic energy of the system is calculated as a function of ρ, according to the Hohenberg–Kohn theorems [11] and the Kohn–Sham (KS) approximation [12].

Atomistic modeling is useful for determining fundamental properties of molecular species such as their electronic structures, which are useful for determining the kinetics and thermodynamics of chemical reactions, i.e., rate of reaction and equilibrium constants [13]. It is also used for determining the effect of the solvents on chemical reactions. It is particularly useful for investigating chemical reactions involving radicals, such as the free-radical polymerization reactions and reversible addition-fragmentation chain transfer (RAFT) polymerization [14–19]; although other types of polymerization reactions have also been considered [20–22]. For polymerization reactions in general, and for radical polymerization in particular, atomistic modeling allows the determination of copolymerization ratios and individual reaction rates for each molecular or macromolecular species [23–26].

3.2 Molecular Scale

At the molecular scale, the behavior of subatomic particles, such as electrons or protons, is no longer relevant, but the interaction between individual atoms and/or molecules is considered as a whole. There are two main types of simulation methods at the molecular scale: (1) stochastic (Monte Carlo) and (2) deterministic (molecular dynamics).

3.2.1 Monte Carlo Simulation

The Monte Carlo (MC) simulation is a purely stochastic method, which generates random configurations of the system for determining the lowest energy configuration, and thus obtaining the most probable thermodynamic properties of the system. The different configurations of the system can be generated in a successive or a parallel way or using a combination of both. Parallel generation allows a fast identification of the low-energy configuration region, whereas successive generation allows a more precise identification of the minimum energy configuration of the system. In the successive generation approach, each new configuration of the system is generated by randomly changing the position of individual atoms or molecules. The total energy of the new configuration (U_{new}) is determined as the sum of the potential interaction energy between all pairs of atoms or molecules. Notice that a prerequisite for performing reliable MC simulations lies in the determination of accurate intermolecular interaction potentials. The choice of the best interaction potential functions depends upon the system investigated and on the quality of the force fields available, such as MM2, MM3, and AMBER [27]. Particularly for polymer systems, different molecular force fields have been proposed, including CHARMM (Chemistry at HARvard Macromolecular Mechanics) [28], and PCFF (Polymer Consistent Force Field) [29–30]. However, other general force fields such as DREIDING [31], COMPASS [32], and OPLS [33] have also been used for modeling polymeric systems [34–37].

The energy of the new configuration is then compared with the energy of the previous configuration (U_{old}) in order to accept or reject the new configuration. A very common acceptance criterion, known as the importance sampling scheme or the Metropolis condition [38] is based on the *Boltzmann factor* (f) (Eq. 3.5). If a uniform random number generated between 0 and 1 is lower than the Boltzmann factor, then the new configuration is accepted; otherwise, the new configuration is rejected, and the previous configuration is retained for the next iteration. This procedure is used as a strategy to overcome local energy minima that may lead to erroneous results.

$$f = e^{-\left(\frac{U_{\mathrm{new}} - U_{\mathrm{old}}}{k_{\mathrm{B}}T}\right)} \tag{3.5}$$

For each configuration accepted during the previous procedure, the value of a certain thermodynamic property A is determined. Then, it is possible to determine the ensemble average as

$$\bar{A} = \frac{1}{m}\sum_{i=1}^{m} A_i \tag{3.6}$$

where m is the total number of configurations considered. The total number of configurations required will depend on the thermodynamic property investigated, since some thermodynamic properties converge more rapidly than others. For example, heat capacities require in general a much larger ensemble sampling than internal energies [27].

MC methods are in general less efficient than molecular dynamic techniques. Therefore, MC methods have been applied only to systems where they are more effective, such as liquids or systems in solution. MC methods have been employed for the investigation of polymerization mechanisms and kinetics [39–45], the determination of thermodynamic properties of polymers [46–49], for the description of interfacial phenomena in polymers [50–53], their structural and morphological properties [54–57], and in heterophase polymerization in general [58–65].

3.2.2 Molecular Dynamics Simulation

Molecular dynamics (MD) simulation is a deterministic method used to follow the trajectories and velocities of an ensemble of atoms or molecules subjected to interatomic or intermolecular forces for a certain period of time. Although the atoms and molecules are composed of quantum subatomic particles, their motion can be satisfactorily described by classical Newton's equations of motion. From this information, it is possible to determine static and dynamic properties of the system, such as thermodynamic properties and transport coefficients.

The basic equations of motion employed are Newton's first law (Eq. 3.7) and the definition of velocity (Eq. 3.8):

$$\frac{dv_i}{dt} = \frac{1}{m_i}\sum_{j \neq i} F_{ij} \tag{3.7}$$

$$\frac{dx_i}{dt} = v_i \tag{3.8}$$

where v_i is the velocity, m_i is the mass and x_i is the position of the i-th molecule, F_{ij} is the interaction force between the i-th and j-th molecules, and t is the time. The interaction force is calculated as the negative gradient of the interaction potential (Eq. 1.1). Additional external or internal (mean field) forces can also be considered.

In conventional MD simulations, a system containing a finite number of particles N (atoms and/or molecules) is placed within a usually cubic cell of fixed volume. A set of velocities, usually drawn from the Maxwell–Boltzmann distribution (Eq. 3.9) at a given temperature T, is also selected and assigned in such a way that the net molecular momentum is equal to zero.

$$f_{\mathrm{MB}}(v_k) = \sqrt{\frac{m}{2\pi k_{\mathrm{B}}T}}\, e^{-\left(\frac{mv_k^2}{2k_{\mathrm{B}}T}\right)} \tag{3.9}$$

f_{MB} is the probability of finding a particle of mass m with a velocity v_k in the k-th direction.

The equations of motion are integrated numerically using very short time steps Δt (normally around 10^{-15} to 10^{-14} s). Larger time steps may lead to instability and erroneous results, whereas shorter time steps increase the computation time without improvements in accuracy. At each step, all the forces acting on each atom and/or molecule are calculated from the position-dependent interaction and external potentials and are used in Eq. 3.7 along with the current particle velocities to determine the new velocities one time step ahead. These new velocities are then used together with the current particle positions to determine the new positions one time step ahead (Eq. 3.8). These new positions and velocities are used again for determining the forces and iterating during the whole simulated time interval τ (typically 10^{-11}–10^{-10} s).

The interaction between atoms and/or molecules can be modeled using different interaction potentials, such as the Lennard–Jones potential, the Buckingham potential, the Morse potential, or even relatively simple interaction potentials such as the hard-disk potential, which present no interaction between particles as long as they are not colliding and an infinite repulsion potential when they are in contact or superposing. The interaction potential is responsible for the attraction of atoms or molecules causing clustering or aggregation, and also for the repulsion resulting in elastic collisions, phase separation, and many other phenomena.

The trajectories obtained by MD simulation can be used to determine thermodynamic properties of the system, by averaging over the dynamic history of the system (by using Eq. 3.10). MD simulation, contrary to MC calculations, also allows the study of time-dependent phenomena.

$$\bar{A} = \frac{1}{\tau}\sum_{j=1}^{\tau/\Delta t} A(t_j)\Delta t \tag{3.10}$$

Numerous algorithms have been proposed to numerically integrate the MD equations of motion [66]. The most popular is the finite-difference method proposed by Verlet [67], which computes the position of the i-th particle (x_i) and its velocity (v_i), one time step ahead from the forces and positions at previous times:

$$x_i(t+\Delta t) = 2x_i(t) - x_i(t-\Delta t) + \frac{\sum_{j\neq i} F_{ij}(t)}{m_i}\Delta t^2 \tag{3.11}$$

$$v_i(t+\Delta t) = \frac{x_i(t+\Delta t) - x_i(t-\Delta t)}{2\Delta t} \tag{3.12}$$

The Verlet method (Eqs. 3.11 and 3.12) is easy to implement, is stable at sufficiently small time steps (10^{-15} s to 10^{-14} s), and is relatively fast. Other numerical integration methods of MD equations include the velocity Verlet [68], the Beeman [69], and the leapfrog [70] algorithms, derived from the Verlet scheme; and predictor–corrector techniques such as the methods presented by Rahman [71] and Gear [72].

During the integration of Newton's equations of motion, the energy is conserved. However, slow temperature drifts may occur as a result of the numerical integration and numerical truncations. The simplest method used to keep the system at a constant temperature T_{ref} is rescaling the velocities at appropriate intervals by a factor of $\sqrt{T_{\text{ref}} / T(t)}$, where $T(t)$ is determined using Eq. 3.13.

$$T(t) = \frac{\frac{1}{N}\sum_{i=1}^{N} m_i v_i^2}{3k_{\text{B}}} \tag{3.13}$$

MD simulations have rapidly emerged as a powerful tool to calculate structural and thermodynamic properties of complex liquids, molten salts, crystals, polymers, and proteins in solution [73]. Relevant applications of MD simulation in the field of polymers include

- Polymer thermodynamics [74–77]
- Phase transitions: Swelling [78], glass transition [79–80], crystallization [81], collapse [82], spinodal decomposition [83]
- Polymer fluid dynamics [84–89]
- Transport properties [90–93]

- Vibrational spectra of macromolecules [94–95]
- Mass transfer in heterophase polymerization. The molecular approach also appears to be a very good alternative for the prediction of non-equilibrium monomer concentration inside polymer particles in emulsion polymerization [96].

Some nice reviews of the application of MD simulation to polymers are currently available [97–99].

3.3 Macromolecular Scale

Molecular modeling techniques are suitable for relatively small molecules, but as the size of the molecules increases, new difficulties and complexities arise. In general, macromolecular chains exhibit dynamics along a wide range of length and time scales. For example, the relevant length scales for a polymer chain include the length of single bonds ($\sim 10^{-10}$ m), the ***persistence length*** [100] ($\sim 10^{-9}$ m), and the coil radius ($\sim 10^{-8}$ m). On the other hand, the relevant time scales for a polymer include the vibration of internal bonds ($\sim 10^{-13}$ s) and the relaxation time of the polymer chain ($\sim 10^{-8}$–10^{-4} s) [97]. This means that a full MD simulation of a macromolecule requires the simulation of a cell containing ~ 1000 times more atoms on each side of the cell than for a small molecule, i.e., a system 10^9 times larger. Using the same time step than for the small molecule, it requires making almost 10^{18} times more calculations, thus making the solution to this problem very difficult with conventional computational capabilities.

Persistence length: Characteristic length of a polymer below which the segment behaves as an elastic rod.

To solve this problem, it is necessary to abandon the chemical detail of single atoms and their bonds and introduce a *coarse-grained model*, which consists on the integration of various monomer units into one segment unit. Thus, a macromolecule will consist of just a few segment units, in the same way as a small molecule consists of just a few atoms (Fig. 3.3). Even though the definition of the segments is arbitrary, it is recommended to consider ***Kuhn's effective segments*** [101, 102].

Kuhn segment: Independent statistical chain element of the polymer whose internal dynamics is fast enough to achieve a relaxed state compared to the whole polymer.

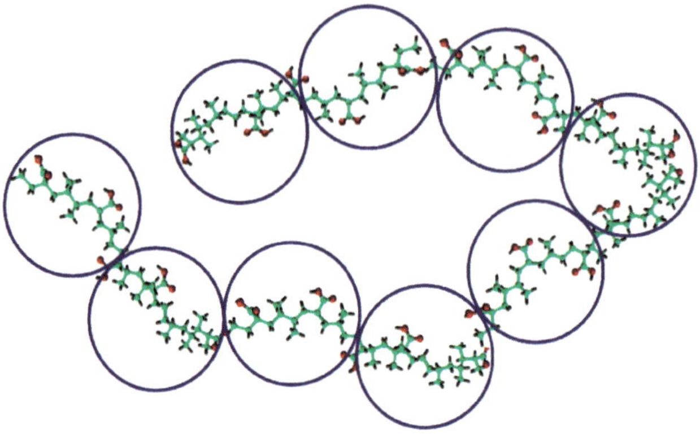

Figure 3.3 Coarse-grained segments of a macromolecular chain (circles).

There are many ways to build coarse-grained models of polymer systems. They can be classified into two general categories: lattice and off-lattice models. In lattice models, the simulation space is represented by a lattice (usually a regular lattice), where the segments of the polymer can only be positioned at the nodes (Fig. 3.4), and the connected segments must always be at neighboring nodes. The segments are allowed to move to adjacent nodes, taking into account that the polymer cannot cross itself at any point (self-avoiding walk).

In off-lattice models, each segment can be positioned anywhere in space, as long as the bond distance between segments is conserved. Some types of off-lattice coarse-graining models include the freely joint chain, the pearl necklace model, and the bead-spring (Rouse) model [103] among others (Fig. 3.5).

The interaction between segments can also be modeled in different ways. The simplest way is to keep constant bond lengths, changing only the rotation angle of the bonds. More realistic models include the Lennard–Jones potential (Eq. 1.11), the harmonic bond potential (Eq. 3.14), and the finite extensible nonlinear elastic (FENE) bond potential (Eq. 3.15).

$$V_{\mathrm{H}}(l)=\begin{cases} k(l-l_0)^2 & l_{\min}<l<l_{\max} \\ \infty & l\le l_{\min} \wedge l\ge l_{\max} \end{cases} \tag{3.14}$$

$$V_{\mathrm{FENE}}(l)=\begin{cases} -\dfrac{kl_{\max}^2}{2}\ln\left[1-\left(\dfrac{l}{l_{\max}}\right)^2\right] & l<l_{\max} \\ \infty & l\ge l_{\max} \end{cases} \tag{3.15}$$

$V_{\mathrm{H}}(l)$ is the harmonic bond potential, $V_{\mathrm{FENE}}(l)$ is the FENE bond potential, l is the bond length between segments, $l_{\min}$ is the minimum bond length, $l_{\max}$ is the maximum bond length, l_0 is the equilibrium bond length ($l_{\min} < l_0 < l_{\max}$), and k is an interaction parameter.

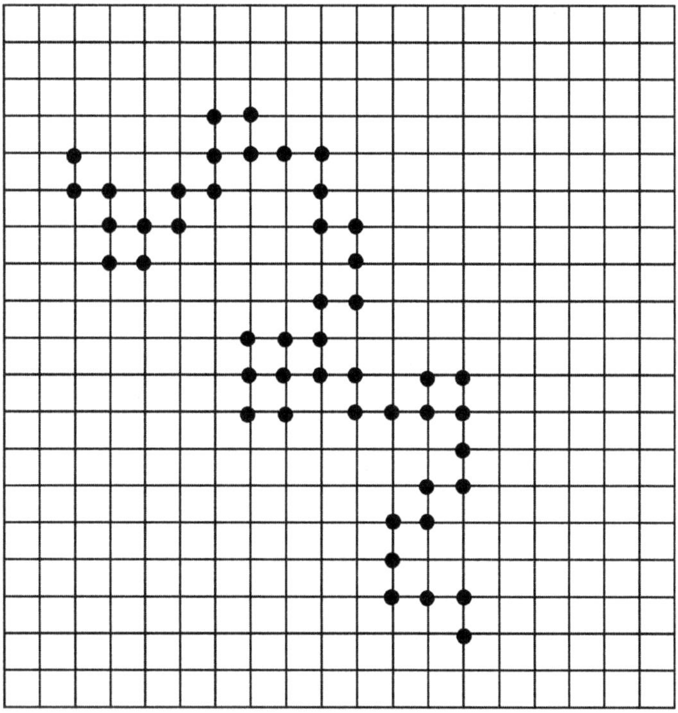

Figure 3.4 Two-dimensional lattice coarse-graining of a macromolecular chain.

The field of computer simulation at the macromolecular scale is a very active area of research [104–108]. Some applications include large-scale simulations of polymer networks [109–111], determination of structural properties of polymers [112], polymer solutions and polyelectrolytes [113–115], polymer brushes [116–117], block copolymer mesophase ordering [118], etc.

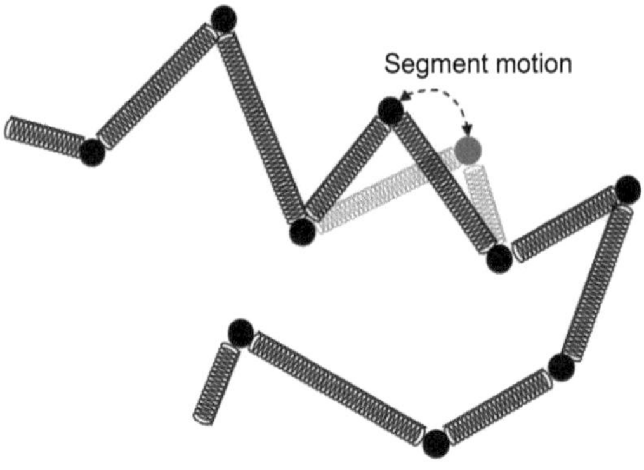

Figure 3.5 Bead-spring off-lattice coarse-graining of a macromolecular chain.

3.4 Supramolecular and Colloidal Scale

The close interaction of molecules and macromolecules gives rise to a higher scale involving both *supramolecular* and *colloidal* phenomena. ***Supramolecularity*** is a property of basically all molecules present in a molecular neighborhood [119]. The idea of *supramolecular chemistry* was first proposed by Jean-Marie Lehn [120] for describing intermolecular forces involved in molecular recognition processes, chemistry of receptors, and catalysis. The original intended difference between the general intermolecular interactions proposed by Johannes van der Waals [121] and supramolecular interactions is the geometrical characteristic of the latter. In other words, general intermolecular interactions occur in all directions, whereas supramolecular interactions only occur in specific directions. The concept of supramolecular interaction has evolved to consider in general any non-covalent interaction between molecules [122]. Thus, it can be stated that as a result of intermolecular/supramolecular interactions, molecular aggregates are formed.

Supramolecularity: Presence of reversible non-covalent extramolecular interactions and associations between chemical compounds.

Molecular aggregates represent very important entities in heterophase polymerization. First of all, polymer particles are aggregates of macromolecules and other small molecules. Thus, in principle, they are molecular aggregates bonded by non-covalent intermolecular interactions between the (macro-)molecules in the particle. Amphiphilic molecules, usually employed for stabilizing the particles, may also aggregate forming self-assembled micelles [123].

Molecular aggregation (such as during nucleation processes within the frame of classical nucleation theory) can be modeled considering the following general step-wise process:

$$A_{s-1} + A_1 \underset{k_s^-}{\overset{k_s^+}{\Leftrightarrow}} A_s \tag{3.16}$$

where A_1 represents an individual aggregatable molecule, A_{s-1} is a molecular aggregate containing $s - 1$ molecules, and k_s^+ and k_s^- are rate coefficients of aggregation/disaggregation for an aggregate of s molecules. This step-wise model considers that molecular aggregates gain or lose only one molecule at a time. Of course, molecular aggregates can combine or split in groups of molecules, but this case will be considered later. Representative examples of the step-wise aggregation model include surfactant micellization [124], and isodesmic supramolecular polymerization proceeding exclusively by monomer association and dissociation events [125].

For this model, the rate of aggregation into a molecular aggregate of size s is

$$\frac{d[A_s]}{dt} = J_{s-1} - J_s \tag{3.17}$$

where

$$J_{s-1} = k_s^+[A_1][A_{s-1}] - k_s^-[A_s] \tag{3.18}$$

J_{s-1} is the flux of aggregates from size $s - 1$ to size s in the space of cluster sizes (the net number of aggregates of size $s - 1$ absorbing one additional molecule per unit time), and $[A_s]$ is the concentration of molecular aggregates of size s. In this approach, the aggregation (or disaggregation) process involves the transfer of an individual molecule from (or to) the continuous phase.

The thermodynamic formulation of the aggregation process completely neglects dynamics and considers that the system reaches equilibrium when the Gibbs free energy is at its minimum [126].

$$\mu_s^0 + k_{\mathrm{B}}T\ln X_s = s\left[\mu_1^0 + k_{\mathrm{B}}T\ln X_1\right] \tag{3.19}$$

where $\mu_s{}^0$ is the standard chemical potential of a molecular aggregate of size s (assuming all molecules identical), and X_s is the mole fraction of these aggregates in the solution. As it can be seen in Eq. 3.19, the transfer of individual molecules toward molecular aggregates is driven by the differences in chemical potential between the aggregates and the individual molecules. Implementation of this model requires a very precise determination of the chemical potentials [127].

In addition, molecular aggregation kinetics has also been modeled using classical or modern nucleation theories [128, 129]. This can be explained by the fact that nucleation can be considered as spontaneous aggregation of individual entities (atoms, molecules, or ions), where only aggregates above a certain size are stable. At this critical size, the Gibbs free energy of the aggregate is maximal. Aggregates with sizes smaller than the critical size have a limited probability of growing because the free energy increases with each aggregation step. However, the growth of subcritical aggregates is favored by the presence of thermal fluctuations. When a molecular aggregate reaches its critical size, its growth becomes spontaneous because from that moment on, each further aggregation step leads to a decrease in the free energy. The flux of aggregates is assumed to be determined by the free energy of formation of an aggregate according to the Boltzmann's distribution:

$$J_s = n_s^+\left[A_s\right]\left(1-\exp\left(\frac{\Delta W_s + \ln\left(\frac{\left[A_{s+1}\right]}{\left[A_s\right]}\right)}{k_{\mathrm{B}}T}\right)\right) \tag{3.20}$$

where $n_s{}^+$ is the number of molecules absorbed from the continuous phase by the aggregate of size s per unit of time, and ΔW_s is the difference in aggregation work between the aggregate of size $s + 1$ and the aggregate of size s, which is defined by

$$W_s = \mu_s - s\mu_1 \tag{3.21}$$

μ_s is the chemical potential of the molecular aggregates of size s.

n_s^+ is usually determined using Smoluchowski's equation (Eq. 2.13), which is basically the solution to Fick's equation for the absorption of molecules by a molecular aggregate in infinite dilution.

Alternatively, MD simulation can also be used for investigating the molecular aggregation process [130]. This method requires defining all different interaction potentials present in the system. Although computationally demanding, this method also allows the description of the dynamics of molecular aggregation as well as of its equilibrium properties.

As it was previously anticipated, aggregation and dissociation may involve smaller aggregates or fragments, instead of individual molecules. Such general aggregation scheme is summarized as follows:

$$A_i + A_j \underset{k_{i,j}^-}{\overset{k_{i,j}^+}{\Leftrightarrow}} A_{i+j} \tag{3.22}$$

where $k_{i,j}^+$ and $k_{i,j}^-$ represents the kinetic rate coefficients for aggregation/fragmentation of a molecular aggregate involving fragments of sizes i and j.

At equilibrium,

$$k_{i,j}^+ = K_{i,j} k_{i,j}^- \tag{3.23}$$

where $K_{i,j}$ is the equilibrium constant involving fragments of sizes i and j.

Mavelli [131] proposed a solution to this aggregation model using Gillespie's stochastic simulation algorithm (SSA) [132], also known as kMC. The equilibrium constants were determined using the following expression:

$$K_{i,j} = \frac{ij}{(i+j)[A]} \frac{P(i+j)}{P(i)P(j)} \tag{3.24}$$

where

$$[A] = \sum_{i=1}^{n_{\max}} i[A_i] \tag{3.25}$$

is the total concentration of aggregatable molecules, $n_{\max}$ is the maximum aggregation number considered, and $P(i)$ is a Boltzmann-type probability distribution of aggregate sizes given by

$$P(i) = \frac{\exp\left(-\frac{E_i}{k_B T}\right)}{\sum_j \exp\left(-\frac{E_j}{k_B T}\right)} \tag{3.26}$$

where

$$E_i = i\left(\frac{i - n_{\max}}{n_{\max}}\right)^2 E_0 \tag{3.27}$$

Furthermore, the fragmentation rate coefficient was determined as follows:

$$k_{i,j}^- = (i+j)\frac{i!\,j!}{(i+j-1)!}k_0^- \tag{3.28}$$

In this case, k_0^- and E_0 are parameters of the model.

Fragmentation and fragment aggregation play an important role in peptide self-assembly, in supramolecular polymerization, and in other types of aggregation sensitive to mechanical agitation [125].

Also for heterophase polymerization, particle nucleation can be considered being the result of molecular aggregation [133, 134]. However, the situation is complicated as both the concentration and the physicochemical properties of the aggregating species change with time. Moreover, the presence of other segregated phases strongly influences the nucleation process due to interactions with the molecular aggregates. This interaction lowers the free energy of nucleation and thus increases the probability of particle formation. There is experimental evidence that this so-called "*heterogeneous nucleation*" is not only crucial for the formation of clouds and rain drops in the atmosphere [135], but also for particle nucleation in heterogeneous polymerization [136].

The molecules conforming a stable molecular aggregate or *nucleus* can no longer be considered dissolved in the continuous phase; they already form a segregated phase dispersed in the continuous phase. As these aggregates grow by the incorporation of additional molecules, their sizes soon reach the ***colloidal*** scale

(10^{-9}–10^{-6} m). Systems consisting of dispersions of colloidal entities are denoted as ***colloids***. The term colloid was first used by Thomas Graham, considered the founder of *colloid chemistry* [137], to denote very high molecular weight materials. He probably borrowed the term that was used at the time to describe gelatin-like substances [138], as he considered gelatin to be the representative of this type of materials [139]. However, 50 years later, this definition was revised by different scientists, who led by Wolfgang Ostwald founded the *Kolloid-Gesellschaft* in 1922 [140]. Their new definition considered colloids based only on their size, irrespective of their molecular weight.

Colloid: A colloid is a system where the interfacial energy significantly contributes to its overall energy. For dispersion colloids, this is the case when there is a large enough number of molecular aggregates (or discontinuities between them) having at least one dimension approximately between 1 nm and 1 μm.

Entities at the colloidal scale are characterized by Brownian motion (cf. Section 1.3.1), which can be mathematically described by Langevin's stochastic equation (Eq. 1.24). For a completely relaxed system, the solution to the equations of Brownian motion can be performed using BD simulation (cf. Section 1.3.2).

BD simulation is a powerful tool for describing processes at the colloidal scale. It has very good predictive capabilities and can be used in complex systems [141–143], especially macromolecules [144–151], interfaces [152–154], and heterophase polymerization [155, 156]. It is, however, limited to a relatively small number of colloidal particles or molecules depending on the available computational power.

Another approach for simulating colloidal systems is DPD [157–161]. In the DPD (Dissipative Particle Dynamics) method, the total force (F_T) acting on a colloidal particle is the result of three forces: a conservative force (F_C), a dissipative force (F_D), and a random force (F_R):

$$F_{T,i} = \sum_{j \neq i} \left(F_{C,ij} + F_{D,ij} + F_{R,ij} \right) \qquad (3.29)$$

where $F_{C,ij}$ is given by the interaction potential between particles, $F_{D,ij}$ is a force opposing the relative motion of particle i with respect to particle j (Eq. 3.30), and $F_{R,ij}$ is a Gaussian white-noise random force, which depends on the relative position between particles i and j (Eq. 3.31).

$$F_{D,ij} = -\gamma w_D(r_{ij})(\hat{r}_{ij} \cdot \vec{v}_{ij})\hat{r}_{ij} \tag{3.30}$$

$$F_{R,ij} = \sigma w_R(r_{ij})\hat{r}_{ij}\xi_{ij} \tag{3.31}$$

where γ is the friction coefficient, $w_D(r_{ij})$ and $w_R(r_{ij})$ are weight functions depending on the interparticle distance r_{ij} (see for example Eq. 3.32), $\hat{r}_{ij}$ is the unit vector describing the position between particles i and j, $\vec{v}_{ij}$ is the relative velocity between particles i and j, σ is the amplitude of the noise, and ξ_{ij} is a normalized Gaussian white-noise random number. Additionally, the dissipative and the random forces must satisfy the fluctuation-dissipation theorem (Eq. 3.33) in order to guarantee the thermal equilibrium of the system.

$$w_R(r) = \sqrt{w_D(r)} = \begin{cases} (1-r)^2 & r < 1 \\ 0 & r \geq 1 \end{cases} \tag{3.32}$$

$$\sigma = \sqrt{2k_B T\gamma} \tag{3.33}$$

The numerical solution for DPD simulation can also be done using the Verlet algorithm. One of the most important features of DPD is that by considering the fluctuation-dissipation theorem, contrary to the other methods, it preserves the hydrodynamics of the system. DPD simulation is also a very popular alternative for simulating macromolecular systems at the colloidal scale [162–169]. Other simulation methods used at the colloidal scale are discussed by Park et al. [170].

3.5 Microscopic Scale

The microscopic scale can be found beyond the colloidal domain but still below the *thermodynamic limit* of a system. At the thermodynamic limit, the number of molecules present in the system is large enough so that the overall behavior of the system becomes deterministic instead of probabilistic. This limiting number of molecules depends on the particular system considered, as well as on the property of interest. For example, the interfacial tension of single colloidal entities, such as water nuclei in air, just after the nucleation event (consisting of much less than 100 molecules), already corresponds to the observed macroscopic value [171].

A graphical explanation of the thermodynamic limit is presented in Fig. 3.6. Let us assume that every single molecule in the system behaves according to an arbitrary probability distribution function PDF (a uniform distribution is used in Fig. 3.6a as an example) for a certain variable *x*. When two molecules are considered, the average PDF is no longer distributed according to the original PDF (Fig. 3.6b). If the effect of several independent molecules is added, according to the central limit theorem [172, 173], a normal or Gaussian probability distribution function will be obtained (Fig. 3.6c). This effect can be easily understood considering the characteristic functions of the probability distributions (and particularly the mathematical nature of the binomial coefficient for large numbers) [173, 174]. If the number of events (or molecules) considered is increased, the width of the Gaussian distribution (characterized by the standard deviation) is reduced (Fig. 3.6d). At a certain critical point, the thermodynamic limit of the system, the number of events considered is so large that the standard deviation of the distribution becomes negligible and the distribution resembles a single line (i.e., Dirac's delta function) (Fig. 3.6e).

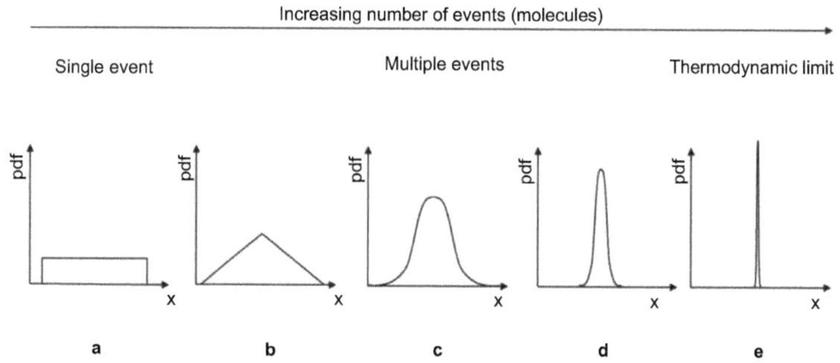

Figure 3.6 Graphical representation of the thermodynamic limit of a system for an arbitrary variable *x*, evolving from single molecular events.

Microscopic stochastic modeling comprises methods with different degrees of complexity: methods with temporal resolution only, such as kMC methods [132, 175–178]; methods with temporal and spatial resolution [179–181]; and methods with single-molecule temporal and spatial resolution [182]. Due to its discrete nature, this microscopic stochastic approach is ideally suited for heterogeneous reaction systems where reactants of very different size and mobility participate.

The kMC method, also known as the stochastic simulation algorithm (SSA), was formally introduced by Gillespie [132] as a method for obtaining singular realizations of processes involving chemical reactions described by the chemical master equation (CME) [174, 183], as an alternative to the deterministic solutions obtained from the rate of reaction equations (RRE). In the thermodynamic limit of a system, the results obtained using both kMC and RRE are identical. However, chemically reacting systems involving infrequent chemical reactions or chemical species in limited amounts are better represented by stochastic than by deterministic methods. The SSA method has been extended to consider not only chemical reactions but also physical processes such as adsorption or desorption from surfaces [181, 184–186].

One main disadvantage of the original SSA formulation is that it is a very inefficient method for simulating chemical reactions described by stiff differential equations. Such stiffness is observed when at least one reaction pathway is orders of magnitude more frequent than the others. Several approaches, including the hybrid stochastic method, have been proposed to overcome stiffness using stochastic algorithms [176, 179, 187].

In the original SSA formulation (also known as *direct method*), the time at which the next stochastic event occurs (τ) can be calculated assuming an exponential probability distribution of time between events, using the following equation:

$$\tau = -\frac{\ln(\xi_{\mathrm{U}})}{\sum_i a_i} \tag{3.34}$$

where ξ_{U} is a uniformly distributed random number between 0 and 1, and a_i is the *propensity function* of the i-th stochastic event (in frequency units). For chemical reactions, the propensity function can be expressed as

$$a_i = c_i f(n_1, n_2, \ldots) \tag{3.35}$$

where c_i is a reaction probability, and $f(n_1, n_2, ...)$ is a function of the number of molecules in the system, which depends on the order of the reaction. For the general case of a bimolecular reaction between the molecules A and B, $c_i \equiv k_i/N_A V$, where k_i is the rate coefficient of the i-th reaction, and the propensity function is then given by

$$a_i = \frac{k_i n_A n_B}{N_A V} \tag{3.36}$$

and can also be expressed as a function the molar concentrations (C):

$$a_i = k_i C_A C_B N_A V \tag{3.37}$$

Proceeding similarly with other types of reactions, it can be found that in general:

$$a_i = N_A V k_i f_i(C) \tag{3.38}$$

The type of event taking place at time τ is determined randomly, where the probability P of choosing an i-th event is

$$P(i) = \frac{a_i}{\sum_j a_j} \tag{3.39}$$

The previous formulation of the SSA is based on the assumption of perfectly mixed reaction volumes; thus, the probability of finding a given single molecule at any position in the system is uniform. This means that a single molecule has the same probability of reacting with every other molecule present in the system, which is not possible in real systems because of mass transfer limitations.

The SSA can be modified in order to consider also imperfectly mixed systems. In the stochastic simulation algorithm of imperfectly mixed systems (SSA-IM) [188], the type of reaction must be determined before calculating the propensity functions. From Eqs. 3.38 and 3.39, the probability of the i-th reaction being the next event is

$$P(i) = \frac{k_i f_i(C)}{\sum_j k_j f_j(C)} \tag{3.40}$$

Notice that it is not necessary to know the volume of the system in order to determine the next event. Once the next reaction has been identified, we need to determine the time at which the next reaction takes place and the perfectly mixed volume for the particular reaction at the particular time.

Stickler [189] proposed the use of Einstein's equation of Brownian motion (Eq. 1.23) for the determination of effective reaction volumes in diffusion-controlled reactions. Einstein's equation describes the diameter of a sphere inside which the probability of finding the diffusing molecule is about 50%. Therefore, we can estimate the diffusion volume for the particular molecule as twice the volume of the sphere described by Eq. 1.23:

$$V_{\text{dif}} = \frac{8\pi}{3}(6D)^{3/2}\,\Delta t^{3/2} \tag{3.41}$$

An estimate of the perfectly mixed volume (V_{pm}) is then given by the diffusion volume of the fastest molecule involved in the next reaction, considering that $\Delta t = \tau$:

$$V_{\text{pm}} = \frac{8\pi}{3}\left[6\tau \max(D)\right]^{3/2} \tag{3.42}$$

The next reaction time is then determined from

$$\tau = \left[-\frac{3}{8\pi N_{\text{Av}}\,(6\max(D))^{3/2}}\frac{\ln(\xi_{\text{U}})}{\sum_i k_i f_i(C)}\right]^{2/5} \tag{3.43}$$

and the perfectly mixed volume is found to be

$$V_{\text{pm}} = \left(1536\pi^2\right)^{1/5}\left[-\frac{\max(D)}{N_{\text{A}}}\frac{\ln(\xi_{\text{U}})}{\sum_i k_i f_i(C)}\right]^{3/5} \tag{3.44}$$

For this perfectly mixed volume, the number of molecules for each species is given by the floor-rounding operation:

$$n_k = \left\lfloor C_k V_{\text{pm}} N_{\text{A}} \right\rfloor \tag{3.45}$$

Using the number of molecules actually present in the perfectly mixed volume, it is now possible to calculate the actual propensity function of the next reaction (Eq. 3.36). If the calculated propensity function is zero, the time is updated but no reaction takes place. If the actual propensity function is greater than zero, the event takes place and the concentrations of the species are updated using the following expression:

$$C_{k,\text{new}} = C_{k,\text{old}} + \frac{v_{k,i}}{N_A V_{\text{pm}}} \tag{3.46}$$

where $v_{k,i}$ is the stoichiometric coefficient for the i-th reaction and k-th component. In this way, it is possible to carry out stochastic simulations of diffusion-controlled systems, such as those where interfacial mass transfer events are involved. It should be noticed that the accuracy of this algorithm relies on the use of an adequate model for the estimation of diffusion coefficients in the reacting mixture.

The kMC methods are very popular for modeling surface phenomena [190–195], complex reaction networks [196–199], transport phenomena [200–202], and diffusion-controlled reactions [203, 204].

The kMC methods are particularly useful for radical polymerization processes. Radicals are present in very low concentrations and take part in different reactions occurring simultaneously; therefore, the simulation of the process using stochastic methods is a better representation of the real radical polymerization system and allows getting a deeper understanding of the polymerization process. kMC has been mainly used to describe molecular mass distributions in radical polymerization [178, 204–208], and heterophase polymerization systems in general [58, 61, 202, 209–213].

3.6 Mesoscopic Scale

Even if the thermodynamic limit of the system is reached, it may still be necessary to describe local macroscopic properties of the system such as temperature, pressure, velocity, density, and composition, as a result of the particular spatial configuration of the system. Such is the case, naming just a few examples, of mixing in stirred tanks, heat transfer in heat exchangers, and hot spots in heterogeneous reactors. The modeling of systems at the mesoscopic scale is done by spatially discretizing the system, and numerically solving the corresponding equations describing the particular phenomenon. If the phenomenon of interest is the fluid mechanics of the system, the numerical solution is known, in general, as computational fluid dynamics (CFD). Thus, CFD involves the solution to the equations governing the transport of mass, momentum, and energy in moving fluids.

The general procedure for mesoscopic modeling is as follows [214]:

1. Pre-processing:
    - Discretization of the system
    - Development of governing equations for the discrete element
    - Assembly of elements, creating a global matrix
    - Incorporation of boundary and initial conditions
2. Solution:
    - Solution to the set of linear or nonlinear algebraic equations
3. Post-processing:
    - Determination of additional properties of the system

Several techniques are available for performing the spatial discretization of the system, including among others the FEM, the finite volume method (FVM), the finite differences method (FDM), and the spectral element method (SEM). From these methods, the FEM is perhaps the most popular because of its improved stability and its flexibility to describe irregular and complex geometries.

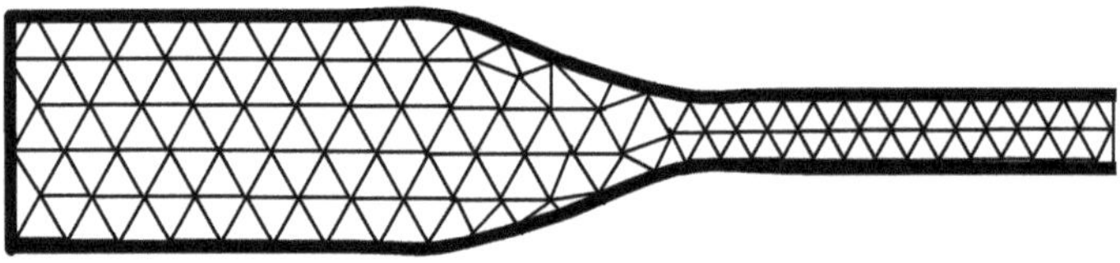

Figure 3.7 Example of domain discretization: 2D discretization of a pipe reduction cross section. Each triangular section represents a discrete element.

Discretization is a vital step of mesoscopic simulation because it transforms the governing equations of the system from a set of nonlinear partial differential equations to a set of linear or nonlinear algebraic equations. This set of algebraic equations can be easily solved even for irregular systems. In addition, within each element, unknown functions can be approximated using interpolation procedures. Additional assumptions can be made in a particular case to further reduce the complexity of the formulation (e.g., constant transport coefficients, Newtonian flow, etc.). Figure 3.7 shows an example of discretization of an irregular domain.

In general, the transformation of partial differential equations into algebraic equations is done using the following expression:

$$\left(\frac{\partial^n y}{\partial x^n}\right)_j \approx \sum_{i=-b}^{f} a_{i,j} y_{j+i} \tag{3.47}$$

where y is a local variable, x is a dimension of the system, j is the position of the grid point where the n-th partial derivative is calculated, b is the number of backward neighbors whereas f is the number of forward neighbors around j, and $a_{i,j}$ are coefficients related to the type (forward, backward, central, etc.) and order of the approximation. The set of algebraic equations thus obtained can be expressed in the matrix form as:

$$AY = B \tag{3.48}$$

where A is the matrix of coefficients $a_{i,j}$ associated to each node i when evaluating the equation at node j, Y is the vector of values of the local variable for each node along the x-dimension, and B is the corresponding vector of independent terms. Thus, the values of the local variable are obtained by solving

$$Y = A^{-1}B \tag{3.49}$$

Mesoscopic modeling techniques are commonly applied in the simulation of industrial polymer processing operations [215], in the investigation of the effect of mixing on polymerization reactions [216–220], in the simulation of certain emulsification and deposition processes [221–225], and in heterophase polymerization processes [226–230].

A very important feature of heterophase polymerization at the mesoscale is the polymer PSD. One of the most common methods for describing the evolution of PSD during heterophase polymerization is the population balance modeling (PBM) [230–234].

A population balance model is a continuity equation for the number of particles sharing a specific property (internal coordinate), specifically particle size for polymer dispersions. It is, therefore, necessary to consider one balance equation for each particle size. Since particle size is a continuous variable, in practice it must be divided into a finite number of particle size intervals. The population balance for each interval is expressed in terms of the population density (ρ) (relative proportion of particles with size within the corresponding interval), as follows:

$$\frac{\partial \rho(r_i,t)}{\partial t} = R_{i,\text{nucl}} + R_{i,\text{coal}} - \frac{\rho(r_i,t)}{V}\frac{dV}{dt} - \frac{\partial\left(\rho(r_i,t)G(r_i,t)\right)}{\partial r_i} \quad (3.50)$$

where r_i represents the mean particle radius of the i-th interval, $R_{i,\text{nucl}}$ is the rate of nucleation of particles with radius in the i-th interval, $R_{i,\text{coal}}$ is the net rate of formation of particles with radius in the i-th interval by coalescence (subtracting the coalescence rate of particles of this size), V is the volume of the system, and $G(r_i,t)$ is the growth rate function. The growth rate function takes into account the net disappearance of particles from the size interval because of growth by polymerization.

The internal coordinate in PBM for heterophase polymerization can also be the polymer particle volume, surface, or mass. It is also possible to consider the average molecular mass, monomer concentration, and number of active molecules as internal coordinates of the particles.

Equation 3.50 is a deterministic expression, which does not take into account the random nature of polymerization processes. It is also possible to formulate a stochastic version of the PBM, and in this case, it corresponds to the Fokker–Planck equation (FPE) [235, 236]. Further advances in the modeling of PSD can be found in the review paper by Sheibat-Othman et al. [237].

3.7 Macroscopic Scale

Following the concept introduced in Fig. 3.6, a macroscopic property of the system can be defined as the average of the probabilities of all the corresponding microscopic states. At the macroscopic scale, the most relevant phenomena involve, in general, the transport and conservation of mass, momentum, and energy. The balance or conservation principles can be applied to a general macroscopic system (Fig. 3.8) as follows:

$$\begin{aligned}\text{Accumulation rate} &= \text{Input flux} - \text{Output flux}\\ &\quad + \text{Generation rate} - \text{Consumption rate}\end{aligned} \quad (3.51)$$

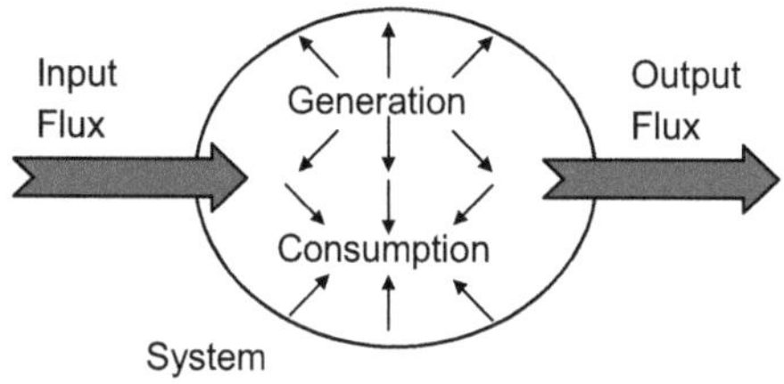

Figure 3.8 Schematic representation of transport in a general macroscopic system.

If the system considered is a differential volume element, Eq. 3.51 evolves naturally into a differential equation or a partial differential equation [238]. A generalized differential representation of transport and conservation equations is presented in Eq. 3.52.

$$\frac{\partial \Phi}{\partial t}+\frac{\partial (u_i \cdot \Phi)}{\partial x_i}=\frac{\partial}{\partial x_i}\left(\Gamma_{\text{eff}}\frac{\partial \Phi}{\partial x_i}\right)+S_{\Phi} \tag{3.52}$$

where Φ is the conservative macroscopic variable of interest (mass, momentum, energy, etc.), u_i is the velocity of matter flowing across the volume element in the i-th direction x_i, Γ_{eff} is an effective exchange coefficient, and S_{Φ} is the net source (generation – consumption). The first term in Eq. 3.52 represents accumulation of the macroscopic conservative variable, the second term is a convective rate of change, and the third term (first term of the right-hand side) is a diffusive rate of change. Table 3.2 summarizes the corresponding terms for the different macroscopic conservative variables.

Table 3.2 General representation of conservation equations

Macroscopic conservative variable	Φ	Γ_{eff}	S_{Φ}
Total mass	ρV	0	0
Moles of species j	$[C_j]V$	D_j	$r_j V$
Momentum	$\rho u_i V$	μ_{eff}	$-V\partial P/\partial x_i$
Energy (enthalpy)	$\rho h V$	$k_{\text{eff}}/\rho C_{\text{p}}$	Q

The variables used in Table 3.2 correspond to the total mass density (ρ), total volume of the system (V), molar concentration of the j-th species ($[\boldsymbol{C_j}]$), effective diffusion coefficient (D_j), net reaction rate of the j-th component (r_j), effective kinematic viscosity (μ_{eff}), pressure (P), specific enthalpy (h), effective thermal conductivity (k_{eff}), specific heat capacity (C_p), and net heat flow (Q).

Additional approximations can be made for each variable at the macroscopic scale in order to reduce the complexity of the formulation and to facilitate the description and/or the parameter identification of the system. Particularly, a brief review of different approaches and approximations used to describe macroscopic mass transfer in heterophase polymerization [96] will be presented in this section. Modeling of heat and momentum transfer can be done using similar expressions, considering the equivalence presented in Table 3.2.

Different types of macroscopic models have been used to describe mass transfer processes in heterophase polymerization systems. One possible classification of these models includes (i) first principles, (ii) semi-empirical, (iii) equilibrium thermodynamics, and (iv) empirical models.

3.7.1 First-Principles Modeling

The first-principles or fundamental models of mass transfer are based on the differential equations obtained by Fick [239]. Fick's results for isotropic materials can be summarized into two laws as follows:

$$J_j = -D_j \nabla \left[C_j \right] \tag{3.53}$$

$$\frac{\partial \left[C_j \right]}{\partial t} = D_j \nabla^2 \left[C_j \right] \tag{3.54}$$

These equations indicate that the total number of molecules of type j crossing a completely permeable section per unit area per unit time (J_j) is related to the diffusion coefficient of the molecules in the medium (D_j) and to the local spatial gradient (∇) of the molecular concentration (C_j); and that the rate of change in the local molecular concentration is related to the diffusion coefficient of the molecules and to the Laplacian (∇^2) of the molecular concentration, i.e., to the second partial derivative of the concentration with respect to the position.

Equations 3.53 and 3.54 are sometimes interpreted by assuming that molecular transfer is caused by the difference in concentration between two regions. However, this is just an apparent interpretation. If the net force acting on the diffusing molecules is zero (for example, in pure components or in ideal mixtures), the random displacement of each molecule is uniformly distributed; therefore, the number of molecules moving on each direction is the same. The net displacement is then proportional to the difference in the number of molecules at each side, which is proportional to the gradient in local concentration. The true cause of molecular diffusion is the frequent random collisions with the neighboring molecules and not the difference in molecular concentration, even though, as the result of the random motion, the net flux of molecules becomes proportional to the difference in molecular concentration. If there is a nonzero force acting on the molecules (for example, at interfaces), Eqs. 3.53 and 3.54 are no longer valid, and chemical potentials should be considered instead of concentrations to describe the diffusion process.

The rigorous application of Fick's equations to heterophase polymerization systems involves the solution to partial differential equations with four independent variables (time and the three spatial dimensions) expressed usually in spherical coordinates centered on a representative segregated phase. However, if the geometry of the system is not perfectly spherical, cylindrical, or planar, the analytical solution to Fick's equations becomes a very difficult task. These complex geometries can be modeled by partitioning the system into smaller regions having simpler geometries [240] but at the expense of an increased number of equations to be solved.

Anisotropy: Characteristic of certain materials that present different properties when measured in different directions.

In anisotropic media, diffusion depends on the direction considered. The diffusion coefficient is, in this case, a function of the local spatial composition around the diffusing molecule. Some common examples of anisotropic media are crystals, textile fibers, and polymer films in which the molecules have a preferential direction of orientation. In these cases, Fick's laws remain valid, but the diffusion coefficient is now a matrix and not a scalar variable. ***Anisotropy*** is especially important at interfaces since diffusion across the interface is different compared to diffusion in any other direction.

For this reason and considering that chemical potentials should be used instead of concentrations, rigorous analytical solutions of molecular transfer across interfaces in real heterogeneous systems using fundamental equations are very difficult if not impossible to achieve. Numerical solutions can be obtained using finite elements approximations, but a detailed knowledge of anisotropic diffusion coefficients for all molecules at interfaces is required. In practice, fundamental modeling is used to describe diffusion at both sides of an interface, whereas some assumptions are introduced in order to obtain analytical solutions describing mass transfer across the interface.

3.7.2 Semi-empirical Modeling

An additional simplification of Fick's equations can be obtained by assuming that the bulk of each phase (both continuous and segregated) is perfectly mixed. In this case, mass transfer of molecular species is expressed in terms of the difference in concentration of this species at both sides of the interface, using semi-empirical parameters denoted as local mass transfer coefficients (k_L) [241]. Net fluxes of molecules across the interface between phases A and B are calculated as follows:

$$J_j = k_{L,j,AB}\left(\left[C_j\right]_A^{AB} - \left[C_j\right]_B^{AB}\right) \tag{3.55}$$

where $[C_j]_A^{AB}$ denotes the concentration of the j-th component on the side of phase A at the interface between phases A and B. Additionally, assuming that there is no accumulation of matter at the interface, the flux of molecules across the interface should be the same as the flux of molecules from the bulk to the interface of the concentrated phase or from the interface to the bulk of the diluted phase. Thus,

$$J_j = k_{L,j,A}\left(\left[C_j\right]_A - \left[C_j\right]_A^{AB}\right) = k_{L,j,B}\left(\left[C_j\right]_B^{AB} - \left[C_j\right]_B\right) \tag{3.56}$$

Comparing Eqs. 3.53 and 3.56, it is possible to conclude that the semi-empirical approach is equivalent to the first-principles model as long as

$$k_{L,j,X} = \frac{D_{j,X}}{\delta_X} \tag{3.57}$$

$D_{j,X}$ being the diffusion coefficient of molecule j in phase X, and δ_X the boundary layer at the side of phase X of the interface. The boundary layer represents the distance from the interface at which the local concentration practically reaches the bulk concentration. The measurement of boundary layers according to this definition is not easy; therefore, boundary layers and local and global mass transfer coefficients are calculated using semi-empirical expressions, which are obtained after analyzing large amounts of experimental concentration data under different geometries and flow regimes [242].

Additionally, empirical partition coefficients can be used together with local coefficients of mass transfer in order to facilitate the modeling of molecular transfer across interfaces. Assuming the net flux of molecules across the interface proportional to the concentration difference between (i) the actual bulk concentration of the j-th component in phase A ($[C_j]_A$) and (ii) the theoretical concentration present in A ($[C_j]^{*B}_A$) in equilibrium with the bulk concentration in B ($[C_j]_B$), then

$$J_j = K_{\mathrm{m},j,AB}\left(\left[C_j\right]_A - \left[C_j\right]_A^{*B}\right) \tag{3.58}$$

where $K_{\mathrm{m},j,AB}$ is another semi-empirical parameter called the global mass transfer coefficient. Such global mass transfer coefficient is related to the local mass transfer coefficients at both sides of the interface (Eq. 3.56), resembling two electrical resistances in series, according to the following expressions [241]:

$$\frac{J_j}{K_{\mathrm{m},j,AB}} = \frac{J_j}{k_{\mathrm{L},j,A}} + \frac{J_j}{k_{\mathrm{L},j,B}} \frac{\left[C_j\right]_A^{AB} - \left[C_j\right]_A^{*B}}{\left[C_j\right]_B^{AB} - \left[C_j\right]_B} \tag{3.59}$$

$$\frac{1}{K_{\mathrm{m},j,AB}} = \frac{1}{k_{\mathrm{L},j,A}} + \frac{m_{j,AB}^{AB}}{k_{\mathrm{L},j,B}} \tag{3.60}$$

where

$$m_{j,AB}^{AB} = \frac{\left[C_j\right]_A^{AB} - \left[C_j\right]_A^{*B}}{\left[C_j\right]_B^{AB} - \left[C_j\right]_B} \tag{3.61}$$

$m_{j,AB}^{AB}$ can be regarded as the slope of a straight line intercepting the equilibrium curve at the concentrations $\left[C_j\right]_B^{AB}$ and $\left[C_j\right]_B$. If the equilibrium curve can be assumed to be linear, then the variable $m_{j,AB}^{AB}$ will be equivalent to the equilibrium partition coefficient ($K_{j,AB}$). This type of modeling is widely used to describe macroscopic mass transfer processes across interfaces. It requires the knowledge of mass transfer coefficients (which can be estimated from semi-empirical correlations) and equilibrium distribution coefficients.

3.7.3 Equilibrium Thermodynamics Modeling

A further simplification of the mass transfer process can be made once steady-state conditions have been reached in the system, i.e., when the *net* flux across all interfaces becomes zero. Under these conditions, even though the individual fluxes of molecules back and forth are not zero, it is possible to assume that the system has reached a *thermodynamic equilibrium*. In ideal mixtures, there are no enthalpic effects and mixing occurs spontaneously as a result of the entropy increase. In other words, there is no significant change in the net average forces acting on the molecules (no enthalpic effect), and complete mixing occurs only because it is the most probable state of the system (entropic effect). Ideal mixtures are always homogeneous systems. In nonideal mixtures, phase separation can take place (due to differences in intermolecular forces) and an equilibrium condition is reached when both entropic and enthalpic contributions to the free energy of the system exactly counteract each other. This is the principle used by Flory and Huggins to describe polymer solutions [243], which can be used in general to describe the thermodynamics of mixtures of species with different molecular weights. A detailed description of this type of modeling was already described in Section 2.2.3.

3.7.4 Empirical Modeling

One final approach for the deterministic modeling of mass transfer is based completely on experimental data. In this case, the concentrations of a certain component in two different phases at steady state are related by means of a partition or distribution coefficient, which in general can be expressed as

$$K_{j,AB} = f\left(\left[C_j\right]_A^*, \left[C_j\right]_B^*\right) \quad (3.62)$$

where $K_{j,AB}$ is the equilibrium partition coefficient for the species j between the phases A and B. Very frequently, the empirical distribution function has the following structure:

$$K_{j,AB} = \left(\frac{\left[C_j\right]_A^*}{\left[C_j\right]_B^*}\right)^n \quad (3.63)$$

where n is a parameter obtained experimentally for the particular system. In an ideal system, the value of n is 1, and thus, the following ideal expression (equivalent to Henry's law) is obtained:

$$K_{j,AB} = \frac{\left[C_j\right]_A^*}{\left[C_j\right]_B^*} \quad (3.64)$$

In this context, the partition or distribution coefficient is a constant parameter used to predict the steady-state concentration of a component in one phase by knowing its concentration in the other. As expected, the value of this constant will depend on the particular external conditions of the system (pressure, temperature, etc.). It is possible to predict the values of the partition coefficients at different conditions using thermodynamic expressions. However, it is more common to find tabulated values of the partition coefficients obtained from experimental measurements. Additional information regarding partition and distribution coefficients was presented by Leo et al. [244].

Empirical models are very useful to describe mass transfer in complex systems using relatively simple expressions. However, their main disadvantage is their very limited predictive capabilities.

Macroscopic models are a basic ingredient of any description of heterophase polymerization [245, 246]. Only after the 1970s, macroscopic models were integrated with the population balance models (mesoscopic scale), as it was found that a purely macroscopic description was inadequate for heterogeneous polymerization since they are only capable of predicting temperature, pressure, and monomer concentration in the reactor [247]. A detailed description of multiscale integration and multiscale models in heterophase polymerization is presented in the next section.

3.8 Multiscale Integration

In the previous sections of this chapter, it was shown that relevant phenomena involved in heterophase polymerization take place at different scales, ranging from the subatomic to the macroscopic scale. A single dynamic model of heterophase polymerization covering all phenomena described at their own scales is possible, but its solution would be inefficient using our current computational capabilities. Thus, efficiently solving multiscale models requires assumptions and compromises.

3.8.1 Model Order Reduction

Perhaps the most common approach for multiscale integration is the *bottom-up model order reduction*. The purpose of this method is finding empirical models that can satisfactorily describe, at a higher scale, the behavior of lower-scale phenomena. Let us consider a system involving only two different scales: an upper scale U and a lower scale L. The vector of lower-scale dynamic variables (x_L) will be described by the following partial differential equation:

$$\frac{\partial \boldsymbol{x}_\mathrm{L}}{\partial t} = \boldsymbol{f}_\mathrm{L}\left(\boldsymbol{x}_\mathrm{L}, \boldsymbol{x}_\mathrm{U}, \boldsymbol{u}_\mathrm{L}\right) \tag{3.65}$$

where f_L is an arbitrary nonlinear vector function describing the lower-scale dynamics, x_U is a vector of upper-scale dynamic variables, and u_L is a vector of lower-scale inputs to the system. Equation 3.65 is a partial differential equation because x_L may also present local spatial changes. Similarly, the vector of upper-scale dynamic variables will be

$$\frac{\partial \boldsymbol{x}_\mathrm{U}}{\partial t} = \boldsymbol{f}_\mathrm{U}\left(\boldsymbol{x}_\mathrm{L}, \boldsymbol{x}_\mathrm{U}, \boldsymbol{u}_\mathrm{U}\right) \tag{3.66}$$

where f_U is an arbitrary nonlinear vector function describing the upper-scale dynamics, and u_U is a vector of upper-scale inputs to the system. Even neglecting local spatial changes, analytical solutions for the simultaneous solution to Eqs. 3.65 and 3.66 are, in general, not available. Therefore, numerical solution methods are required. Since the time and length scales of x_L are below those of x_U, the integration step of the numerical solution must be smaller than the characteristic time at the lower scale. If the model could be solved considering the dynamics of the upper scale, then the solution would be more efficient. A general reduced-order model approximation of the lower scale is

$$\mathbf{x}_{\mathrm{L}}(t) \approx \mathbf{g}_{\mathrm{L}}\left(\mathbf{x}_{\mathrm{U}}, \mathbf{u}_{\mathrm{U}}, t\right) \tag{3.67}$$

where $\mathbf{g}_{\mathrm{L}}$ is an arbitrary nonlinear algebraic function, which may also involve time *t*. Thus, Eq. 3.66 becomes

$$\frac{\partial \mathbf{x}_{\mathrm{U}}}{\partial t} \approx \mathbf{f}_{\mathrm{U}}\left(\mathbf{x}_{\mathrm{U}}, \mathbf{u}_{\mathrm{U}}, \mathbf{g}_{\mathrm{L}}\right) \tag{3.68}$$

which can then be efficiently solved at the upper scale.

The use of kinetic expressions is an example of this bottom-up model order reduction. The rate of a chemical reaction depends on factors such as molecular kinetic energy, molecular orientation, and electronic configuration [248], relevant at the atomistic and molecular scales. However, at the macroscopic scale, the rate of reaction is expressed for a component C_i as

$$r_i = v_i k(T) \prod_j \left[C_j\right]^{a_j} \tag{3.69}$$

where v_i is a stoichiometric coefficient, $k(T)$ is a temperature-dependent rate coefficient, $[C_j]$ denotes the concentration of all components involved in the reaction, and a_j are empirical coefficients. Equation 3.69 incorporates all dynamical effects involved at the lower scale, into a macroscopic term.

Empirical expressions for reducing the model order are also normally used for describing heat transfer phenomena (in terms of global heat transfer coefficients), diffusion (e.g., Smoluchowski's equation), particle formation (e.g., classical nucleation theory), etc. Almost all mathematical models of heterophase polymerization include one or more model order reductions by means of empirical expressions.

3.8.2 Sequential Multiscale Simulation

In the model order reduction approach, the empirical expressions and their parameters are usually obtained from previous experimental data. In other words, they are obtained "*off-line.*" However, process conditions may differ between those used for identification, and those used for prediction, resulting in larger prediction errors. To overcome this difficulty, a more recent approach consists of determining "*on-line*" the parameters of reduced-order models, by performing lower-scale simulations of the process. This is the *sequential* (*bottom-up*) *multiscale simulation*. Thus, for example, kinetic polymerization coefficients can be obtained from quantum mechanics calculations and then used in macroscopic mass balance models [17]. It is also possible using BD simulation at the colloidal scale to describe the rate of radical capture in emulsion polymerization, which can then be used at the microscopic scale to simulate competitive chemical reactions inside the particles using kMC simulation [210].

Mathematically, this method consists of solving Eq. 3.65 assuming constant upper-scale variables, and approximating to a reduced-order model to be used at the upper scale:

$$\boldsymbol{x}_{\mathrm{L}}(t)=\int_{t_0}^{t}\left(\frac{\partial \boldsymbol{x}_{\mathrm{L}}}{\partial t}\right)_{\boldsymbol{x}_{\mathrm{U}},\boldsymbol{u}_{\mathrm{U}}} dt \approx \boldsymbol{g}_{\mathrm{L}}\left(\boldsymbol{x}_{\mathrm{U}},\boldsymbol{u}_{\mathrm{U}},t\right) \tag{3.70}$$

The main disadvantage of this method is that local and transient behavior at the lower scale may have a significant effect on the predictions at the upper scale. In order to reduce these effects, not the result of a single simulation but the average of multiple simulations at the lower scale (at different local conditions) must be used for estimating the reduced-order model parameters. This is done, for example, when estimating the average number of radicals per particle in emulsion polymerization by kMC simulation [213], required for the macroscopic kinetic expression of polymerization rate.

It is also possible to define particular conditions for triggering the update of empirical parameters by lower-scale simulation, during upper-scale simulation. In heterophase polymerization, for example, a significant change in average polymer particle size, in polymer volume fraction, in monomer composition, or in process temperature may trigger an update of empirical parameters under the new conditions.

3.8.3 Stochastic Transformation across Scales

Figure 3.9 shows a general multiscale system as perceived by an observer. Such observer is only able to detect the corresponding observation scale of such system. Both the infra-scale (lower scale) and the supra-scale (upper scale) are unknown to the observer. However, the variables at the supra- and infra-scales might be related to variables at the observation scale. Such relationships are denoted as *restrictions* when infra-scale variables are transformed into observable variables, or as *lifting* when supra-scale variables are transformed into observable variables [249].

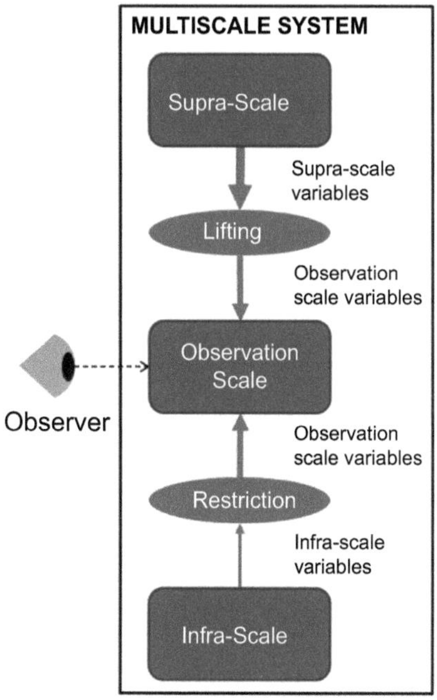

Figure 3.9 General representation of a multiscale system composed of three scales: supra-scale, observation scale, and infra-scale. The observer can only reach the observation scale. Arrow thickness is related to the order of magnitude of the dimension considered [249].

Restrictions always involve reducing the amount of information available to the observer as the resolution of the observation is limited in certain dimensions (e.g., time, length, mass, etc.). Some examples of restrictions include the following:

- **Direct sampling**: The relevant variable at the infra-scale is the same, but only a fraction of the values of the variable is observed, depending on the resolution unit.

- **Cumulative sampling**: The variable observed corresponds to the sum of individual effects at the infra-scale composing one resolution unit.
- **Averaging sampling**: The variable observed corresponds to the average of individual effects at the infra-scale composing one resolution unit.
- **Modal sampling**: The variable observed corresponds to the maximum or minimum value among the individual effects at the infra-scale composing one resolution unit.
- **Functional sampling**: The variable observed corresponds to any other function of the individual values of the infra-scale variable composing one resolution unit.

All these restrictions can be considered stochastic transformations [250, 251] of the infra-scale variables, because the missing information can only be seen as a random variable. Furthermore, each multiscale dimension provides an additional dimension of randomness [252].

Lifting is the opposite transformation of *restriction*. A supra-scale variable may have a significant effect on the observed system depending on the relative situation of the system. In most cases, such relative situation of the system is difficult to identify, resulting in unpredictability. Thus, it is possible to consider lifting as a reverse stochastic transformation. That is, similar systems available at the observation scale may present different behaviors although all of them are described by a certain statistical distribution at the supra-scale. Each observable system then becomes a single realization of the supra-scale distribution.

Figure 3.10 shows a graphical representation of these direct and reverse stochastic transformations, considering the particle size of polymer dispersions as an example. Information from the smaller scale (individual particle size) can be represented by a random variable at the larger scale (PSD). Identifying the random variable representing the lower-scale information corresponds to a probability distribution *reconstruction* process [253].

Random variables with different characteristics (probability density functions) can be used for representing the smaller-scale information, depending on the amount and quality of the information available, and on the identification or transformation method employed. The reverse transformation of a random variable at a larger scale (e.g., PSD) will inevitably lead to a practically infinite number of possible sets of reconstructed information (realizations) at the smaller scale (i.e., individual particle sizes in the system), which exactly or approximately satisfy the behavior described by the large-scale random variable. The reverse transformation process is also denoted, in general, as *sampling* (or *discretization*). Sampling of a random variable can be done using random, deterministic, or *randomistic* methods [254]. Random sampling methods are commonly used to reproduce the variability observed at lower scales. However, they may not preserve the properties of the upper scale variable. Deterministic sampling methods preserve those properties but neglect the natural variability between samples. Randomistic sampling methods, on the other hand, represent a compromise between variability and preservation of the properties of the random variable.

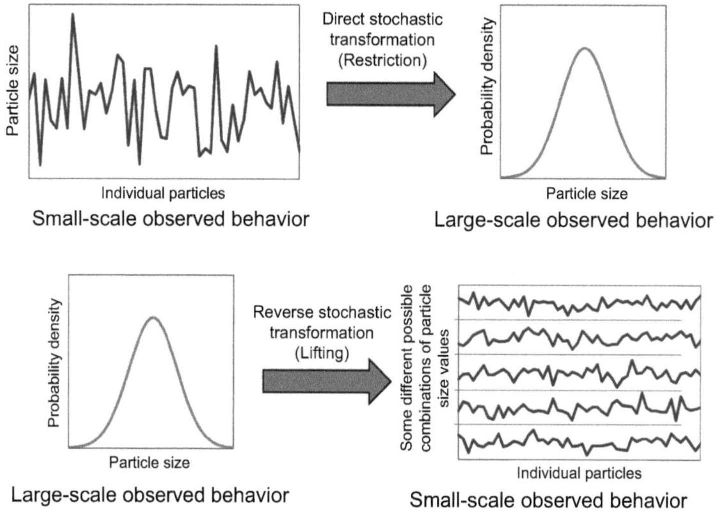

Figure 3.10 Direct (top) and reverse (bottom) stochastic transformation between scales, using the particle size of polymer dispersions as an example.

Understanding the behavior of a multiscale system relying only on the variables available at the observation scale is usually difficult. Therefore, mathematical models describing the behavior of multiscale systems based solely on the variables at the observation scale may be deficient. Multiscale models consist of two or more coupled models, each representing a different scale. In principle, such models should be able to provide a superior performance for describing the behavior of multiscale systems, compared to isolated, single-scale models [255].

Hoekstra et al. [256] suggested that some of the open questions remaining in the field of multiscale modeling and simulation include formulating a general theory or calculus of multiscale modeling (involving scale bridging), dealing with the validation and verification of such models, and considering numerical error propagation. Since scale bridging always involves missing information, multidimensional stochastic transformations should become a key component of multiscale modeling. Furthermore, considering stochastic variables will also help quantify the uncertainty propagation in general (not only numerical errors). Thus, a multiscale modeling theory based on the mathematics and transformations of multidimensional random variables is a promising topic for improving the modeling and simulation of multiscale systems, including of course, heterophase polymerization systems.

3.8.4 Full *ab initio* Multiscale Simulation

Ab initio (from the beginning) simulation refers to performing the simulation of a process without any empirical (previously obtained) parameter. Only known values of physical constants and elemental properties of matter are required [257].

The term *ab initio* in simulation was first used by Parr, Craig, and Ross in 1950 [258]. In 1969, the first mention of *ab initio* simulation in polymers was done by Del Bene and Pople [259]. Particularly, they were considering the theoretical behavior of small supramolecular polymers formed by water (inspired by the polywater hype that happened during the late 1960s [260], a quite controversial topic, which was ultimately found to be caused by organic impurities [261]). Karpfen devoted, in the 1980s, to the *ab initio* simulation of different types of organic polymers [262–265]. These calculations only reached up to the macromolecular scale. Further scales have been reached in order to understand and describe selective adsorption of polymers on metal surfaces [266], the mechanical properties of nanocomposites [267], radical stability in RAFT polymerization [268], charge in conjugated polymers [269], lignin depolymerization [270], and conformational fluctuations of DNA [271], just to mention a few.

While still a challenge, with the continuous increase in human computational power, plus the improved knowledge of physical and chemical processes taking place at different scales, performing a full *ab initio* multiscale simulation of heterophase polymerization is closer to becoming reality every day.

References

1. Fermeglia, M. and Pricl, S. (2007). Multiscale modeling for polymer systems of industrial interest. *Progress in Organic Coatings*, **58**(2–3), pp. 187–199.
2. Gooneie, A., Schuschnigg, S., and Holzer, C. (2017). A review of multiscale computational methods in polymeric materials. *Polymers*, **9**(1), pp. 16.
3. Maroudas, D. (2000). Multiscale modeling of hard materials: Challenges and opportunities for chemical engineering. *AIChE Journal*, **46**(5), pp. 878–882.
4. Chatterjee, A. and Vlachos, D. G. (2006). Multiscale spatial Monte Carlo simulations: Multigriding, computational singular perturbation, and hierarchical stochastic closures. *The Journal of Chemical Physics*, **124**(6), 064110.
5. Broughton, J. Q., Abraham, F. F., Bernstein, N., and Kaxiras, E. (1999). Concurrent coupling of length scales: Methodology and application. *Physical Review B*, **60**(4), pp. 2391–2403.
6. Mills, R. L. (2000). The hydrogen atom revisited. *International Journal of Hydrogen Energy*, **25**(12), pp. 1171–1183.
7. Combes, J. M., Duclos, P., and Seiler, R. (1981). The Born–Oppenheimer approximation. In: Velo, G. and Wightman, A. S., *Rigorous Atomic and Molecular Physics*. NATO Advanced Study Institutes Series, Vol 74. (Springer, USA). pp. 185–213.
8. Hartree, D. R. (1928). The wave mechanics of an atom with a non-Coulomb central field. Part I. Theory and methods. *Mathematical Proceedings of the Cambridge Philosophical Society*, **24**(1), pp. 89–110.
9. Le Bris, C. (2003). *Handbook of Numerical Analysis*, Vol. 10. Special Volume Computational Chemistry. (Elsevier Science, The Netherlands).

10. Levine, I. N. (2008). *Quantum Chemistry*, 6th Ed. (Prentice Hall, USA).
11. Hohenberg, P. and Kohn, W. (1964). Inhomogeneous electron gas. *Physical Review*, **136**(3B), pp. B864–B871.
12. Kohn, W. and Sham, L. J. (1965). Self-consistent equations including exchange and correlation effects. *Physical Review*, **140**(4A), pp. A1133–A1138.
13. McQuarrie, D. A. (1976). *Statistical Mechanics* (HarperCollins Publishers, USA).
14. O'driscoll, K. F. and Yonezawa, T. (1966). Application of molecular orbital theory to vinyl polymerization. *Polymer Reviews*, **1**(1), pp. 1–37.
15. Davis, T. P. and Rogers, S. C. (1994). Ab initio molecular orbital calculations on the transition state for the addition of a methyl radical to vinyl monomers. *Macromolecular Theory and Simulations*, **3**(5), pp. 905–913.
16. Ferguson, C. J., Hughes, R. J., Nguyen, D., Pham, B. T., Gilbert, R. G., Serelis, A. K., Such, C. H., and Hawkett, B. S. (2005). Ab initio emulsion polymerization by RAFT-controlled self-assembly. *Macromolecules*, **38**(6), pp. 2191–2204.
17. Coote, M. L. (2009). Quantum-chemical modeling of free-radical polymerization. *Macromolecular Theory and Simulations*, **18**(7–8), pp. 388–400.
18. Cuccato, D., Dossi, M., Moscatelli, D., and Storti, G. (2011). A density functional theory study of poly (vinyl chloride)(PVC) free radical polymerization. *Macromolecular Symposia*, **302**(1), pp. 100–109.
19. Mavroudakis, E., Cuccato, D., and Moscatelli, D. (2015). On the use of quantum chemistry for the determination of propagation, copolymerization, and secondary reaction kinetics in free radical polymerization. *Polymers*, **7**(9), pp. 1789–1819.
20. Nikitin, S. V., Nikitin, V. V., Oleynik, I. I., Oleynik, I. V., and Bagryanskaya, E. G. (2016). Activity of phenoxy-imine titanium catalysts in ethylene polymerization: A quantum chemical approach. *Journal of Molecular Catalysis A: Chemical*, **423**, pp. 285–292.
21. Gu, H., Xu, W., Zhang, J., Qi, Z., Zhang, T., and Song, L. (2018). Quantum chemical calculations of amine-catalyzed polymerization of silanol. *IOP Conference Series: Materials Science and Engineering*, **322**(2), 022063.
22. Nifant'ev, I. E., Shlyakhtin, A. V., Kosarev, M. A., Komarov, P. D., Karchevsky, S. G., and Ivchenko, P. V. (2019). Data for quantum-chemical modeling of the mechanisms of ring-opening polymerization of methyl ethylene phosphate. *Data in Brief*, **26**, 104431.

23. Mark, H. F. (2004). *Encyclopedia of Polymer Science and Technology*, 3rd Ed. (Wiley-Interscience, USA).
24. Davis, T. P. and Rogers, S. C. (1993). Application of ab initio MO calculations to methods for predicting reactivity in free radical copolymerization. *European Polymer Journal*, **29**(10), pp. 1311–1317.
25. Bebe, S., Yu, X., Hutchinson, R. A., and Broadbelt, L. J. (2006). Estimation of free radical polymerization rate coefficients using computational chemistry. *Macromolecular Symposia*, **243** (1), pp. 179–189.
26. Dossi, M. and Moscatelli, D. (2012). A QM approach to the calculation of reactivity ratios in free-radical copolymerization. *Macromolecular Reaction Engineering*, **6**(2–3), pp. 74–84.
27. Doucet, J. P. and Weber, J. (1996). *Computer-Aided Molecular Design: Theory and Applications* (Academic Press, USA), pp. 124–170.
28. Brooks, B. R., Bruccoleri, R. E., Olafson, B. D., States, D. J., Swaminathan, S. A., and Karplus, M. (1983). CHARMM: A program for macromolecular energy, minimization, and dynamics calculations. *Journal of Computational Chemistry*, **4**(2), pp. 187–217.
29. Sun, H., Mumby, S. J., Maple, J. R., and Hagler, A. T. (1994). An ab initio CFF93 all-atom force field for polycarbonates. *Journal of the American Chemical Society*, **116**(7), pp. 2978–2987.
30. Sun, H. (1995). Ab initio calculations and force field development for computer simulation of polysilanes. *Macromolecules*, **28**(3), pp. 701–712.
31. Mayo, S. L., Olafson, B. D., and Goddard, W. A. (1990). DREIDING: A generic force field for molecular simulations. *Journal of Physical Chemistry*, **94**(26), pp. 8897–8909.
32. Sun, H. (1998). COMPASS: An ab initio force-field optimized for condensed-phase applications overview with details on alkane and benzene compounds. *The Journal of Physical Chemistry B*, **102**(38), pp. 7338–7364.
33. Jorgensen, W. L., Maxwell, D. S., and Tirado-Rives, J. (1996). Development and testing of the OPLS all-atom force field on conformational energetics and properties of organic liquids. *Journal of the American Chemical Society*, **118**(45), pp. 11225–11236.
34. Costache, A. D., Ghosh, J., Knight, D. D., and Kohn, J. (2010). Computational methods for the development of polymeric biomaterials. *Advanced Engineering Materials*, **12**(1–2), pp. B3–B17.
35. Li, C. and Strachan, A. (2015). Molecular scale simulations on thermoset polymers: A review. *Journal of Polymer Science Part B: Polymer Physics*, **53**(2), pp. 103–122.

36. Sahputra, I. H., Alexiadis, A., and Adams, M. J. (2018). Temperature and configurational effects on the Young's modulus of poly (methyl methacrylate): A molecular dynamics study comparing the DREIDING, AMBER and OPLS force fields. *Molecular Simulation*, **44**(9), pp. 774–780.

37. Rukmani, S. J., Kupgan, G., Anstine, D. M., and Colina, C. M. (2019). A molecular dynamics study of water-soluble polymers: Analysis of force fields from atomistic simulations. *Molecular Simulation*, **45**(4–5), pp. 310–321.

38. Metropolis, N., Rosenbluth, A. W., Rosenbluth, M. N., Teller, A. H., and Teller, E. (1953). Equation of state calculations by fast computing machines. *The Journal of Chemical Physics*, **21**(6), pp. 1087–1092.

39. He, J., Zhang, H., Chen, J., and Yang, Y. (1997). Monte Carlo simulation of kinetics and chain length distributions in living free-radical polymerization. *Macromolecules*, **30**(25), pp. 8010–8018.

40. Platkowski, K. and Reichert, K. H. (1999). Application of Monte Carlo methods for modelling of polymerization reactions. *Polymer*, **40**(4), pp. 1057–1066.

41. Schmidt-Naake, G., Drache, M., and Koppers, F. (2007). Kinetische Studien und Monte Carlo-Simulation der radikalischen Ter-und Tetrapolymerisation. *Chemie Ingenieur Technik*, **79**(8), pp. 1155–1164.

42. Chaffey-Millar, H., Stewart, D., Chakravarty, M. M., Keller, G., and Barner-Kowollik, C. (2007). A parallelised high performance Monte Carlo simulation approach for complex polymerisation kinetics. *Macromolecular Theory and Simulations*, **16**(6), pp. 575–592.

43. Najafi, M., Roghani-Mamaqani, H., Salami-Kalajahi, M., and Haddadi-Asl, V. (2010). A comprehensive Monte Carlo simulation of styrene atom transfer radical polymerization. *Chinese Journal of Polymer Science*, **28**(4), pp. 483–497.

44. Wang, L. and Broadbelt, L. J. (2010). Kinetics of segment formation in nitroxide-mediated controlled radical polymerization: Comparison with classic theory. *Macromolecules*, **43**(5), pp. 2228–2235.

45. Shere, I. and Malani, A. (2018). Polymerization kinetics of a multi-functional silica precursor studied using a novel Monte Carlo simulation technique. *Physical Chemistry Chemical Physics*, **20**(5), pp. 3554–3570.

46. Mazur, J. and McCrackin, F. L. (1968). Monte Carlo studies of configurational and thermodynamic properties of self-interacting linear polymer chains. *The Journal of Chemical Physics*, **49**(2), pp. 648–665.

47. Munk, P. and Gutierrez, B. O. (1979). The use of Monte-Carlo studies for the evaluation of thermodynamic properties of polymer solutions from their intrinsic viscosity. *Macromolecules*, **12**(3), pp. 467–473.

48. Lepage, T., Képes, F., and Junier, I. (2015). Thermodynamics of long supercoiled molecules: Insights from highly efficient Monte Carlo simulations. *Biophysical Journal*, **109**(1), pp. 135–143.

49. Janke, W. and Paul, W. (2016). Thermodynamics and structure of macromolecules from flat-histogram Monte Carlo simulations. *Soft Matter*, **12**(3), pp. 642–657.

50. Laguecir, A. and Stoll, S. (2005). Adsorption of a weakly charged polymer on an oppositely charged colloidal particle: Monte Carlo simulations investigation. *Polymer*, **46**(4), pp. 1359–1372.

51. John, A., Nagel, J., and Heinrich, G. (2007). Monte Carlo simulation of polymer reactions at interfaces. *Macromolecular Theory and Simulations*, **16**(4), pp. 430–440.

52. Möddel, M., Janke, W., and Bachmann, M. (2011). Adsorption of finite polymers in different thermodynamic ensembles. *Computer Physics Communications*, **182**(9), pp. 1961–1965.

53. Feng, J., Venna, S. R., and Hopkinson, D. P. (2016). Interactions at the interface of polymer matrix-filler particle composites. *Polymer*, **103**, pp. 189–195.

54. Iedema, P. D. and Hoefsloot, H. C. (2006). Conditional Monte Carlo sampling to find branching architectures of polymers from radical polymerizations with transfer to polymer. *Macromolecules*, **39**(8), pp. 3081–3088.

55. Meimaroglou, D. and Kiparissides, C. (2014). Review of Monte Carlo methods for the prediction of distributed molecular and morphological polymer properties. *Industrial and Engineering Chemistry Research*, **53**(22), pp. 8963–8979.

56. Brandão, A. L., Soares, J. B., Pinto, J. C., and Alberton, A. L. (2015). When polymer reaction engineers play dice: Applications of Monte Carlo models in PRE. *Macromolecular Reaction Engineering*, **9**(3), pp. 141–185.

57. Lemos, T., Melo, P. A., and Pinto, J. C. (2015). Stochastic modeling of polymer microstructure from residence time distribution. *Macromolecular Reaction Engineering*, **9**(3), pp. 259–270.

58. Tobita, H. (1995). Monte Carlo simulation of emulsion polymerization: Linear, branched, and crosslinked polymers. *Acta Polymerica*, **46**(3), pp. 185–203.

59. Nie, L., Yang, W., Zhang, H., and Fu, S. (2005). Monte Carlo simulation of microemulsion polymerization. *Polymer*, **46**(9), pp. 3175–3184.
60. Tobita, H. and Yanase, F. (2007). Monte Carlo simulation of controlled/living radical polymerization in emulsified systems. *Macromolecular Theory and Simulations*, **16**(4), pp. 476–488.
61. Stubbs, J., Carrier, R., and Sundberg, D. C. (2008). Monte Carlo simulation of emulsion polymerization kinetics and the evolution of latex particle morphology and polymer chain architecture. *Macromolecular Theory and Simulations*, **17**(4–5), pp. 147–162.
62. Goikoetxea, M., Reyes, Y., Carolina, M., Minari, R. J., Beristain, I., Paulis, M., Barandiaran, M. J., Keddie, J. L., and Asua, J. M. (2012). Transformation of waterborne hybrid polymer particles into films: Morphology development and modeling. *Polymer*, **53**(5), pp. 1098–1108.
63. Hamzehlou, S., Reyes, Y., and Leiza, J. R. (2013). Modeling the mini-emulsion copolymerization of N-butyl acrylate with a water-soluble monomer: A Monte Carlo approach. *Industrial & Engineering Chemistry Research*, **53**(22), pp. 8996–9003.
64. Mahjub, A. (2016). Monte Carlo simulation of the dispersion polymerization of styrene. *RSC Advances*, **6**(54), pp. 48973–48984.
65. Chaloupka, T., Zubov, A., and Kosek, J. (2017). Real-time hybrid Monte Carlo method for modelling of 4 monomer semi-batch emulsion copolymerization. *Computer Aided Chemical Engineering*, **40**, pp. 259–264.
66. Paquet, E. and Viktor, H. L. (2015). Molecular dynamics, Monte Carlo simulations, and Langevin dynamics: A computational review. *BioMed Research International*, 183918.
67. Verlet, L. (1967). Computer "experiments" on classical fluids. I. Thermodynamical properties of Lennard-Jones molecules. *Physical Review*, **159**(1), pp. 98–103.
68. Hinchliffe, A. (2003). *Molecular Modelling for Beginners* (John Wiley & Sons, UK).
69. Beeman, D. (1976). Some multistep methods for use in molecular dynamics calculations. *Journal of Computational Physics*, **20**(2), pp. 130–139.
70. Hockney, R. W. and Eastwood, J. W. (1981). *Computer Simulation Using Particles* (McGraw-Hill Book Company, USA).
71. Rahman, A. (1964). Correlations in the motion of atoms in liquid argon. *Physical Review*, **136**(2A), pp. A405–A411.

72. Gear, C. W. (1971). *Numerical Initial Value Problems in Ordinary Differential Equations* (Prentice Hall, USA).

73. Rapaport, D. C. and Rapaport, D. C. R. (2004). *The Art of Molecular Dynamics Simulation* (Cambridge University Press, UK).

74. Brostow, W., Macip, M. A., and Sochanski, J. S. (1982). Macromolecular conformations in solutions. II. Thermodynamics of interactions. *Journal of Statistical Physics*, **29**(4), pp. 865–878.

75. Fermeglia, M. and Pricl, S. (1999). Equation-of-state parameters for pure polymers by molecular dynamics simulations. *AIChE Journal*, **45**(12), pp. 2619–2627.

76. Indrakanti, A., Maranas, J. K., and Kumar, S. K. (2000). Intramolecular effects on the thermodynamics of polymers. *Macromolecules*, **33**(23), pp. 8865–8869.

77. Kozuch, D., Zhang, W., and Milner, S. (2016). Predicting the Flory–Huggins χ parameter for polymers with stiffness mismatch from molecular dynamics simulations. *Polymers*, **8**(6), 241.

78. Aydt, E. M. and Hentschke, R. (2000). Swelling of a model network: A Gibbs-ensemble molecular dynamics study. *The Journal of Chemical Physics*, **112**(12), pp. 5480–5487.

79. Yu, K. Q., Li, Z. S., and Sun, J. (2001). Polymer structures and glass transition: A molecular dynamics simulation study. *Macromolecular Theory and Simulations*, **10**(6), pp. 624–633.

80. Carbone, P., Rapallo, A., Ragazzi, M., Tritto, I., and Ferro, D. R. (2006). Glass transition temperature and chain flexibility of ethylene-norbornene copolymers from molecular dynamics simulations. *Macromolecular Theory and Simulations*, **15**(6), pp. 457–468.

81. Zhang, D. and Meyer, H. (2007). Molecular dynamics study of polymer crystallization in the presence of a particle. *Journal of Polymer Science Part B: Polymer Physics*, **45**(16), pp. 2161–2166.

82. Michel, A. and Kreitmeier, S. (1997). Molecular dynamics simulation of the collapse of a single polymer chain. *Computational and Theoretical Polymer Science*, **7**(2), pp. 113–120.

83. Koch, S. W. and Liebmann, R. (1983). Comparison of molecular dynamics and Monte Carlo computer simulations of spinodal decomposition. *Journal of Statistical Physics*, **33**(1), pp. 31–41.

84. Gotlib, Y. Y., Balabaev, N. K., Darinskii, A. A., and Neelov, I. M. (1980). Investigation of local motions in polymers by the method of molecular dynamics. *Macromolecules*, **13**(3), pp. 602–608.

85. Kremer, K. and Grest, G. S. (1990). Dynamics of entangled linear polymer melts: A molecular-dynamics simulation. *The Journal of Chemical Physics*, **92**(8), pp. 5057–5086.

86. Paul, W., Smith, G. D., Yoon, D. Y., Farago, B., Rathgeber, S., Zirkel, A., Willner, L., and Richter, D. (1998). Chain motion in an unentangled polyethylene melt: A critical test of the rouse model by molecular dynamics simulations and neutron spin echo spectroscopy. *Physical Review Letters*, **80**(11), pp. 2346–2349.

87. Smith, G. D., Bedrov, D., Li, L., and Byutner, O. (2002). A molecular dynamics simulation study of the viscoelastic properties of polymer nanocomposites. *The Journal of Chemical Physics*, **117**(20), pp. 9478–9489.

88. Kim, J. M. and Baig, C. (2016). Precise analysis of polymer rotational dynamics. *Scientific Reports*, **6**, 19127.

89. Takahashi, K., Nishimura, R., Yasuoka, K., and Masubuchi, Y. (2017). Molecular dynamics simulations for resolving scaling laws of polyethylene melts. *Polymers*, **9**(1), 24.

90. Pant, P. K. and Boyd, R. H. (1993). Molecular-dynamics simulation of diffusion of small penetrants in polymers. *Macromolecules*, **26**(4), pp. 679–686.

91. Richard, J. and Wong, K. (1995). Interdiffusion of polymer chains and molecular dynamics in dried latex films. *Journal of Polymer Science Part B: Polymer Physics*, **33**(9), pp. 1395–1407.

92. Alentiev, A., Economou, I. G., Finkelshtein, E., Petrou, J., Raptis, V. E., Sanopoulou, M., Soloviev, S., Ushakov, N., and Yampolskii, Y. (2004). Transport properties of silmethylene homo-polymers and random copolymers: Experimental measurements and molecular simulation. *Polymer*, **45**(20), pp. 6933–6944.

93. Chatterjee, R., Bisoi, S., Kumar, A. G., Padmanabhan, V., and Banerjee, S. (2018). Polyimides containing phosphaphenanthrene skeleton: Gas-transport properties and molecular dynamics simulations. *ACS Omega*, **3**(10), pp. 13510–13523.

94. Wozny, C. E., Sumpter, B. G., and Noid, D. W. (1994). A molecular dynamics method for obtaining the vibrational spectra of macromolecules. *The Journal of Chemical Physics*, **100**(5), pp. 3520–3531.

95. Tuzun, R. E., Noid, D. W., Sumpter, B. G., and Wozny, C. E. (1997). Recent advances in polymer molecular dynamics simulation and data analysis. *Macromolecular Theory and Simulations*, **6**(5), pp. 855–880.

96. Hernández, H. F. and Tauer, K. (2009). Modeling of molecular transfer in heterophase polymerization. *Macromolecular Reaction Engineering*, **3**(7), pp. 375–397.
97. Binder, K. (Ed.). (1995). *Monte Carlo and Molecular Dynamics Simulations in Polymer Science* (Oxford University Press, USA).
98. Dionisio, M., Alves, N. M., and Mano, J. F. (2004). Molecular dynamics in polymeric systems. *e-Polymers*, **4**(1).
99. Barrat, J. L., Baschnagel, J., and Lyulin, A. (2010). Molecular dynamics simulations of glassy polymers. *Soft Matter*, **6**(15), pp. 3430–3446.
100. Flory, P. J. (1969). *Statistical Mechanics of Chain Molecules* (John Wiley & Sons, USA).
101. Kuhn, W. (1934). Über die gestalt fadenförmiger moleküle in lösungen. *Kolloid-Zeitschrift*, **68**(1), pp. 2–15.
102. Brodin, A. (2008). Segmental versus chain dynamics of linear polymers. *The Journal of Chemical Physics*, **128**(10), 104901.
103. Rouse Jr, P. E. (1953). A theory of the linear viscoelastic properties of dilute solutions of coiling polymers. *The Journal of Chemical Physics*, **21**(7), pp. 1272–1280.
104. Müller-Plathe, F. (2002). Coarse-graining in polymer simulation: From the atomistic to the mesoscopic scale and back. *ChemPhysChem*, **3**(9), pp. 754–769.
105. Sun, Q. and Faller, R. (2005). Systematic coarse-graining of atomistic models for simulation of polymeric systems. *Computers & Chemical Engineering*, **29**(11–12), pp. 2380–2385.
106. Detcheverry, F. A., Pike, D. Q., Nealey, P. F., Müller, M., and de Pablo, J. J. (2009). Monte Carlo simulation of coarse grain polymeric systems. *Physical Review Letters*, **102**(19), 197801.
107. Salerno, K. M., Agrawal, A., Perahia, D., and Grest, G. S. (2016). Resolving dynamic properties of polymers through coarse-grained computational studies. *Physical Review Letters*, **116**(5), 058302.
108. Salerno, K. M. and Bernstein, N. (2018). Persistence length, end-to-end distance, and structure of coarse-grained polymers. *Journal of Chemical Theory and Computation*, **14**(4), pp. 2219–2229.
109. Tschöp, W., Kremer, K., Hahn, O., Batoulis, J., and Bürger, T. (1998). Simulation of polymer melts. II. From coarse-grained models back to atomistic description. *Acta Polymerica*, 49(2–3), pp. 75–79.
110. Uhlherr, A. (2000). A multiple chain Monte Carlo method for atomistic simulation of high molecular weight polymer melts. *Computational and Theoretical Polymer Science*, **10**(1-2), pp. 29–41.

111. Tito, N. B., Storm, C., and Ellenbroek, W. G. (2017). Self-consistent field lattice model for polymer networks. *Macromolecules*, **50**(24), pp. 9788–9795.

112. Reith, D., Meyer, H., and Müller-Plathe, F. (2001). Mapping atomistic to coarse-grained polymer models using automatic simplex optimization to fit structural properties. *Macromolecules*, **34**(7), pp. 2335–2345.

113. Shie, S. C., Lee, C. K., Hua, C. C., and Chen, S. A. (2010). A predictive coarse-grained model for semiflexible polymers in specific solvents. *Macromolecular Theory and Simulations*, **19**(4), pp. 179–189.

114. Andreev, M., Chremos, A., de Pablo, J., and Douglas, J. F. (2017). Coarse-grained model of the dynamics of electrolyte solutions. *The Journal of Physical Chemistry B*, 121(34), pp. 8195–8202.

115. Raubenolt, B., Gyawali, G., Tang, W., Wong, K., and Rick, S. (2018). Coarse-grained simulations of aqueous thermoresponsive polyethers. *Polymers*, 10(5), 475.

116. Freire, J. J. (1999). Conformational properties of branched polymers: Theory and simulations. In: Roovers, J., *Branched Polymers II* (Springer, Germany), pp. 35–112.

117. Kumar, A. (2006). *Molecular Dynamics Simulations of Polyelectrolyte Brushes*. Doctoral Dissertation (Universität Potsdam, Germany).

118. Binder, K. and Milchev, A. (2002). Off-lattice Monte Carlo methods for coarse-grained models of polymeric materials and selected applications. *Journal of Computer-Aided Materials Design*, **9**(1), pp. 33–74.

119. Dance, I. G. (1996). Supramolecular inorganic chemistry. In: Desiraju, G. R., *The Crystal as a Supramolecular Entity* (John Wiley, USA), pp. 137–233.

120. Lehn, J. M. (1978). Cryptates: The chemistry of macropolycyclic inclusion complexes. *Accounts of Chemical Research*, **11**(2), pp. 49–57.

121. van der Waals, J. D. (1873). *Over de Continuiteit van den Gas- en Vloeistoftoestand*. Doctoral Dissertation (Hoogeschool te Leiden, The Netherlands).

122. Mattia, E. and Otto, S. (2015). Supramolecular systems chemistry. *Nature Nanotechnology*, **10**(2), pp. 111–119.

123. Israelachvili, J. N., Mitchell, D. J., and Ninham, B. W. (1976). Theory of self-assembly of hydrocarbon amphiphiles into micelles and bilayers. *Journal of the Chemical Society, Faraday Transactions 2: Molecular and Chemical Physics*, **72**, pp. 1525–1568.

124. Aniansson, E. A. G. and Wall, S. N. (1975). Kinetics of step-wise micelle association and dissociation. In: Wyn-Jones, E., *Chemical and Biological Applications of Relaxation Spectrometry* (Springer, Germany), pp. 223–238.

125. Markvoort, A. J., Eikelder, H. M. T., Hilbers, P. A., and de Greef, T. F. (2016). Fragmentation and coagulation in supramolecular (Co) polymerization kinetics. *ACS Central Science*, **2**(4), pp. 232–241.

126. Tanford, C. (1980). *The Hydrophobic Effect: Formation of Micelles and Biological Membranes*, 2nd Ed. (Wiley-Interscience, USA).

127. Nagarajan, R. and Ruckenstein, E. (1991). Theory of surfactant self-assembly: A predictive molecular thermodynamic approach. *Langmuir*, **7**(12), pp. 2934–2969.

128. Shchekin, A. K., Kuni, F. M., Grinin, A. P., and Rusanov, A. I. (2005). Nucleation in micellization processes. In: Schmelzer, J. W. P., *Nucleation Theory and Applications* (Wiley-VCH, Germany), pp. 312–374.

129. Kashchiev, D. (2000). *Nucleation* (Butterworth-Heinemann, UK).

130. Smit, B., Hilbers, P. A. J., and Esselink, K. (1993). Computer simulations of surfactant self-assembly. *International Journal of Modern Physics C*, **4**(02), pp. 393–400.

131. Mavelli, F. (1997). Stochastic simulations of surfactant aggregation kinetics. *Progress in Colloid and Polymer Science*, **103**, pp. 155–159.

132. Gillespie, D. T. (1976). A general method for numerically simulating the stochastic time evolution of coupled chemical reactions. *Journal of Computational Physics*, **22**(4), pp. 403–434.

133. Kühn, I. (1996) *Untersuchungen zum Mechanismus der Teilchenbildung während der Emulsionspolymerisation*, Doctoral Dissertation (Universität Potsdam, Germany).

134. Kühn, I. and Tauer, K. (1995). Nucleation in emulsion polymerization: A new experimental study 1. Surfactant-free emulsion polymerization of styrene. *Macromolecules*, **28**, pp. 8122–8128.

135. Laaksonen, A., Talanquer, V., and Oxtoby, D. W. (1995). Nucleation: Measurements, theory, and atmospheric applications. *Annual Review of Physical Chemistry*, **46**(1), pp. 489–524.

136. Tauer, K., Hernandez, H., Kozempel, S., Lazareva, O., and Nazaran, P. (2008). Towards a consistent mechanism of emulsion polymerization: New experimental details. *Colloid and Polymer Science*, **286**(5), pp. 499–515.

137. Mokrushin, S. G. (1962). Thomas Graham and the definition of colloids. *Nature*, **195**(4844), pp. 861.

138. Warren, J. C. (1844). Peculiar case of gelatiniform cancer, in which nearly all the organs of the body contained colloid tumours; with the appearances on dissection. *Medico-Chirurgical Transactions*, **27**, pp. 385–397.

139. Graham, T. (1861). X. Liquid diffusion applied to analysis. *Philosophical Transactions of the Royal Society of London*, **151**, pp. 183–224.

140. Ostwald, W. (1975). Zur Gründung der Kolloid-Gesellschaft. In: Steinkopff, J., *Konzepte der Kolloidchemie* (Springer-Verlag, Germany), pp. 1–2.

141. Neelov, I. M. and Binder, K. (1995). Brownian dynamics simulation of grafted polymer brushes. *Macromolecular Theory and Simulations*, **4**(1), pp. 119–136.

142. Lodge, J. M. and Heyes, D. (1999). Transient colloidal gels by Brownian dynamics computer simulation. *Physical Chemistry Chemical Physics*, **1**(9), pp. 2119–2130.

143. Doyle, P. S. and Underhill, P. T. (2005). Brownian dynamics simulations of polymers and soft matter. In: Yip, S., *Handbook of Materials Modeling* (Springer, Germany), pp. 2619–2630.

144. Edwards, S. F. and Muthukumar, M. (1984). Brownian dynamics of polymer solutions. *Macromolecules*, **17**(4), pp. 586–596.

145. Brostow, W., Drewniak, M., and Medvedev, N. N. (1995). Chain overlap and entanglements in dilute polymer solutions: Brownian dynamics simulation. *Macromolecular Theory and Simulations*, **4**(4), pp. 745–758.

146. Li, L. and Larson, R. G. (2000). Comparison of Brownian dynamics simulations with microscopic and light-scattering measurements of polymer deformation under flow. *Macromolecules*, **33**(4), pp. 1411–1415.

147. Kumar, K. S. and Prakash, J. R. (2003). Equilibrium swelling and universal ratios in dilute polymer solutions: Exact Brownian dynamics simulations for a delta function excluded volume potential. *Macromolecules*, **36**(20), pp. 7842–7856.

148. Pamies, R., Cifre, J. H., and De La Torre, J. G. (2007). Brownian dynamics simulation of polyelectrolyte dilute solutions: Relaxation time and elongational flow. *Journal of Polymer Science Part B: Polymer Physics*, **45**(6), pp. 714–722.

149. Pham, T. T., Bajaj, M., and Prakash, J. R. (2008). Brownian dynamics simulation of polymer collapse in a poor solvent: Influence of implicit hydrodynamic interactions. *Soft Matter*, **4**(6), pp. 1196–1207.

150. Saadat, A. and Khomami, B. (2015). Matrix-free Brownian dynamics simulation technique for semidilute polymeric solutions. *Physical Review E*, **92**(3), 033307.

151. Young, C. D., Marvin, M., and Sing, C. E. (2018). Conformationally averaged iterative Brownian dynamics simulations of semidilute polymer solutions. *The Journal of Chemical Physics*, **149**(17), 174904.

152. Ravichandran, S., Madura, J. D., and Talbot, J. (2001). A Brownian dynamics study of the initial stages of hen egg-white lysozyme adsorption at a solid interface. *The Journal of Physical Chemistry B*, **105**(17), pp. 3610–3613.

153. Cerbelaud, M., Videcoq, A., Alison, L., Tervoort, E., and Studart, A. R. (2017). Early dynamics and stabilization mechanisms of oil-in-water emulsions containing colloidal particles modified with short amphiphiles: A numerical study. *Langmuir*, **33**(50), pp. 14347–14357.

154. Schuss, Z. (2015). *Brownian Dynamics at Boundaries and Interfaces* (Springer-Verlag, USA).

155. Hernández, H. F. and Tauer, K. (2007). Brownian Dynamics simulation of the capture of primary radicals in dispersions of colloidal polymer particles. *Industrial & Engineering Chemistry Research*, **46**(13), pp. 4480–4485.

156. Hernández, H. F. and Tauer, K. (2007). Brownian dynamics simulation studies on radical capture in emulsion polymerization. *Macromolecular Symposia*, **259**(1), pp. 274–283.

157. Hoogerbrugge, P. J. and Koelman, J. M. V. A. (1992). Simulating microscopic hydrodynamic phenomena with dissipative particle dynamics. *Europhysics Letters*, **19**(3), pp. 155–160.

158. Espanol, P. and Warren, P. (1995). Statistical mechanics of dissipative particle dynamics. *Europhysics Letters*, **30**(4), pp. 191–196.

159. Groot, R. D. and Warren, P. B. (1997). Dissipative particle dynamics: Bridging the gap between atomistic and mesoscopic simulation. *The Journal of Chemical Physics*, **107**(11), pp. 4423–4435.

160. Liu, M. B., Liu, G. R., Zhou, L. W., and Chang, J. Z. (2015). Dissipative particle dynamics (DPD): An overview and recent developments. *Archives of Computational Methods in Engineering*, **22**(4), pp. 529–556.

161. Sokhan, V. P. and Todorov, I. T. (2019). Dissipative particle dynamics: Dissipative forces from atomistic simulation. *Molecular Simulation*, Published online. doi: 10.1080/08927022.2019.1578353.

162. Gibson, J. B., Zhang, K., Chen, K., Chynoweth, S., and Manke, C. W. (1999). Simulation of colloid-polymer systems using dissipative particle dynamics. *Molecular Simulation*, **23**(1), pp. 1–41.

163. van Vliet, R. E., Hoefsloot, H. C., Hamersma, P. J., and Iedema, P. D. (2000). Pressure-induced phase separation of polymer-solvent systems with dissipative particle dynamics. *Macromolecular Theory and Simulations*, **9**(9), pp. 698–702.

164. Goujon, F., Malfreyt, P., and Tildesley, D. J. (2004). Dissipative particle dynamics simulations in the grand canonical ensemble: Applications to polymer brushes. *ChemPhysChem*, **5**(4), pp. 457–464.

165. Ortiz, V., Nielsen, S. O., Discher, D. E., Klein, M. L., Lipowsky, R., and Shillcock, J. (2005). Dissipative particle dynamics simulations of polymersomes. *The Journal of Physical Chemistry B*, **109**(37), pp. 17708–17714.

166. Zhang, J. M., Li, H., Liu, H., and Sun, C. C. (2011). Dissipative particle dynamics simulation study on controlling molecular weight distribution in emulsion polymerization. *Journal of Theoretical and Computational Chemistry*, **10**(5), pp. 615–628.

167. Masubuchi, Y., Langeloth, M., Böhm, M. C., Inoue, T., and Müller-Plathe, F. (2016). A multichain slip-spring dissipative particle dynamics simulation method for entangled polymer solutions. *Macromolecules*, **49**(23), pp. 9186–9191.

168. Sindelka, K., Limpouchová, Z., Stpánek, M., and Procházka, K. (2017). Stabilization of coated inorganic nanoparticles by amphiphilic copolymers in aqueous media. Dissipative particle dynamics study. *Colloid & Polymer Science*, **295**(8), pp. 1429–1441.

169. Lemos, T., Abreu, C., and Pinto, J. C. (2020). Mesoscopic simulation of dispersed copolymers: Effects of chain length, chemical composition, and block length distributions on self-assembly. *Macromolecular Theory and Simulations*, **29**(1), 1900042.

170. Park, J. D., Myung, J. S., and Ahn, K. H. (2016). A review on particle dynamics simulation techniques for colloidal dispersions: Methods and applications. *Korean Journal of Chemical Engineering*, **33**(11), pp. 3069–3078.

171. Galimzyanov, B. N. and Mokshin, A. V. (2017). Surface tension of water droplets upon homogeneous droplet nucleation in water vapor. *Colloid Journal*, **79**(1), pp. 26–34.

172. Gardiner, C. W. (1990). *Handbook of Stochastic Methods*, 2nd Ed. (Springer-Verlag, Germany), pp. 37.

173. Hernandez, H. (2019). Sums and averages of large samples using standard transformations: The central limit theorem and the law of large numbers. *ForsChem Research Reports*, **4**(1), FRR 2019-01. doi: 10.13140/RG.2.2.32429.33767.

174. van Kampen, N. G. (1992). *Stochastic Processes in Physics and Chemistry* (Elsevier, The Netherlands), pp. 26–29.

175. Gillespie, D. T. (2007). Stochastic simulation of chemical kinetics. *Annual Review of Physical Chemistry*, **58**, pp. 35–55.

176. Cao, Y., Gillespie, D., and Petzold, L. (2005). Multiscale stochastic simulation algorithm with stochastic partial equilibrium assumption for chemically reacting systems. *Journal of Computational Physics*, **206**(2), pp. 395–411.

177. Saidel, G. M. and Katz, S. (1969). Emulsion polymerization: A stochastic approach to the polymer size distribution. *Journal of Polymer Science Part C: Polymer Symposia*, **27**(1), pp. 149–169.

178. Drache, M., Schmidt-Naake, G., Buback, M., and Vana, P. (2005). Modeling RAFT polymerization kinetics via Monte Carlo methods: Cumyl dithiobenzoate mediated methyl acrylate polymerization. *Polymer*, **46**(19), pp. 8483–8493.

179. Salis, H. and Kaznessis, Y. (2005). Accurate hybrid stochastic simulation of a system of coupled chemical or biochemical reactions. *The Journal of Chemical Physics*, **122**(5), 054103.

180. Marquez-Lago, T. T. and Burrage, K. (2007). Binomial tau-leap spatial stochastic simulation algorithm for applications in chemical kinetics. *The Journal of Chemical Physics*, **127**(10), 09B603.

181. Chatterjee, A. and Vlachos, D. G. (2007). An overview of spatial microscopic and accelerated kinetic Monte Carlo methods. *Journal of Computer-Aided Materials Design*, **14**(2), pp. 253–308.

182. Andrews, S. S. and Bray, D. (2004). Stochastic simulation of chemical reactions with spatial resolution and single molecule detail. *Physical Biology*, **1**(3), pp. 137–151.

183. Gillespie, D. T. (1977). Exact stochastic simulation of coupled chemical reactions. *The Journal of Physical Chemistry*, **81**(25), pp. 2340–2361.

184. Evans, J. W. (2005). Kinetic Monte Carlo simulation of non-equilibrium lattice-gas models: Basic and refined algorithms applied to surface adsorption processes. In: Yip, S., *Handbook of Materials Modeling* (Springer, Germany), pp. 1753–1767.

185. Erban, R. and Chapman, S. J. (2007). Time scale of random sequential adsorption. *Physical Review E*, **75**(4), 041116.

186. Schneider, D., Kapteijn, F., and Valiullin, R. (2019). Transport properties of mixed-matrix membranes: A kinetic Monte Carlo study. *Physical Review Applied*, **12**(4), 044034.

187. Marchetti, L., Priami, C., and Thanh, V. H. (2017). Hybrid simulation algorithms. In: Marchetti, L., Priami, C., and Thanh, V. H., *Simulation Algorithms for Computational Systems Biology* (Springer, Switzerland), pp. 181–205.

188. Hernandez, H. F. and Tauer, K. (2008). Stochastic simulation of imperfect mixing in free radical polymerization. *Macromolecular Symposia*, **271**(1), pp. 64–74.

189. Stickler, M. (1983). Free-radical polymerization kinetics of methyl methacrylate at very high conversions. *Die Makromolekulare Chemie: Macromolecular Chemistry and Physics*, **184**(12), pp. 2563–2579.

190. Takeuchi, H. (1999). Monte Carlo simulations of the exchange kinetics of polymers adsorbed on a solid surface. *Macromolecular Theory and Simulations*, **8**(4), pp. 391–401.

191. Hansen, E. W. and Neurock, M. (2000). First-principles-based Monte Carlo simulation of ethylene hydrogenation kinetics on Pd. *Journal of Catalysis*, **196**(2), pp. 241–252.

192. Mei, D., Du, J., and Neurock, M. (2010). First-principles-based kinetic Monte Carlo simulation of nitric oxide reduction over platinum nanoparticles under lean-burn conditions. *Industrial & Engineering Chemistry Research*, **49**(21), pp. 10364–10373.

193. Cheng, F., He, X., Chen, Z. X., and Huang, Y. G. (2015). Kinetic Monte Carlo simulation of surface segregation in Pd–Cu alloys. *Journal of Alloys and Compounds*, **648**, pp. 1090–1096.

194. Darby, M. T., Piccinin, S., and Stamatakis, M. (2016). First principles-based kinetic Monte Carlo simulation in catalysis. In: Kasai, H. and Sison, M. C., *Physics of Surface, Interface and Cluster Catalysis* (IOP Publishing, UK).

195. Andersen, M., Panosetti, C., and Reuter, K. (2019). A practical guide to surface kinetic Monte Carlo simulations. *Frontiers in Chemistry*, **7**, 202.

196. Chaffey-Millar, H., Stewart, D., Chakravarty, M. M., Keller, G., and Barner-Kowollik, C. (2007). A parallelised high performance Monte Carlo simulation approach for complex polymerisation kinetics. *Macromolecular Theory and Simulations*, **16**(6), pp. 575–592.

197. Galina, H. and Lechowicz, J. B. (2002). Kinetic and Monte-Carlo modelling of hyperbranched polymerisation. *e-Polymers*, **2**(1).

198. Slepoy, A., Thompson, A. P., and Plimpton, S. J. (2008). A constant-time kinetic Monte Carlo algorithm for simulation of large biochemical reaction networks. *The Journal of Chemical Physics*, **128**(20), 05B618.

199. Wang, R., Lin, T. S., Johnson, J. A., and Olsen, B. D. (2017). Kinetic Monte Carlo simulation for quantification of the gel point of polymer networks. *ACS Macro Letters*, **6**(12), pp. 1414–1419.

200. Katsoulakis, M. A. and Vlachos, D. G. (2003). Coarse-grained stochastic processes and kinetic Monte Carlo simulators for the diffusion of interacting particles. *The Journal of Chemical Physics*, **119**(18), pp. 9412–9427.

201. Tien, W. J. and Chiu, C. C. (2018). Generic parameters of trajectory-extending kinetic Monte Carlo for calculating diffusion coefficients. *AIP Advances*, **8**(6), 065311.

202. Marien, Y. W., van Steenberge, P. H., D'hooge, D. R., and Marin, G. B. (2019). Particle by particle kinetic Monte Carlo tracking of reaction and mass transfer events in miniemulsion free radical polymerization. *Macromolecules*, **52**(4), pp. 1408–1423.

203. Martins, J., Naqvi, K. R., and Melo, E. (2000). Kinetics of two-dimensional diffusion-controlled reactions: A Monte Carlo simulation of hard-disk reactants undergoing a Pearson-type random walk. *The Journal of Physical Chemistry B*, **104**(20), pp. 4986–4991.

204. Gao, H., Broadbelt, L. J., Konstantinov, I. A., and Arturo, S. G. (2017). Acceleration of kinetic Monte Carlo simulations of free radical copolymerization: A hybrid approach with scaling. *AIChE Journal*, **63**(9), pp. 4013–4021.

205. Lu, J., Zhang, H., and Yang, Y. (1993). Monte Carlo simulation of kinetics and chain-length distribution in radical polymerization. *Macromolecular Theory and Simulations*, **2**(5), pp. 747–760.

206. He, J., Zhang, H., Chen, J., and Yang, Y. (1997). Monte Carlo simulation of kinetics and chain length distributions in living free-radical polymerization. *Macromolecules*, **30**(25), pp. 8010–8018.

207. Habibi, A. and Vasheghani-Farahani, E. (2007). Bayesian modeling and Markov chain Monte Carlo simulations for a kinetic study of homo- and co-polymerization systems. *Macromolecular Theory and Simulations*, **16**(3), pp. 269–294.

208. Serajian, S., Najafi, M., and Amiri, A. (2019). The impacts of feed fraction on copolymer microstructure in ATRcP: A kinetic Monte Carlo simulation. *International Journal of Modelling and Simulation*, **39**(2), pp. 96–109.

209. Tobita, H., Takada, Y., and Nomura, M. (1994). Molecular weight distribution in emulsion polymerization. *Macromolecules*, **27**(14), pp. 3804–3811.

210. Hernandez, H. F. and Tauer, K. (2008). Brownian dynamics and kinetic Monte Carlo simulation in emulsion polymerization. *Computer Aided Chemical Engineering*, **25**, pp. 769–774.

211. Drache, M., Brandl, K., Reinhardt, R., and Beuermann, S. (2018). Ab initio kinetic Monte Carlo simulation of seeded emulsion polymerizations of styrene. *Physical Chemistry Chemical Physics*, **20**(16), pp. 10796–10805.

212. Brandl, F., Drache, M., and Beuermann, S. (2018). Kinetic Monte Carlo simulation based detailed understanding of the transfer processes in semi-batch iodine transfer emulsion polymerizations of vinylidene fluoride. *Polymers*, **10**(9), 1008.

213. Urrea-Quintero, J. H., Ochoa, S., and Hernández, H. (2019). A reduced-order multiscale model of a free-radical semibatch emulsion polymerization process. *Computers & Chemical Engineering*, **127**, pp. 11–24.

214. Moaveni, S. (1999). *Finite Element Analysis Theory and Application with ANSYS* (Prentice Hall, USA).

215. Nassehi, V. (2002). *Practical Aspects of Finite Element Modelling of Polymer Processing* (John Wiley & Sons, USA).

216. Kolhapure, N. H. and Fox, R. O. (1999). CFD analysis of micromixing effects on polymerization in tubular low-density polyethylene reactors. *Chemical Engineering Science*, **54**(15–16), pp. 3233–3242.

217. Meszéna, Z. G. and Johnson, A. F. (2001). Prediction of the spatial distribution of the average molecular weights in living polymerisation reactors using CFD methods. *Macromolecular Theory and Simulations*, **10**(2), pp. 123–135.

218. Elgebrandt, R. C., Fletcher, D. F., Gomes, V. G., and Romagnoli, J. A. (2006). A framework for modeling particle size effects in emulsion polymerization systems using computational fluid dynamics linked to a detailed population balance model. *Computer Aided Chemical Engineering*, **21**, pp. 551–556.

219. Patel, H., Ein-Mozaffari, F., and Dhib, R. (2010). CFD analysis of mixing in thermal polymerization of styrene. *Computers & Chemical Engineering*, **34**(4), pp. 421–429.

220. Xu, C. Z., Wang, J. J., Gu, X. P., and Feng, L. F. (2017). CFD modeling of styrene polymerization in a CSTR. *Chemical Engineering Research and Design*, **125**, pp. 46–56.

221. Abrahamse, A. J., van der Padt, A., Boom, R. M., and De Heij, W. B. C. (2001). Process fundamentals of membrane emulsification: Simulation with CFD. *AIChE Journal*, **47**(6), pp. 1285–1291.

222. Do Amaral, M., Arevalillo, A., Santos, J. L., and Asua, J. M. (2004). Novel insight into the miniemulsification process: CFD applied to ultrasonication. In: Tauer, K., *Aqueous Polymer Dispersions* (Springer, Germany), pp. 103–106.

223. Fogliati, M., Fontana, D., Garbero, M., Vanni, M., Baldi, G., and Donde, R. (2006). CFD simulation of paint deposition in an air spray process. *JCT Research*, **3**(2), pp. 117–125.

224. Keshani, S., Montazeri, M. H., Daud, W. R. W., and Nourouzi, M. M. (2015). CFD modeling of air flow on wall deposition in different spray dryer geometries. *Drying Technology*, **33**(7), pp. 784–795.

225. Janssen, J. and Mayer, R. (2016). Computational fluid dynamics (CFD)-based droplet size estimates in emulsification equipment. *Processes*, **4**(4), 50.

226. Pohn, J., Cunningham, M., and McKenna, T. F. (2013). Scale-up of emulsion polymerization reactors Part II–Simulations and interpretations. *Macromolecular Reaction Engineering*, **7**(8), pp. 393–408.

227. Roudsari, S. F. (2015). *Experimental and CFD Investigation of the Mixing of MMA Emulsion Polymerization in a Stirred Tank Reactor*. Doctoral Dissertation (Ryerson University, Canada).

228. Fathi Roudsari, S., Dhib, R., and Ein-Mozaffari, F. (2016). Using a novel CFD model to assess the effect of mixing parameters on emulsion polymerization. *Macromolecular Reaction Engineering*, **10**(2), pp. 108–122.

229. Nogueira, A. L., Quadri, M. B., Machado, R. A., and Claumann, C. A. (2017). Experimental and CFD study of a vertically stirred tubular reactor designed for suspension polymerization reactions. *Macromolecular Reaction Engineering*, **11**(3), 1600040.

230. Aryafar, S., Sheibat-Othman, N., and McKenna, T. F. (2017). Coupling of CFD simulations and population balance modeling to predict brownian coagulation in an emulsion polymerization reactor. *Macromolecular Reaction Engineering*, **11**(5), 1600054.

231. Ramkrishna, D. (1985). The status of population balances. *Reviews in Chemical Engineering*, **3**(1), pp. 49–95.

232. Immanuel, C. D. and Doyle III, F. J. (2003). Computationally efficient solution of population balance models incorporating nucleation, growth and coagulation: Application to emulsion polymerization. *Chemical Engineering Science*, **58**(16), pp. 3681–3698.

233. Kiparissides, C. (2006). Challenges in particulate polymerization reactor modeling and optimization: A population balance perspective. *Journal of Process Control*, **16**(3), pp. 205–224.

234. Hosseini, A., Bouaswaig, A. E., and Engell, S. (2012). Comparison of classical population balance models of emulsion polymerization with experimental results and a stochastic extension. *Chemical Engineering Science*, **72**, pp. 179–194.

235. Hosseini, A., Bouaswaig, A. E., and Engell, S. (2013). Novel approaches to improve the particle size distribution prediction of a classical emulsion polymerization model. *Chemical Engineering Science*, **88**, pp. 108–120.

236. Urrea-Quintero, J. H., Marino, M., Hernandez, H., and Ochoa, S. (2019). Multiscale modeling of a free-radical emulsion polymerization process: Numerical approximation by the Finite Element Method. Submitted to *Computers and Chemical Engineering*.

237. Sheibat-Othman, N., Vale, H. M., Pohn, J. M., and McKenna, T. F. (2017). Is modeling the PSD in emulsion polymerization a finished problem? An overview. *Macromolecular Reaction Engineering*, **11**(5), 1600059.

238. Rice, R. G. and Do, D. D. (2012). *Applied Mathematics and Modeling for Chemical Engineers*, 2nd Ed. (Wiley, USA), pp. 3–30.

239. Fick, A. (1855). Ueber diffusion. *Annalen der Physik*, **170**(1), pp. 59–86.

240. Gómez, H., Colominas, I., Navarrina, F., and Casteleiro, M. (2007). A finite element formulation for a convection–diffusion equation based on Cattaneo's law. *Computer Methods in Applied Mechanics and Engineering*, **196**(9–12), pp. 1757–1766.

241. Treybal, R. E. (1981). *Mass Transfer Operations*, 3rd Ed. International Edition (McGraw-Hill, USA), pp. 45–87.

242. Knudsen, J. G., Hottel, H. C., Sarofim, A. F., Wankat, P. C., and Knaebel, K. S. (1997). Heat and mass transfer. In: Perry, R. H., Green, D. W., and Maloney, J. O., *Perry's Chemical Engineer's Handbook*, 7th Ed. (McGraw-Hill, USA), pp. 5.1–5.79.

243. Flory, P. J. (1953). *Principles of Polymer Chemistry* (Cornell University Press, USA).

244. Leo, A., Hansch, C., and Elkins, D. (1971). Partition coefficients and their uses. *Chemical Reviews*, **71**(6), pp. 525–616.

245. Schork, F. J., Deshpande, P. B., and Leffew, K. W. (1993). *Control of Polymerization Reactors* (Marcel-Dekker, USA), pp. 31–39.

246. Min, K. W. and Ray, W. H. (1974). On the mathematical modeling of emulsion polymerization reactors. *Journal of Macromolecular Science, Part C: Polymer Reviews*, **11**(2), pp. 177–255

247. Penlidis, A., MacGregor, J. F., and Hamielec, A. E. (1985). Dynamic modeling of emulsion polymerization reactors. *AIChE Journal*, **31**(6), pp. 881–889.

248. Hernandez, H. (2019). Collision energy between Maxwell–Boltzmann molecules: An alternative derivation of Arrhenius equation. *ForsChem Research Reports*, **4**, 2019-13. doi: 10.13140/RG.2.2.21596.33926.

249. Hernandez, H. (2018). On the stochastic nature of multiscale systems: Research perspectives. *ForsChem Research Reports*, **3**, 2018-03. doi: 10.13140/RG.2.2.18642.04803.

250. Hernandez, H. (2017). Ergodic-stochastic transformations. *ForsChem Research Reports*, **3**, 2017-12. doi: 10.13140/RG.2.2.20325.70881.

251. Hernandez, H. (2017). Multivariate probability theory: Determination of probability density functions. *ForsChem Research Reports*, **2**, 2017-13. doi: 10.13140/RG.2.2.28214.60481.

252. Hernandez, H. (2018). Multidimensional randomness, standard random variables and variance algebra. *ForsChem Research Reports*, **3**, 2018-02. doi: 10.13140/RG.2.2.11902.48966.

253. Hernandez, H. (2018). Comparison of methods for the reconstruction of probability density functions from data samples. *ForsChem Research Reports*, **3**, 2018-12. doi: 10.13140/RG.2.2.30177.35686.

254. Hernandez, H. (2019). Discretization of probability distributions: Random, deterministic and randomistic sampling. *ForsChem Research Reports*, **4**, 2019-11. doi: 10.13140/RG.2.2.11389.92643.

255. Yang, A. (2013). On the common conceptual and computational frameworks for multiscale modeling. *Industrial & Engineering Chemistry Research*, **52**(33), pp. 11451–11462.

256. Hoekstra, A., Chopard, B., and Coveney, P. (2014). Multiscale modelling and simulation: A position paper. *Philosophical Transactions of the Royal Society A*, **372**(2021), 20130377.

257. Leach, A. R. (2001). *Molecular Modeling: Principles and Applications*, 2nd Ed. (Prentice Hall, UK), p. 65.

258. Parr, R. G., Craig, D. P., and Ross, I. G. (1950). Molecular orbital calculations of the lower excited electronic levels of benzene, configuration interaction included. *The Journal of Chemical Physics*, **18**(12), pp. 1561–1563.

259. Del Bene, J. and Pople, J. A. (1969). Intermolecular energies of small water polymers. *Chemical Physics Letters*, **4**(7), pp. 426–428.

260. Derjaguin, B. (1983). Polywater reviewed. *Nature*, **301**(5895), pp. 9–10.

261. Derjaguin, B. V., Zorin, Z. M., Rabinovich, Y. I., and Churaev, N. V. (1974). Results of analytical investigation of the composition of "anomalous" water. *Journal of Colloid and Interface Science*, **46**(3), pp. 437–441.

262. Karpfen, A. (1979). Ab initio studies on polymers. I. The linear infinite polyyne. *Journal of Physics C: Solid State Physics*, **12**(16), 3227.

263. Karpfen, A. (1982). Ab Initio calculations on the ground state properties of polymers. *Physica Scripta*, **1982**(T1), 79.

264. Karpfen, A. and Beyer, A. (1984). Ab initio studies on polymers. VII. Polyoxymethylene. *Journal of Computational Chemistry*, **5**(1), pp. 19–23.

265. Karpfen, A. (1987). Ab initio studies on polymers: Torsional potential in helical sulfur chains. *Chemical Physics Letters*, **136**(6), pp. 571–574.

266. Delle Site, L., Abrams, C. F., Alavi, A., and Kremer, K. (2002). Polymers near metal surfaces: Selective adsorption and global conformations. *Physical Review Letters*, **89**(15), 156103.

267. Valavala, P. K. and Odegard, G. M. (2005). Modeling techniques for determination of mechanical properties of polymer nanocomposites. *Reviews on Advanced Materials Science*, **9**(34), e44.

268. Coote, M. L. and Henry, D. J. (2005). Effect of substituents on radical stability in reversible addition fragmentation chain transfer polymerization: An ab initio study. *Macromolecules*, **38**(4), pp. 1415–1433.

269. Vukmirovic, N. and Wang, L. W. (2009). Charge carrier motion in disordered conjugated polymers: A multiscale ab initio study. *Nano Letters*, **9**(12), pp. 3996–4000.

270. Mar, B. D., Qi, H. W., Liu, F., and Kulik, H. J. (2015). Ab initio screening approach for the discovery of lignin polymer breaking pathways. *The Journal of Physical Chemistry A*, **119**(24), pp. 6551–6562.

271. Machado, M. R., Zeida, A., Darré, L., and Pantano, S. (2019). From quantum to subcellular scales: Multi-scale simulation approaches and the SIRAH force field. *Interface Focus*, **9**(3), 20180085.

Chapter 4

Recent Advances and Future Perspectives in Heterophase Polymerization

Heterophase polymerization is a scientific and industrial field with a long history of more than a century of discoveries and theoretical developments. The high complexity of heterophase polymerization processes, resulting from their multiscale and stochastic nature, has probably slowed down the progress in the basic understanding of all the mechanisms and phenomena involved. However, significant contributions have been continuously made for more than 100 years. Before reviewing the most recent advances in this field occurring in the last decade, it is also important to briefly reference the firm experimental and scientific ground supporting these developments.

The first interest in heterophase polymerization resulted from the search of synthetic substitutes for natural rubber. These activities where initially carried out in industry and the cradle of heterophase polymerization stood in Elberfeld, Germany [1, 2]. The first industrial innovations were summarized by Frederick Marchionna and Royce Noble in the 1930s [3–5].

In 1953, Hermann Staudinger, the pioneer of macromolecular chemistry [6, 7] received the Nobel Prize in Chemistry acknowledging the development of the ***macromolecular theory*** [8]. His theory had a significant influence in all areas involving polymers and

Heterophase Polymerization: Basic Concepts and Principles
Hugo Hernandez and Klaus Tauer

ISBN 978-981-4877-32-9 (Hardcover), 978-1-003-11929-6 (eBook)
www.jennystanford.com

greatly contributed to a better understanding of the chemistry of macromolecules. This included, of course, the rubber industry [9–11].

Macromolecular theory: Proposed by Hermann Staudinger in the 1920s, it assumed that polymers were macromolecules obtained by covalent bonds between low-molecular-weight compounds. The theory accepted at the time was that polymers were noncovalently bonded aggregates of small molecules. However, later the *aggregate* theory has been found to be valid only for *supramolecular polymers.*

Polymer latex (or just **Latex**): Stable colloidal dispersion of a polymeric substance in an aqueous medium. The term *latex* emerged because of the milky appearance of polymer dispersions, similar to certain plant juices denoted as *latex* by botanists.

The knowledge gained in synthetic rubber was soon expanded to similar products, giving rise to the field of ***polymer latices*** (or ***polymer latexes***) utilizing adventitiously water as polymerization medium. In 1966, D. C. Blackley published a two-volume comprehensive survey of the science and technology of polymer latices at the time [12]. Some years later, Robert Fitch edited the most relevant advances presented in the 1970 Symposium on ***Polymer Colloids*** (a more general concept that includes polymer latices), resulting in another landmark book on this topic [13].

Polymer colloid: Dispersion of fine polymer particles, preferentially in fluid media, and characterized by presenting considerably lower viscosity compared with polymer solutions at a given polymer content.

Even though the most important scientific progress was made in aqueous polymer dispersions, the science of solvent-based polymer dispersions was also nicely summarized by Barrett in 1975 [14].

The scientific progress made in the 1960s and 1970s in the field of polymer colloids revitalized the interest in the topic and promoted a better understanding of heterophase polymerization and, particularly, polymer dispersions [15] and emulsion polymerization processes [16].

The most important breakthroughs in the science and technology of heterophase polymerization took place during the last decade of the past millennium. New and revised books in the field summarized the newly gained knowledge in the mechanisms of heterophase polymerization [17–21].

At the beginning of the new millennium, Urban and Takamura edited a nice summary of the technology of polymer dispersions, from an industrial perspective [22]. Some years later, two graduate-level textbooks on emulsion polymerization were published [23, 24], including the most recent progress in the field.

Finally, some important publications in the field of heterophase polymerization during the last decade include a book exclusively devoted to miniemulsion polymerization [25], several chapters of a recent encyclopedia on polymers [26], and an updated edition of the textbook on emulsion polymerization edited by van Herk [27].

It is also relevant mentioning the most recent review on advances in heterophase polymerization presented by Jenjob, Seidi, and Crespy [28]. They highlight the progress of heterophase polymerization beyond the classical free-radical emulsion polymerization, reaching other reaction mechanisms, including ionic, metal-catalyzed, step-growth polymerization, and particularly, ***click-chemistry*** reactions. The formation of hybrid structures from nanoparticles (carbon nanotubes, quantum dots, graphene, etc.) and the use of different dispersion media (supercritical carbon dioxide, fluorinated solvents, etc.) are also mentioned. Particular interest is also placed on biomedical applications and the synthesis of precise macromolecular architectures. Such precise control of polymer architecture is possible thanks to a unique property of heterophase polymerization systems: The *segregation effect*.

Click chemistry: Any stereospecific chemical reaction occurring in high yields, favoring a single reaction product (may result in easily removable inoffensive byproducts), insensitive to most environmental conditions, and easy to perform. For example: Thiol-ene reaction, Diels–Alder reaction.

Segregation, compartmentalization, or confinement of reactive species in finely dispersed phases has a very important effect on the kinetics of polymerization, on the molecular mass distribution, on the particle size distribution, and on the particle morphology [29].

Each growing particle in heterophase polymerization behaves as an individual *nanoreactor* [30], with its own composition, kinetics, and properties. The overall properties of the dispersion will be the result of the distribution of properties obtained inside each nanoreactor. A better understanding of the unique characteristics of heterophase polymerization will clearly allow additional progress in this exciting field.

The purpose of this chapter is continuing describing the recent history of heterophase polymerization, by reviewing some of the most relevant advances reported in the last decade in the scientific literature, not covered in previous reviews.

4.1 Surfactants and Dispersants

Surfactants, dispersants, and stabilizers, in general, are one of the key ingredients in most heterophase polymerization formulations. Although they were already included in the recipes for the first emulsion polymerization, there is still much room for improvement in this area. The number of possible stabilizers used in heterophase polymerization increases every day. Currently there are more than 20,000 different industrial surfactants available in the market, including at least 5000 different ingredients [31]. Considering that most heterophase polymerization formulations commonly use blends of commercial surfactants in different proportions, the number of possibilities increases dramatically. According to Steven Abbott, "*single surfactants are generally unsatisfactory because they are in a useless part of surfactant space*"; however, two useless surfactants become useful upon blending [32].

Although surfactants are mainly added for stabilizing the polymer dispersion, they may also have a negative effect during the final use of the product, as a result of migration [22]. In the late 1980s, an interesting idea emerged, consisting of binding the surfactants to the polymer chains, thus avoiding surfactant migration. This type of surfactants was denoted as ***reactive surfactants*** [33–38]. The remaining challenge, however, is to effectively achieve a complete reaction between the reactive surfactants and the polymer, as their reactivity and selectivity are usually low and depend on the particular polymerization system used.

Reactive (or polymerizable) surfactant: Surface active agent incorporating a reactive group, which can form covalent bonds with polymer chains. The reactive group can be an initiator (*inisurf*), a transfer agent (*transurf*), or a monomer (*surfmer*).

Maleic anhydride is one of the preferred modifiers of surfactants because the anhydride ring easily reacts with different functional groups; it provides a polymerizable double bond to the surfactant, and it also leaves one ionizable carboxyl group, which contributes to the surface activity of the molecule. An additional advantage of maleic acid derivatives as surfmers is that they only hardly homopolymerize and have only a low tendency to copolymerize compared with the corresponding fumarates, which are easily used up in alternating copolymerization [39, 40]. Thus, fumarates will, to a larger extent, be buried inside the latex particles and lost for the stabilization of latex particles. Mekki et al. [41], for example, synthesized reactive surfactants using phenyl-modified linear alcohols of different length and maleic anhydride and used them for the emulsion polymerization of styrene. The general structure of these reactive surfactants is presented in Fig. 4.1. It was shown, however, that the overall polymer conversion was reduced (60–83%) when short alkyl chains (i.e., benzyl maleate) are used.

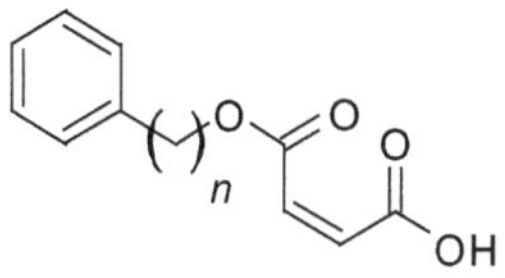

Figure 4.1 Maleate esters of phenyl-modified linear alcohols used as reactive surfactants [41]. n = 1, 4, 7.

Since maleate reactive surfactants usually present reactivity limitations, different chemistries have been explored. One alternative is polyoxyethylene alkylphenyl ether ammonium sulfate with an allyl group attached to the phenyl ring in meta position to the alkyl chain (Fig. 4.2), successfully commercialized by Dai-Ichi Kogyo Seiyaku (Japan) as Hitenol™ BC. Ovando-Medina et al. [42, 43] used Hitenol™ BC10 (10 units of ethylene oxide) in semicontinuous heterophase polymerization of methyl methacrylate. They found that Hitenol BC10 successfully incorporated into the polymer, as evidenced by differential scanning calorimetry and ^{1}H nuclear magnetic resonance. As it can be seen, the ethylenic double bond of this surfmer is not hindered by oxygen atoms as in the case of maleate esters.

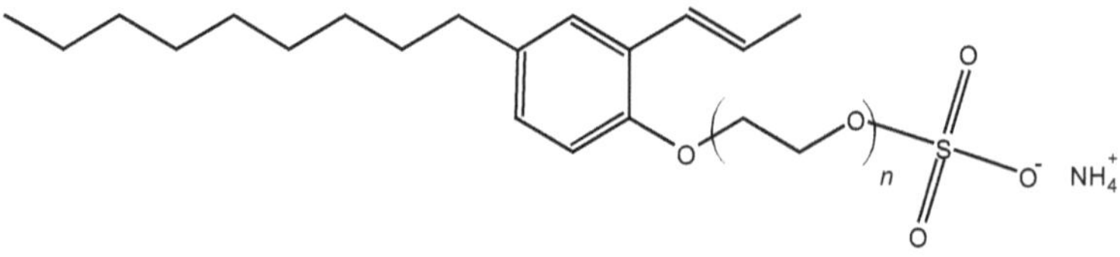

Figure 4.2 Polyoxyethylene alkylphenyl ether ammonium sulfate (Hitenol™ BC10) used as reactive surfactant in heterophase polymerization [42]. $n = 10$.

Another difficulty commonly found in reactive surfactants is that they can become buried inside the polymer particles, reducing their efficiency as colloidal stabilizers. A possible solution to this problem was presented by Herold et al. [44]. The proposed cationic surfmer, p-(11-acrylamido)undecanoyloxyphenyl dimethylsulfonium methyl sulfate (AUPDS) is shown in Fig. 4.3. The acrylamide reactive group is not hindered and provides good reactivity too. The main disadvantage of this cationic surfactant is that it is not compatible with initiators generating anionic end groups in the polymer (e.g., persulfates).

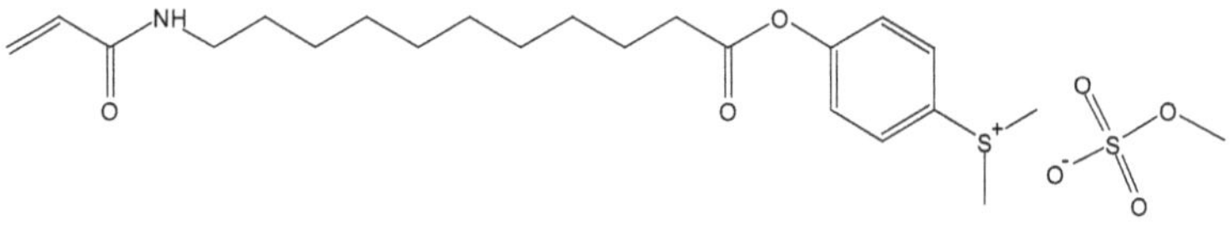

Figure 4.3 p-(11-acrylamido)undecanoyloxyphenyl dimethylsulfonium methyl sulfate (AUPDS) cationic surfmer [44].

Albernaz et al. [45] tested this AUPDS surfmer in miniemulsion polymerizations initiated with AIBN, and confirmed by zeta potential measurements that the surfmer remains in the particle surface for poly(methyl methacrylate) and polystyrene as well as crosslinked copolymers.

The stability of miniemulsions is usually improved by hydrophobic compounds (e.g., hexadecane) sometimes in an incorrect way (cf. below, p. 47) also denoted as co-stabilizers. Even though they are not surfactants, they help improve the stability of the miniemulsion, particularly against Ostwald ripening. Thus, Asua [46] proposed that the concept of reactive surfactants can also be extended to reactive co-stabilizers, (i.e., hydrophobic monomers, initiators, or chain transfer agents), thus ensuring incorporation of those compounds into the polymer and avoiding unwanted side effects such as residual emissions of volatile organic compounds (VOCs) or reduced performance. He found that an optimal reactivity of the co-stabilizer is required for each miniemulsion system, in order to achieve good stability without degrading the final performance of the product.

Returning to reactive surfactants, Itoh et al. [47] proposed a surface-active anionic macromonomer based on poly(L-glutamic acid), as illustrated in Fig. 4.4. The styrene-like end group guarantees good surfmer reactivity. The pH-responsive behavior of the polyelectrolyte chain is used to verify the presence of the surfmer at the surface of the particles.

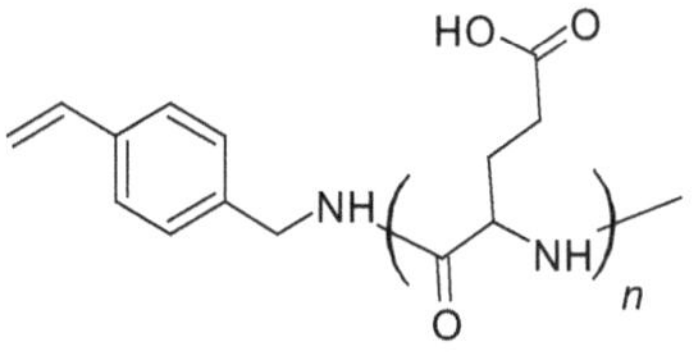

Figure 4.4 Poly(L-glutamic acid)-based macromonomer [47].

Using high-molecular-weight polymer surfactants and stabilizers is also a good approach for reducing migration and performance deterioration. In this case, since there are no chemical reactions involved, there is no need to worry about reactivity or surfactant burial. The most representative polymeric stabilizers include polyelectrolytes and amphiphilic copolymers. In the case of block copolymers, for example, two different blocks in the polymer chain are required, one compatible with the polymer particle and the other compatible with the continuous phase. Although block copolymers are already a mature technology [48], innovations in this field are still possible. Atanase and Riess [49] summarize some of the most recent developments in the field, particularly their use as stabilizers for oil-in-oil emulsions used, for example, when the use of moisture-sensitive monomers or catalysts is required, or when each oil phase must be polymerized by different polymerization mechanisms. Multiple emulsions and other elaborate systems can also be stabilized using block copolymers. More recently, Sanders et al. [50] investigated the stability of *block-random copolymers* (one homopolymer block and one random copolymer block). According to the authors, this type of stabilizer shows a novel nucleation behavior during the emulsion polymerization of styrene, which is currently under further investigation. In addition, they concluded that these *hybrid* copolymer chains are capable of self-folding into single-chain nanoparticles, and small aggregates, resulting in stable dispersions. Thus, amphiphilic copolymers are able to stabilize polymer dispersions in the form of extended, partially solvated chains, or as collapsed amphiphilic particles [51]. The latter stabilization mechanism is also known as ***Pickering stabilization***. Although Pickering stabilization is even older than heterophase polymerization itself [52, 53], its effect on the mechanisms and properties in heterophase polymerization still remains an active subject of research [54–56].

Pickering stabilizers: Colloidal particles capable of stabilizing a heterogeneous system by adsorbing at the interface between the dispersed and the continuous phases.

The technology of amphiphilic block copolymers has greatly improved with the emergence of controlled polymerization mechanisms, as well as *click-chemistry* reactions. They allow a better control of topological structure of the block copolymer, resulting in linear-, cyclic-, and multiblock structures, which can also provide a better control on particle formation, polymer particle size, and stability in heterophase polymerization [57].

4.2 Controlled Radical Polymerization

Controlled radical polymerization has become very important for heterophase polymerization, not only because it allows the synthesis of a wide variety of novel surfactants and stabilizers, but also because it allows a precise control of the architecture and molecular mass of the final polymer. From all types of controlled polymerizations, the *reversible addition–fragmentation chain transfer* (RAFT) process is perhaps the most promising for being used in heterophase polymerization [58]. The mechanism of RAFT polymerization is similar to a conventional radical polymerization, but additional equilibrium reactions (depicted in Fig. 4.5) are present. P_n represents a growing polymer with chain length n, and X, Z, A, and R represent certain particular molecular groups (e.g., methylene or sulfur for both X and A). A radical is represented by a dot, and k denotes the rate coefficient of reaction.

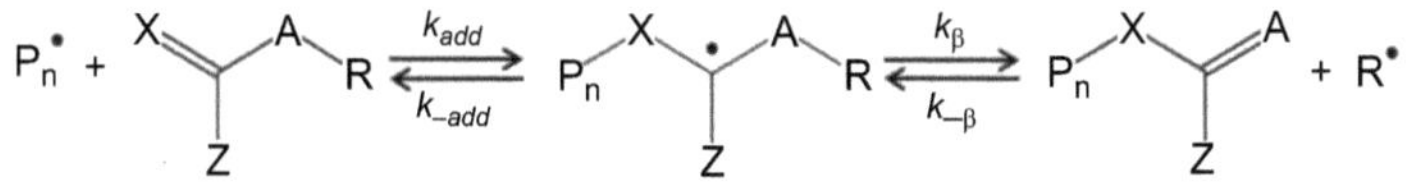

Figure 4.5 RAFT mechanism.

The balance between the forward and reverse reactions determines the degree of control of the polymerization. In one extreme, for example, if $k_{\text{add}} << k_{-\text{add}}$, the system behaves as a conventional free-radical polymerization. On the opposite end, if $k_{\text{add}} >> k_{-\text{add}}$, polymer chains will be unable to grow. For intermediate relationships, controlled chain growth will occur. Also, the balance between k_β and $k_{-\beta}$ determines the degree of chain transfer for starting new growing chains.

While successful implementations of RAFT heterophase polymerizations have been reported since the early 2000s, a great research effort is devoted to improving the understanding of the process, and the range of possible applications.

Polymerization-induced self-assembly (PISA): Phenomenon that takes place when the growth of a second block, insoluble in the polymerization medium, on a soluble chain leads to the formation of block copolymers that self-assemble into nanoparticles. PISA can be performed in both emulsion (insoluble monomers) and dispersion (soluble monomers) polymerization.

One of the most interesting phenomena observed during RAFT heterophase polymerization is ***polymerization-induced self-assembly*** (PISA). This approach was developed in 2002 by Ferguson et al. [59] while trying to eliminate various difficulties observed in RAFT emulsion polymerization. For the sake of completeness, it should be mentioned that a similar approach is the application of water-soluble polymeric radicals (e.g., poly(ethylene glycol) azoinitiators) to initiate surfactant-free emulsion polymerization of hydrophobic monomers as described in Refs. [60–65]. This procedure can be used to make block copolymer as well as nanocomposite particles and can be considered preceding the PISA method.

During the last decade, PISA via RAFT heterophase polymerization has awakened the curiosity of polymer scientists [66–69]. Brotherton et al. [70] used *in situ* small-angle X-ray scattering for observing, for the first time, the morphology evolution of block copolymer aggregates during RAFT emulsion polymerization. They evidenced the evolution of block copolymer aggregates from spheres to worms to vesicles in the case of RAFT aqueous emulsion polymerization of 2-methoxyethyl methacrylate (MOEMA) on poly(glycerol monomethacrylate) (PGMA). Chernikova et al. [71] and Wang and An [72] published two recent reviews on this topic.

RAFT polymerization has been used for synthesizing different types of polymer nanoparticle dispersions, using emulsion [59, 70, 73], dispersion [66, 67, 69, 73, 74], and miniemulsion [73, 75] polymerization. Even reactive surfactants, mentioned in the previous topic, have also been used with RAFT heterophase polymerization [76]. Palmiero et al. [77] summarizes the best practices for performing RAFT-controlled polymerization in aqueous heterophase polymerization.

Nitroxide-mediated polymerization (NMP) is another type of controlled polymerization mechanism, which has been explored in heterogeneous systems. NMP polymerization uses nitroxide structures, as radical traps due to the relatively stable (persistent) radicals formed by nitroxides (such as those depicted in Fig. 4.6). Their operating principle is similar to RAFT in the sense that an equilibrium reaction (reversible deactivation) controls the polymerization (Fig. 4.7).

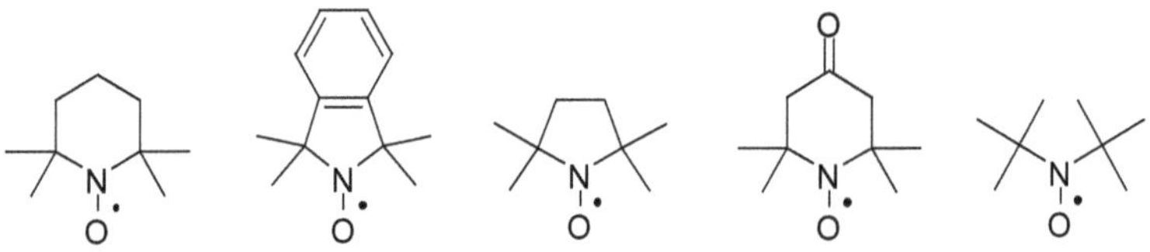

Figure 4.6 TEMPO radical (first structure from the left), and other stable nitroxide radicals commonly found in NMP.

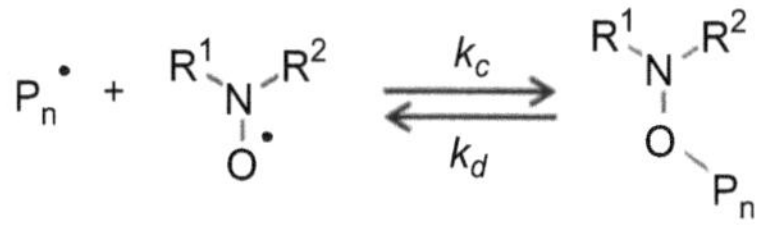

Figure 4.7 Controlling equilibrium reaction in NMP.

Heterophase NMP has been successfully implemented in heterogeneous systems where radical transfer from the dispersed phase is negligible, and particle nucleation is avoided. Ballard et al. [78, 79], for example, have used NMP in both miniemulsion and suspension polymerization processes. NMP emulsion polymerization has been challenging because of colloidal stability and polymerization control issues, although the introduction of water-soluble alkoxamines as initiators has been found to be a promising route [80]. Additional examples of heterophase NMP, including miniemulsion, microemulsion, dispersion, and suspension polymerization, are reviewed by Delaittre [81].

It is also possible to control the architecture of the polymer using *macroinimers* (macromonomer initiators) [82]. These compounds allow synthesizing branched, hyperbranched, star, arborescent polymers, and polymer brushes and other types of crosslinked polymer networks, using either conventional or controlled polymerization mechanisms.

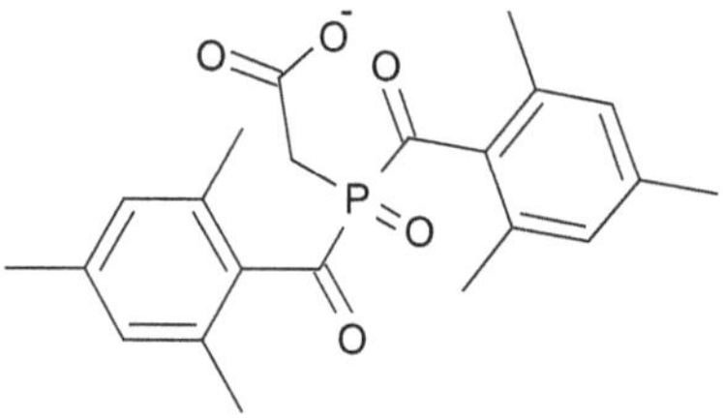

Figure 4.8 Photoinitiator, 2-(bis(2,4,6-trimethylbenzoyl)phosphoryl) acetic acid (BAPO-AA), leading to snowballing radical generation [75].

On the other side of the spectrum, it is also possible to release the control on the radical polymerization process leading to ultrahigh-molecular-weight (UHMW) polymers. Laurino et al. [83] found a particular type of photoinitiator (Fig. 4.8) capable of producing biradical polymer chains, resulting in a *snowballing radical generation*, leading to the formation of polymer chains with UHMW, beyond 10^7 Da.

UHMW polymers (>10^6 Da) have also been reported by Carmean et al. [84] using RAFT aqueous polymerization. The key to UHMW polymers is the use of photosensitive *iniferters* (initiators with chain transfer activity), such as the trithiocarbonate presented in Fig. 4.9.

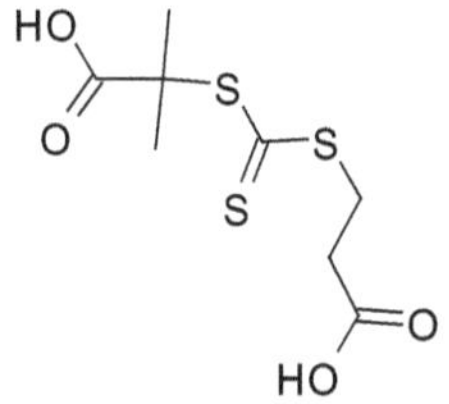

Figure 4.9 Photosensitive iniferter, 2-(2-carboxyethylsulfanylthiocarbonylsulfanyl)-2-methylpropionic acid, used in the synthesis of UHMW polymers [84].

Liu et al. [85] review the latest developments in heterophase-controlled polymerization, particularly when it is performed at room temperature.

4.3 Kinetics and Mechanisms

Despite the huge progress achieved so far in understanding heterophase polymerization kinetics and its underlying mechanisms, experimental information on particular systems is still needed. The number of factors involved in heterogeneous polymerization systems is usually larger compared to any homophase polymerization process. Not only is the number of components increased (by including one or more stabilizers), but also the polymerization system becomes more sensitive to process conditions (e.g., temperature, pH, ionic strength, stirring, and mechanical strain). Thus, there are still continuous efforts toward a better comprehension of heterophase polymerization kinetics and mechanisms for particular polymerization systems [86–94].

Another aspect of ongoing interest in heterophase polymerization is swelling of polymer particles. Ferapontov et al. [95] proposed different models for describing the swelling of hydrophilic polymer gels in electrolyte solutions. According to their findings, the internal structure of the polymer particle is not homogeneous as usually assumed but may present particular arrangements. In their example considered, a layered, sandwich-type structure seems to better describe their experimental results. Tauer et al. [96–98] recently questioned the relevance of molecular diffusion in the swelling of polymer particles. They found that monomer droplet–polymer particle collisions are probably the main driver of particle swelling in emulsion polymerization. They also remarked the importance of *spontaneous emulsification* in the formation of polymer particles. These results are consistent with the dynamics of molecular aggregates described in the previous chapter.

Another assumption recently questioned by Tauer et al. [99] is the effect of oxygen and other gases on the kinetics of heterophase free-radical polymerization. Oxygen, on one hand, may accelerate or decelerate the polymerization depending on the polymerization conditions and on the phase where radicals are generated. Furthermore, oxygen as well as any other gas present in the system may dissolve in the continuous phase affecting mass transfer phenomena. A very special influence of air dissolved in the aqueous phase was observed during particle nucleation experiments. The reproducibility of the experimental data could be drastically improved when the aqueous phase was degassed prior to the experiments [100]. A similar effect was observed when styrene quiescently was placed on top of water and the equilibration of styrene in water was followed with multi-angle laser light scattering [101]. During the equilibration period, the scattering intensity increased for degassed water much stronger than for water saturated with air. The increase in the scattering intensity goes along with the formation of styrene drops (cf. Section III). The origin of both effects is not clear yet. At the moment, the most plausible explanation is the influence, in the sense of heterogeneous nucleation, of tiny associates of air molecules (nanobubbles) on the formation of particles and droplets in the first and second examples, respectively.

As it can be seen, any material present even in small amounts in a heterophase polymerization system may have a significant, unexpected effect, either on the kinetics of reaction or on the final product properties. For example, Wei et al. [102] found an interesting effect of the type of polymerization (bulk or heterophase), and the initiation mechanism on the thermal stability (against aging and degradation) of different polymers. The exact underlying mechanism is still a matter of investigation.

4.4 Process Engineering

It has been frequently said that "emulsion polymers are products by process" [103]. However, this is also true for all other types of heterophase polymers. Relatively small changes in reactor design, reactor operation, and polymerization conditions may result in completely different products. Even changes in the fluctuation of certain process variables result in completely different molecular mass and/or particle size distributions, affecting product quality. Thus, improvements in online monitoring and control of heterophase polymerization processes are required for reducing off-spec products and optimizing process performance. Frauendorfer et al. [104], Reed and Alb [105], as well as Haven and Tunkers [106] recently compiled the state of the art of the most important methods for characterizing and monitoring polymerization reactions, including spectroscopic methods, calorimetry, rheology, light scattering, chromatographic methods, and flow-through detectors in general. Monitoring heterophase polymerization systems represent an additional degree of complexity because of their heterogeneous nature. Scattering methods, for example, may fail for the usually high concentration of scatterers present in industrial processes, and thus specialized methodologies are sometimes required. Backscattering inline turbidity has been found to be a good and versatile alternative for different heterophase polymerization systems [107]. Feng et al. [108] used a fluorescence spectroscopy method for monitoring the kinetics of methyl methacrylate microemulsion polymerization, employing N-(2-anthracene) methacrylamide as a fluorescent probe. Ghasemi et al. [109] monitored the kinetics and particle size distribution in the emulsion polymerization of styrene using ***electrical impedance spectroscopy***. They found that polymerization kinetics can be monitored at high frequencies (>1 kHz), whereas particle size is observed at low frequencies (~1 Hz). They also tested this method for monitoring the kinetics of the inverse-phase emulsion polymerization of acrylamide [110]. In this case, particle size was monitored by measuring electrical conductance, and their results were validated using offline dynamic light scattering particle size determinations.

Electrical impedance spectroscopy (EIS): Spectroscopic technique that determines the electrical impedance (resistance or opposition to alternating electrical current) of a material as a function of frequency.

Alternatively, Houben et al. [111] explored the possibility of simultaneously determining monomer concentration and particle size using *in situ* ***Raman spectroscopy***. This is possible when two different comonomers are used, but when the number of comonomers increases, only the overall conversion can be determined by this method.

Raman spectroscopy: Spectroscopic technique that determines vibrational and other low-frequency modes of molecules, by inelastic (non-conservative) scattering of photons.

Yamamoto et al. [112] explored the use of laser ionization time-of-flight mass spectrometry for determining oligomers present in the local microenvironment of emulsion polymerization systems and concluded that this method can be useful for monitoring the early stages of the process. Indeed, this was already earlier shown in a study of surfactant-free emulsion polymerization of styrene initiated with potassium peroxodisulfate [113]. These results support the existence of at least ten different combination of end groups. These are three non-ionic (hydrogen, hydroxyl, and the combination), three ionic (sulfate, carboxylate, and the combination), and the four combinations of the ionic with the nonionic end groups. The results prove the importance of peroxodisulfate side reactions with water and/or chain ends and are particularly important for particle nucleation because the aggregation of oligomers is strongly influenced by the nature of the end groups. A more precise identification of the oligomers present in the continuous phase during heterophase polymerization can be achieved by coupling the MALDI-TOF MS (***matrix-assisted laser desorption ionization time-of-flight mass spectrometry***) technique with SEC (***size exclusion chromatography***) [114].

Matrix-assisted laser desorption ionization time-of-flight mass spectrometry (MALDI-TOF MS): Spectrometric technique that determines the mass of macromolecules by ionizing a sample with a laser, resulting in minimal fragmentation of the macromolecules thanks to the effect of the absorbing matrix (e.g., sinapinic acid).

Size exclusion chromatography (SEC): Chromatographic technique that separates a mixture of molecules according to their hydrodynamic volume, by retaining smaller molecules in the pores of solid beads packed in a column. When the molecules are transported by a liquid mobile phase, it is called *gel permeation chromatography* (GPC).

An alternative strategy for monitoring heterophase polymerization processes consists of implementing mathematical estimation models from readily available process measurements. This, of course, requires the formulation of simple and approximate but reliable mathematical models of the process. Those models are usually denoted as meta-models, surrogate models, reduced-order models, estimators, observers, or soft-sensors [115]. Meta-models are useful also for process optimization and control tasks. Madhuranthakam and Penlidis [116], for example, developed a surrogate model based on ***artificial neural networks*** (ANN) for the emulsion copolymerization of nitrile butadiene rubber, which was then used for minimizing off-spec product generation. This ANN estimates different variables in the polymerization process, including monomer conversion, copolymer composition, average molecular weight, branching, and number of particles. In another example, Kang et al. [117–119] have been developing soft-sensors for estimating the time evolution of the molecular weight distribution in gas-phase polyolefin polymerization. These models were also used for minimizing off-spec product during grade transitions. Irzhak et al. [120] reviewed the estimation of molecular weight distributions and branching in crosslinked polymers using rheological and relaxation properties of the polymer. Adequate physical models of the polymers are required for obtaining accurate structure-properties predictions.

Artificial neural network (ANN): Mathematical model inspired by biological neural networks, based on the interaction between different units of nonlinear mathematical functions (artificial neurons).

The main requirement of meta-models is fast and robust computation. For that reason, more precise but complex models are uncommon for process monitoring, optimization, and control. However, progress in computational capabilities (including parallel computation, for example) and innovations in multiscale model integration allow using better models in process systems engineering. Chaloupka et al. [121] successfully implemented a multiscale hybrid Monte Carlo model for predicting the molecular architecture of different copolymers, solving the model within just a few seconds. These levels of computational efficiency allow using multiscale models in demanding tasks such as ***real-time dynamic optimization*** (RT-DO) and ***nonlinear model predictive control*** (NMPC). Marien et al. [122] also used kinetic Monte Carlo simulation of samples of particles for estimating the evolution of both particle size and chain length distribution in miniemulsion polymerization. Urrea-Quintero et al. [123] developed a reduced-order multiscale model for predicting secondary particle nucleation, particle size distribution, and polymerization kinetics in emulsion polymerization. Even though the proposed model is a computationally efficient approximation, it incorporates models at three different scales: microscopic, mesoscopic, and macroscopic. This model is expected to be used for a better control of particle size and morphology.

Real-time dynamic optimization (RT-DO): Model-based mathematical procedure continuously evaluates and manipulates process conditions to maximize the economic productivity of an operating plant.

Nonlinear model predictive control (NMPC): Advanced process control method used to regulate certain process variables while satisfying a set of optimization constraints, verified by means of a nonlinear model of the plant.

In terms of process control, Zubov et al. [124] presented an example of a successful NMPC implementation for a semi-batch emulsion copolymerization reactor at a pilot-plant scale, demonstrating the feasibility and advantages of this approach. Colegrove et al. [125] presented different examples for the control of copolymer composition manipulating the feed rates of the different monomers and using online spectroscopy. Particularly for heterophase polymerization, they recommend using inline fiber optic Raman spectroscopy for determining solid contents. Gerlinger et al. [126] also used inline Raman spectroscopy in combination with online calorimetry and mechanistic model-based soft-sensors for an NMPC of particle morphology in a multistage semi-batch emulsion polymerization. The authors also explored using online ***transmission electron microscopy*** (TEM). Although not successful, online TEM is a promising technology for particle control that requires further development.

Transmission electron microscopy (TEM): Microscopy technique that uses the intensity of a beam of electrons transmitted through the sample, to create an image.

Cho et al. [127] observed that the particle size obtained by dispersion polymerization can be controlled during polymerization by adjusting the reaction temperature and the composition of the continuous phase. By feeding water to the original ethanol medium, the polymerization may eventually switch from dispersion to emulsion polymerization. They also found that larger clusters of particles can be obtained by emulsifying the dispersion in hexadecane. A different alternative for controlling particle size was proposed by Nauman et al. [128] for the miniemulsion polymerization of methyl methacrylate. They used a porous glass membrane with different pore sizes for adjusting the final polymer particle size. Hydrophobic initiators are required for minimizing secondary particle nucleation in the continuous phase.

4.5 Green Chemistry and Engineering

Heterophase polymerization (aqueous, supercritical fluids, or inert media) emerged as an environmentally friendly alternative to the organic-solvent-based synthesis of polymers [129]. Thus, from the beginning it has been a ***green chemistry*** approach for making polymers. Generally, not only the process and the products should be considered when discussing the "*greenness*" of a certain technology but much more the whole procedure from the cradle to the grave. Accordingly, it is required considering the footprints of all ingredients, over their use in the production process to fabricate the product, up to the disposal of the applied products at the end of their service life.

Green chemistry (or sustainable chemistry): Design of chemical products and processes reducing or eliminating the use or generation of substances hazardous to living beings and the environment.

Green chemistry was formally proposed in the 1990s as a science-based strategy for minimizing the environmental impact of industrial chemistry. Anastas and Warner [130] summarized the main concepts of this strategy in a practical guiding list called the ***12 principles of green chemistry***.

12 Principles of Green Chemistry:

1. Waste prevention
2. Atom economy
3. Less hazardous chemical syntheses
4. Designing safer chemicals
5. Safer solvents and auxiliaries
6. Design for energy efficiency
7. Use of renewable feedstocks
8. Reduce derivatives
9. Catalysis
10. Design for degradation
11. Real-time analysis for pollution prevention
12. Inherently safer chemistry for accident prevention

12 Principles of Green Engineering:

1. Inherently nonhazardous material and energy inputs
2. Prevention instead of treatment
3. Design for separation
4. Maximize mass, energy, space, and time efficiency
5. Output-pulled rather than input-pushed
6. Conserve complexity
7. Targeted durability, not immortality
8. Meet needs, minimize excess
9. Minimize material diversity
10. Energy and material integration
11. Design for commercial "afterlife"
12. Renewable rather than depleting

Some years later, Anastas and Zimmerman [131] proposed an equivalent list of ***12 principles of green engineering***, focusing more on the industrial plant rather than on the R&D lab. Zhang and Dubé [132] recently presented a nice review of the applications of the principles of green chemistry and engineering on emulsion polymerization processes. They considered different relevant aspects of sustainable polymer chemistry, including *greener* raw materials, safer and more efficient processes, and negative-impact prevention strategies.

Raw materials for heterophase polymerization include monomers, dispersing media, stabilizers, catalysts, and other additives. Novel materials used in heterophase polymerization should preferably be sustainable and inherently safe along its life cycle (from cradle to grave). While renewability is desirable (7th green chemistry and 12th green engineering principles), unfortunately it does not guarantee sustainability. A raw material can be considered truly sustainable only when: (i) the local and global demand rate of the material is smaller than the rate of renewal; (ii) renewing the material does not compromise the sustainability of any species or community; and (iii) the overall cost of renewing the material is competitive. Employing materials whose rate of renewal is very low compared to its demand, requiring several years for renewing its monthly consumption for example, is not sustainable in the long term. A vegetable-based material requiring very large cultivation areas, compromising human food security, or putting in risk other native vegetal or animal species cannot be considered sustainable either. Finally, the overall cost (including both economic and environmental costs) should be low in order to become an attractive alternative for the manufacturers. In the case of surfactants, stabilizers, and additives in general, the low demand and high cost of current alternatives allow finding interesting sustainable alternatives for heterophase polymerization. Abundant renewable by-products from other industries are also relatively cheap attractive sources of materials, as it was shown by Schmidt et al. [133], who synthesized effective stabilizers for heterophase polymerization using lignin residues from the paper industry. In the case of monomers, the large demand of material represents a challenge for sustainability. The most important sources of renewable monomers include [132] terpenes, vegetable oils and derivatives (including glycerol), and sugars. Terpenes provide a rich, diverse chemistry, usually containing unsaturations. However, they may not be sufficiently reactive for free-radical polymerization. Furthermore, they are usually present in relatively small amounts in plants, requiring large cultivation areas for high-volume production. Thus, terpenes should be considered an interesting source of specialty monomers, but probably not as main backbone monomers. Vegetable oils have been used as raw materials in polymerization processes for a long time. Their main disadvantage is their limited capability for replicating the polymer properties obtained with the most important monomers used in heterophase polymerization. Glycerol-derived monomers are very promising, but their availability is limited by the production of biodiesel and fatty acids obtained from vegetable oils. Just to get an idea of the issue, the worldwide production of crude glycerol from biodiesel in 2008 was estimated in 0.19 MTon/year [134], whereas the production of styrere-butadiene rubber alone in 2014 required more than 7 MTon/year of monomers [135]. On the other hand, while worldwide sugar production is larger, care must be taken not to compromise human food security. Thus, the search for truly sustainable monomers is still active. One potential approach is synthesizing monomers from residual biomass. According to the European Commission [136], 440 MTon of agricultural residues were produced in Europe by 2017, whereas the worldwide polymer production (excluding certain synthetic fabrics) reached 348 MTon in the same year [137]. In this direction, Vobecka et al. [138] reported the synthesis of the renewable monomer α-methylene-γ-valerolactone, suitable for radical heterophase polymerization, obtained from γ-valerolactone produced from cellulosic biomasses. Recently, Hatton [139] summarized the current developments in the synthesis of monomers useful for RAFT-controlled polymerization, from renewable resources.

Ionic liquids: Salts with relatively low melting point. Thus, at typical processing temperatures, these salts are found in the liquid state.

Green products must also be inherently safe along its full life cycle. For heterophase polymerization, it implies that (i) all raw materials (especially monomers and dispersing media) should be nontoxic and nonhazardous, (ii) no hazardous by-products are obtained as a result of the polymerization, (iii) the polymer dispersion itself is inherently safe during storage and use, and (iv) the polymer has no negative environmental impact at the end of its service life. While water, supercritical CO2, ***ionic liquids***, and low-molecular-weight poly(ethylene glycol)s are green alternatives for the dispersing media [140], inherently safe monomers are challenging, particularly for radical polymerization processes. Any radical-polymerizable molecule is usually highly reactive inside living cells because free radicals are produced during mitochondrial respiration [141]. For that reason, those monomers tend to be toxic in any way. It is, therefore, desirable finding alternative non-radical polymerization mechanisms if inherently safe monomers must be used. Otherwise, rigorous controls and careful manipulation of these materials will always be required in order to avoid negative environmental and health issues due to heterophase polymerization. In addition, most petroleum-based monomers are considered ***volatile organic compounds*** (VOCs) and are tightly regulated because of their health and environmental problems. Another challenge for finding alternative inherently safe monomers is cost, as they must be competitive in order to successfully replace industrially common non-inherently safe monomers.

Volatile organic compounds (VOCs): Any organic compound with a low boiling point (≤250°C) at standard atmospheric pressure.

Polymers, on the other hand, are in general safer than their building blocks. For that reason, at the end of a polymerization process, the most common hazardous by-product residue is free monomer (and VOCs in general), since 100% monomer conversion is unfeasible in industrial processes. De San Luis et al. [142] presents a comprehensive review of the different methods employed to remove VOCs from the final polymer, thus reducing exposure risks during its intended end use. Some methods covered include devolatilization (using a stripping agent such as steam, nitrogen, or air), post-polymerization (addition of initiators at the end of the process), use of highly reactive comonomers, adsorption (using ion-exchange resins, activated carbon, or molecular sieves), and supercritical CO_2 extraction. Another source of VOCs in dispersion polymerization and inverse emulsion polymerization is the dispersing medium itself. When the nature of the dispersing medium cannot be changed, it should at least be reduced. This can be achieved by designing higher solid content polymer dispersions [143, 144]. Even aqueous-based polymer dispersions benefit from increasing the solid content due to lower transportation costs, reduced water footprint, and also performance improvement for certain applications.

Ionic liquids have been explored as dispersing media, but also they have been considered monomers [145], surfactants [146, 147], and catalysts (initiators) [148] in heterophase polymerization. Care must be taken when using ionic liquids since some of them, although nonvolatile, may be highly toxic [149] and thus, not inherently safe.

Persistent organic pollutants (POPs): Any organic compound resistant to environmental degradation (chemical, biological, or photolytic).

Another environmental concern with polymers is the possibility of migration of their additives, ultimately impacting the environment [150]. Salts of perfuorooctanoic acid (SPFOA) are optimum stabilizers for tetrafluoroethylene (TFE) emulsion polymerization to produce the polymer PTFE. However, there exist massive environmental, health, and safety concerns connected with this stabilizer due to its long-time stability if released to the environment. SPFOAs are neither by natural reactions nor biologically degradable; they are omnipresent and have been detected in water, animals (fishes, birds, and even in polar bears), and human blood. In Europe, these surfactants are forbidden. Since fluoropolymers have unique properties and are used for very special applications, the search for alternative surfactants is an active topic [151–154]. Thus, it is desirable using non-***persistent***, inherently safe additives. This means that additives must be nonhazardous and degradable, and also the degradation products must be safe. This is particularly important for the surfactants and stabilizers used in heterophase polymerization, which can easily migrate toward the environment. Istratov et al. [155], for example, investigated the synthesis of polystyrene by heterophase polymerization using block copolymers of polylactide and poly(ethylene glycol) as biodegradable surfactants. Alternatively, reactive dispersants [156] and reactive surfactants, such as those mentioned in Section 4.1.1, are environmentally friendly solutions. Furthermore, the ideal solution would be completely removing the surfactant from the polymer dispersion, as it is the case of surfactant- or emulsifier-free polymer dispersions [157–160]. The stability of surfactant-free dispersions is achieved thanks to the presence of ionic blocks in the polymer chains, coming from ionic/ionizable comonomers, chain transfer agents, or initiator fragments.

Finally, it is important to produce polymer dispersions where the polymer chains are environmentally friendly. One of the most critical issues for the sustainability of the polymer industry is perhaps the persistence of polymers in the environment, particularly when dealing with micro- and nanoparticles [161], as is the case of polymer dispersions. It is, therefore, important to design polymer materials where their lifetime matches their expected service life [162]. For example, a disposable single-use material such as food packaging should not last longer than the food product contained. Fine-tuning the environmental lifetime of polymers to their intended application is an important challenge for our immediate future. For instance, microplastic in the oceans is human made in the sense that the material at the end of its service life is not properly disposed but released to the environment in an uncontrolled way. Collecting and recycling of polymeric materials should be consequently enforced or endorsed, thus avoiding unwanted and harmful effects for Mother Nature. Recent reviews on the synthesis of biodegradable and sustainable polymers were published by Lendlein and Sisson [163] and Thakur and Thakur [164]. Waste prevention can also be achieved using any of the following techniques: reducing, repairing, regenerating, reusing, recycling, and recovering. Reducing material in a product is a design challenge, involving using higher-performance components allowing meeting the product specifications with less material. Repairing and regenerating, when possible, is another important method for reducing the generation of waste. Particularly, ***self-repairing*** or ***self-healing materials*** may extend the service lifetime of the products, by responding to a certain damage stimulus to recover the material properties [165]. Reusing (without further processing) and recycling (processing required) are also important strategies for reducing polymer wastes [166]. Their main disadvantage is the irreversible cumulative deterioration of polymer performance. In that sense, supramolecular polymers may provide a technological solution to self-healing and recycling while preserving the original polymer properties [167, 168]. Finally, when all other strategies fail, mass and energy can be recovered by thermal and combustion processes. In order to improve this last alternative, polymers should avoid (or at least reduce) the use of additives that may release hazardous materials during combustion (such as heavy metals, for example).

Self-healing (self-repairing or self-mendable) materials: Any material with the ability to transform physical energy into a chemical and/or physical response to heal the damage (recovering its initial properties), as a response to a particular external stimulus.

4.6 Product Innovation in Polymer Dispersions

Self-healing, mentioned in the previous section, requires the response of the material to an external stimulus related to damage. Any material responding to an external stimulus, in general, is denoted as a ***smart material***. Smart polymers, in particular, have been greatly explored [169]. Smart polymers may respond to different types of stimuli, including temperature, pH, ionic strength, light, ultrasound, mechanical stress, magnetic fields, voltage, concentration of specific compounds (e.g., enzymes, glucose, solvents, etc.), and even fire [170]. The most common response of smart polymers is phase change (from polymer solution to polymer dispersion and back). Smart polymer dispersions can be prepared using different types of heterophase polymerization processes [171–177]. Certain *multi-stimuli responsive* smart polymers may also change as a result of different stimuli, such as temperature and pH [178], temperature and ionic strength [179, 180], and many other combinations [181–183]. Smart materials, as already mentioned, can be used for self-healing applications but also for light-induced self-cleaning [184, 185], sensors [173, 186], and controlled chemical-delivery applications [187–190]. Some of these applications benefit from carefully controlled particle morphologies [191].

Smart (intelligent or responsive) materials: Any material whose properties change significantly as a result of an external stimulus.

The morphology of structured polymer particles can be controlled either kinetically (via multistage monomer-starved heterophase polymerization), or thermodynamically (swelling and phase separation), or by a combination of both [192]. Vonka and Kosek [193] present a thermodynamic model for describing reaction-induced phase separation and phase inversion in heterophase polymers. More recently, Hamzehlou et al. [194] proposed an integrative model of particle morphology distribution considering both kinetic and thermodynamic effects. Different types of particle morphologies can be obtained, including core–shell structures [179, 195–197], raspberry-like particles [198–201], confetti-like particles [201, 202], ***Janus*** particles [202–205], hollow particles [206], mesoporous particles [207, 208], and others [202, 209–211]. Particle morphology may also be controlled by adjusting the surface chemistry of the seed particles, producing particular topologies ranging from Janus particles to heterogeneous nanoporous particles [212]. It is also possible to control morphology with the assistance of pre-determined templates, which are removed after polymerization [213, 214].

Janus particles: Inspired by the two-faced roman god Janus, the surface of this type of particles has two clearly distinguishable regions with significantly different properties.

Most of these morphologies involve non-spherical particles. Typically, polymer particles made by heterophase polymerizations are of spherical shape as the interfacial free energy tends to be minimized as long as the mobility of the matter inside the particles during polymerization stays high enough. At the end of normal heterophase polymerizations, the spherical shape of hydrophobic polymers is frozen in as long as the temperature is lower than the glass transition temperature of the polymer. However, particles composed of amphiphilic polymers (amphiphilic particles) may, despite the glass transition temperature of the polymer, change their shape in dependence on the solids content. These particles made of gradient-like copolymers (e.g., polystyrene-co-poly(styrene sulfonate)) reversibly change their shape from spheres (e.g., at solids content above 1%) to rod-like, ring-like, and web-like structure upon dilution [215].

The production of non-spherical particles requires special measures during the polymerization. The popular way to produce non-spherical, particularly ellipsoidal, particles is a multistage, typically three-stage batch heterophase polymerization procedure using classical techniques. These techniques require in the first and second stage the preparation of composite particles containing both a linear and a crosslinked fraction of polymers, respectively, as spherically shaped seed. The formation of anisotropic particles happens during swelling in the third polymerization and is controlled by the entropy gain of the linear fraction in the semi-interpenetrating network [216–219].

Another direct synthetic route to control the shape of polymeric particles is using microfluidic devices for monomer droplet formation or deformation, respectively, followed by fast photo-polymerization in the capillary channels [220, 221].

There is still a third, non-polymerization route to non-spherical particles, which is the mechanical deformation/stretching of preformed particles embedded in rubbery films [222–227].

A great variety of polymer particles with various unconventional shapes and morphologies have been developed by the research group of Masayoshi Okubo in Kobe, Japan. An incomplete list of his landmark studies [228–231] illustrates the possibilities.

Potential application possibilities for non-spherical or anisotropic polymeric particles are biomedical as well as electronic areas [232], and more possible fields are mentioned in the cited papers.

Structured polymer particles in general can be considered *composite* particles, since they incorporate multiple solid phases within each dispersed entity. Furthermore, when the particles or their internal structures present at least one characteristic dimensions below 100 nm, they are denoted as *nanocomposites* [233–239]. Nanocomposites can be synthesized incorporating different types of nanoparticles, including carbon nanotubes [235], inorganic oxides [240, 241], silica [237, 242–244], clays [233, 239, 245], fibrillated cellulose and whiskers [234, 246], carboxymethyl cellulose [247], and others. When the composite particles include a polymer phase and an inorganic phase, they are usually denoted as hybrid particles [248, 249]. An innovative approach for incorporating inorganic components to polymer particles is using inorganic molecules as surfactants or stabilizers in heterophase polymerization [250, 251]. Jensen et al. [252] reviews the recent developments in the synthesis of hybrid particles by heterophase polymerization. Hybrid particles can be used for specific applications such as sensors [253] or nanocapsules [254], but in general they are used for improving specific properties of the polymer, such as electronic, dielectric and magnetic properties, surface properties (adhesion, wettability, catalytic, antibacterial, etc.), optical properties, mechanical properties, and thermal and degradation-resistance properties just to mention a few [255–257].

Conjugated polymers: Macromolecular chains whose backbone is composed of alternating single and double bonds. This configuration facilitates electron motion along the chain, providing electrical conductivity, photoluminescence, and magnetic properties.

While electro-optical properties of polymers can be significantly improved using hybrid composites, all-polymer systems can also provide the desired performance, particularly for electronics and energy applications. This is achieved by means of ***conjugated polymers***. Conjugated polymers have received recent considerable attention [258], particularly in the form of nanoparticles [259] obtained by microemulsion polymerization [260, 261], miniemulsion polymerization [262, 263], dispersion polymerization [264], and emulsion polymerization [265–267]. It has also been found that heterophase polymerization of conjugated polymers can be influenced by the presence of short-chain alcohol "co-surfactants" [260, 265], affecting monomer conversion, polymer particle size and morphology, polymer structure, and polymer conductivity.

Furthermore, preparing nanocomposites of inorganic materials with conjugated polymers emerges as an interesting alternative for improved conductive and photoactive properties. Castillo-Reyes et al. [268], for example, synthesized TiO_2/polypyrrole nanomposites as photocatalysts for the visible light degradation of methylene blue. Photoactive compounds, such as photocatalysts [269] and light stabilizers [270], have also been incorporated to polymer dispersions by miniemulsion polymerization. Ghasimi et al. [271] also prepared heterophase photocatalysts for the degradation of organic dyes and chromium (VI) reduction using conjugated polymers modified with ionic-liquid species. Hu et al. [272] found that polystyrene particles decorated with polypyrrole prepared by heterophase polymerization can also be used to remove chromium (VI) by ion adsorption on the surface of the particles.

Dou et al. [273] prepared magnetic hybrid nanocomposite particles incorporating magnetite (Fe_3O_4) into the polymer by photo-initiated miniemulsion polymerization. Bonnefond et al. [274] were also able to obtain magnetic particles by emulsion polymerization using magnetic ionic-liquid surfmers derived from 1-vinyl-3-dodecyl-imidazolium. Paramagnetic behavior was observed even using only 2% magnetic *surfmer* in the formulation.

4.7 Novel Applications of Heterophase Polymerization

Heterophase polymerization has been traditionally employed to prepare polymer dispersions for paper surface sizing and coating, printing inks, protective and decorative coatings, automotive coatings, adhesives, carpet backing and laminating, leather finishing, bonding non-wovens, modification of asphalts and plastic materials, and for dipping gloves and other rubber products [22, 27]. Advances in heterophase polymerization technology result in the continuous improvement of current applications [275, 276] and the development of new markets.

Bioassay: Analytical method used to determine, qualitatively or quantitatively, the concentration, activity, or relative potency of a specific compound in a sample, by observing its effect on living organisms (*in vivo*) or cell/tissue cultures (*in vitro*).

Perhaps the most significant innovations in the use of heterophase polymerization products can be found in the biomedical field. Biomedical applications of polymer dispersions include diagnostic markers, bioimaging, analytical ***bioassays,*** excipients and carriers of drugs and biomolecules, and tissue engineering and regenerative medicine [277–280].

Generalova et al. [281] recently reviewed the use of polymer dispersions in biomedical assays and cell labeling. The general principle consists of functionalizing the polymer particle with complementary stereospecific molecules for the corresponding biological analytes (e.g., antibodies, enzymes, DNA, etc.). Such high-affine host–guest interactions include antigen–antibody, carbohydrate–lectin, and receptor protein–transport protein, just to mention a few. Conjugation of biomolecules can be done simply by adsorption on the particle surface, by covalent bond formation or by

molecular imprinting [282, 283]. Quantitative determination of the analyte can be done by different methods, including turbidimetry (when latex agglutination occurs), photoluminescence (e.g., fluorescence, when fluorescence-labeled biomolecules are used), and bioseparation (filtration, centrifugation, or magnetic separation). The technology of diagnostic bioassays has been evolving toward a personalized medicinal therapy concept, denoted as ***theranostics***. A novel approach for theranostics is the use of upconversion (sequential photon absorption) nanoparticles [284]. Upconversion nanoparticles (UCNP) are usually lanthanide-based cores, which can be encapsulated, for example, by using heterophase polymerization [285], improving its water dispersibility and biocompatibility.

Molecularly imprinted polymer: Polymer material presenting stereospecific functionalized cavities left behind by the removal of a certain template molecule. Such cavity will then behave as a lock-and-key system with the original molecule.

Theranostics: Integration of bioimaging and specific targeted therapy, allowing a personalized detection and follow-up of the treatment.

Polymer dispersions have also been used as delivery systems for a wide range of pharmaceutical products, including ibuprofen [286, 287], rifampicin [288], lorazepam [289, 290], doxycycline [291], and also for cancer chemotherapy [292], just to mention a few. Polymer delivery mechanisms are particularly useful in the case of insoluble drugs [293], specific-organ delivery requirements [294], or controlled and sustained release [290, 295, 296].

Biocompatibility: Characteristic of a material that does not have toxic or injurious effects when used with biological systems. However, biocompatibility does not imply biodegradability, and vice versa.

Polymers used in biomedical applications should be ***biocompatible***, biodegradable, nontoxic, non-mutagenic, and non-immunogenic [297] in order to avoid undesirable side effects in the human body. Before using them in biomedical applications, the polymers must first be approved by the competent authority. For example, Saade et al. [298] indicate that the copolymer poly(methyl methacrylate-co-methacrylic acid) p(MMA-co-MAA) in a 2:1 molar ratio is a synthetic polymer approved by the FDA. Mansour et al. [299] report other FDA-approved synthetic polymeric inactive ingredients, including poly(lactic acid) (PLA), poly(lactic-co-glycolic acid) (PLGA), poly(ethylene glycol) (PEG), poly(vinyl acetate) (PVA), and poly(vinyl pyrrolidone) (PVP).

Conducting polymers also have interesting biomedical applications. They can be used as intelligent electric field-responsive materials interacting with different types of biological tissues (muscles, nerves, etc.), and in tissue interfaces of biomedical devices for enhancing their electrical sensitivity and stability [300]. The challenge in this field is obtaining high-performance conductive polymers with excellent mechanical properties, processability, and most importantly biocompatibility [301]. Since conventional conjugated polymers have usually poor biocompatibility, incorporation of conducting nanoparticles (e.g., carbon nanotubes, graphene, metal oxides) to non-conducting biocompatible polymers seems a promising route. Semiconducting polymers may also contribute to the improvement of bioimaging and cellular labeling methods. This can be achieved by the synthesis of stable, functionalized, fluorescent ***polymer dots*** [302–304]. The emission wavelength of polymer dots strongly depends on the polymer composition but also on the particle size and shape, variables that can be controlled in heterophase polymerization by adjusting the process conditions [305].

Polymer dots: Nanoparticles (<100 nm in size) made of conjugated polymers exhibiting special properties such as high brightness, tunable excitation and emission wavelengths, and excellent photostability. They can be considered the polymer equivalents of *quantum dots.*

Photoluminescence of polymer particles is an interesting property for applications beyond the biomedical field. Meng et al. [306], for example, prepared luminescent electrophoretic crosslinked polystyrene particles copolymerized with a fluorescent comonomer ((aminoethyl) methacrylate hydrochloride-fluorescein isothiocyanate, AEMH-FITC) via miniemulsion polymerization over pigment cores. The resulting material might be used for night-vision electrophoretic displays. The chemical structure of the fluorescent comonomer AEMH-FITC is presented in Fig. 4.10.

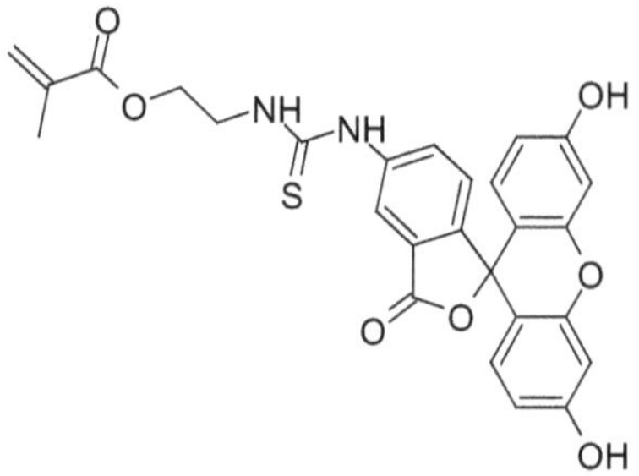

Figure 4.10 Fluorescent comonomer AEMH-FITC [306].

4.8 Future Perspectives in Heterophase Polymerization

Heterophase polymerization, despite being a century-old field, is full of hidden surprises waiting to be found. The history of heterophase polymerization has shown that new developments particularly in polymer and colloid chemistry, but also in other scientific fields, have been adopted and incorporated for the production of novel polymeric particles, thus expanding the envelop of this fascinating polymerization technique.

There are still many open research questions, many raw materials and combinations for exploring, many opportunities for continuous improvement (especially in industry), and many promising novel applications to be further developed.

Novel and innovative types of surfactants will always be welcome, since they have a critical impact on end-use performance (particle size, polymer mass, surface properties, etc.). Reducing undesired migration and minimizing the environmental impact of these materials are clearly a priority.

Similarly, greener (sustainable and inherently safe) monomers and polymers are needed. The current environmental impact of polymers is a result of misuse and should be quickly reduced, or even completely suppressed (e.g., microplastics in oceans). Green and sustainable chemistry and engineering for the design and improvement of products, processes, and applications is probably the best route for achieving this goal. Fine-tuning the lifetime of a polymer to its service life, while promoting polymer recycling, is also important. Supramolecular chemistry is a promising technology for achieving it.

Precise control of polymer chain length and architecture, of polymer particle size and morphology, and of polymer performance is the dreamland of industry. Many opportunities are waiting for innovative technology breakthroughs in this field.

Further developments in biomedical and electronic applications of heterophase polymer dispersions are expected in the near future. Careful attention must be placed to every emerging technology field, as they will provide additional opportunities for heterophase polymerization.

Finally, a better understanding of the mechanisms of heterophase polymerization will eventually allow building accurate and reliable full *ab initio* multiscale models. And if those models can be quickly solved, they could also contribute to better real-time industrial applications. Since the frontiers between the different types of heterophase polymerization techniques are becoming fuzzier, general (rather than specific) models of heterophase polymerization should be preferred.

References

1. Farbenfabriken former F. Bayer & Co. in Elberfeld (1909). Manufacture and Production of a Caoutchouc Substance. German Patent DE000000250690.
2. Farbenfabriken former F. Bayer & Co. in Elberfeld (1912). Process for Producing Artificial Caoutchouc. German Patent DE000000254672.
3. Marchionna, F. (1933). *Latex and Its Industrial Applications*, Vol. 1. (Rubber Age, USA).
4. Noble, R. J. (1936). *Latex in Industry* (Rubber Age, USA).
5. Marchionna, F. (1937). *Latex and Rubber Derivatives and Their Industrial Applications*, Vol. 2 and 3. (Rubber Age, USA).
6. Staudinger, H. (1920). Über polymerisation. *Berichte der deutschen chemischen Gesellschaft (A and B Series)*, **53**(6), pp. 1073–1085.
7. Schubert, U. S. and Zechel, S. (2020). The year of polymers: 100 years of macromolecular chemistry. *Macromolecular Chemistry and Physics*, **221**(1), 1900530.

8. Mülhaupt, R. (2004). Hermann Staudinger and the origin of macromolecular chemistry. *Angewandte Chemie International Edition*, **43**(9), pp. 1054–1063.
9. Cook, P. G. (1956). *Latex, Natural and Synthetic* (Reinhold, USA).
10. Fisher, H. L. (1957). *Chemistry of Natural and Synthetic Rubbers* (Reinhold, USA).
11. Huke, D. W. (1961). *Introduction to Natural and Synthetic Rubbers* (Chemical Publishing, USA).
12. Blackley, D. C. (1966). *High Polymer Latices: Their Science and Technology*, Vol. 1 and 2. (Maclaren, UK).
13. Fitch, R. M. (1971). *Polymer Colloids* (Plenum Press, USA).
14. Barrett, K. E. (1975). *Dispersion Polymerization in Organic Media* (John Wiley & Sons, UK).
15. Yeliseeva, V. I. (1980). *Polimernije Dispersii* (Izd. Chimija, USSR).
16. Piirma, I. (1982). *Emulsion Polymerization* (Academic Press, USA).
17. Gilbert, R. G. (1995). *Emulsion Polymerization: A Mechanistic Approach* (Academic Press, USA).
18. Blackley, D. C. (1997). *Polymer Latices*, Vol. 1, 2, and 3. (Chapman & Hall, UK).
19. Fitch, R. M. (1997). *Polymer Colloids: A Comprehensive Introduction* (Academic Press, USA).
20. Lovell, P. (1997). *Emulsion Polymerization and Emulsion Polymers* (Wiley, USA).
21. Asua, J. M. (1997). *Polymeric Dispersions: Principles and Applications* (Kluwer Academics, The Netherlands).
22. Urban, D. and Takamura, K. (2002). *Polymer Dispersions and Their Industrial Applications* (Wiley-VCH, Germany).
23. van Herk, A. M. (2005). *Chemistry and Technology of Emulsion Polymerisation* (Blackwell, UK).
24. Chern, C. S. (2008). *Principles and Applications of Emulsion Polymerization* (John Wiley & Sons, USA).
25. Mittal, V. (Ed.) (2011). *Miniemulsion Polymerization Technology* (Wiley, Germany).
26. Schlüter, D. A., Hawker, C., and Sakamoto, J. (2012). *Synthesis of Polymers: New Structures and Methods* (Wiley-VCH, Germany).
27. van Herk, A. M. (2013). *Chemistry and Technology of Emulsion Polymerisation*, 2nd Ed. (John Wiley & Sons, USA).

28. Jenjob, R., Seidi, F., and Crespy, D. (2018). Recent advances in polymerizations in dispersed media. *Advances in Colloid and Interface Science*, **260**, pp. 24–31.

29. Krüger, K., Hernandez, H., and Tauer, K. (2014). Particle size-dependent effects in hydrophobically initiated emulsion polymerization. *Macromolecular Theory and Simulations*, **23**(3), pp. 125–135.

30. Zetterlund, P. B. and D'hooge, D. R. (2019). The nanoreactor concept: Kinetic features of compartmentalization in dispersed phase polymerization. *Macromolecules*, **52**(21), pp. 7963–7976.

31. Ash, M. and Ash, I. (2010). *Handbook of Industrial Surfactants*, Vol. 1, 5th Ed. (Synapse Information Resources, USA).

32. Abbott, S. (2017). *Surfactant Science: Principles & Practice* (DEStech Publications, USA).

33. Pavljuchenko, V. N., Ivanchev, S. S., Birdina, N. A., Alekseeva, S. M., and Lesnikova, N. N. (1981). Peculiarities of emulsion polymerization of styrene with radical generation acts located in the adsorption layer of the emulsifier (in Russian). *Dokl. Akad. Nauk SSSR*, **259**(2), pp. 641–645.

34. Ivanchev, S. S., Pavljuchenko, V. N., and Byrdina, N. A. (1987). Elementary reactions of the emulsions polymerization of styrene with the localization of radical formation acts at the interface. *Journal of Polymer Science Part A: Polymer Chemistry*, **25**(1), pp. 47–62.

35. Guyot, A. and Tauer, K. (1994). Reactive surfactants in emulsion polymerization. *Advances in Polymer Science*, **111**, pp. 43–65.

36. Pavljuchenko, V. N., Lesnikova, N. N., Birdina, N. A., and Ivanchev, S. S. (1995). Kinetic peculiarities of emulsion polymerization of styrene with surface-active initiators (in Russian). *Dokl. Akad. Nauk SSSR*, **342**(1), pp. 70–72.

37. Asua, J. M. and Schoonbrood, H. A. S. (1998). Reactive surfactants in heterophase polymerization. *Acta Polymerica*, **49**(12), pp. 671–686.

38. Bognolo, G. (2006). Polymerizable surfactants. In: Farn, R. J., *Chemistry and Technology of Surfactants* (Blackwell, UK), pp. 204–226.

39. Bader, A. R., Buckley, R. P., Leavitt, F., and Szwarc, M. (1957). Addition of methyl radical to cis and trans isomers. *Journal of the American Chemical Society*, **79**(21), pp. 5621–5625.

40. Otsu, T., Shiraishi, K., and Matsumoto, A. (1993). Radical polymerization and copolymerization reactivities of fumarates bearing different alkyl ester groups. *Journal of Polymer Science Part A: Polymer Chemistry*, **31**(10), pp. 2523–2529.

41. Mekki, S., Saïdi-Besbes, S., Elaïssari, A., Valour, J. P., and Derdour, A. (2010). Novel polymerizable surfactants: Synthesis and application in the emulsion polymerization of styrene. *Polymer Journal*, **42**(5), pp. 401–405.

42. Ovando-Medina, V. M., Piña-García, P. S., Corona-Rivera, M. A., Díaz-Flores, P. E., Peralta, R. D., Martínez-Gutiérrez, H., and Lara-Ceniceros, T. E. (2012). Semicontinuous heterophase polymerization of methyl methacrylate in the presence of reactive surfactant HITENOL BC10. *Polymer Bulletin*, **68**(9), pp. 2313–2322.

43. Vargas-Salazar, C. Y., Ovando-Medina, V. M., Ledezma-Rodríguez, R., Peralta, R. D., and Martínez-Gutiérrez, H. (2015). Ultrasound-assisted polymerization of methyl methacrylate using the reactive surfactant Hitenol BC10 in a semicontinuous heterophase process. *Iranian Polymer Journal*, **24**(1), pp. 41–50.

44. Herold, M., Brunner, H., and Tovar, G. E. (2003). Polymer nanoparticles with activated ester surface by using functional surfmers. *Macromolecular Chemistry and Physics*, **204**(5–6), pp. 770–778.

45. Albernaz, V., Bach, M., Weber, A., Southan, A., and Tovar, G. (2018). Active ester containing surfmer for one-stage polymer nanoparticle surface functionalization in mini-emulsion polymerization. *Polymers*, **10**(4), 408.

46. Asua, J. M. (2018). Ostwald ripening of reactive costabilizers in miniemulsion polymerization. *European Polymer Journal*, **106**, pp. 30–41.

47. Itoh, T., Kojima, K., Shimomoto, H., and Ihara, E. (2018). Control of lengths and densities of surface-attached chains on polymer particles prepared by dispersion polymerization using macromonomer stabilizer. *Polymer*, **158**, pp. 158–165.

48. Noshay, A. and McGrath, J. E. (1977). *Block Copolymers: Overview and Critical Survey* (Academic Press, USA).

49. Atanase, L. I. and Riess, G. (2011). Block copolymers as polymeric stabilizers in non-aqueous emulsion polymerization. *Polymer International*, **60**(11), pp. 1563–1573.

50. Sanders, C. A., George, S. R., Deeter, G. A., Campbell, J. D., Reck, B., and Cunningham, M. F. (2019). Amphiphilic block-random copolymers: Self-folding behavior and stabilizers in emulsion polymerization. *Macromolecules*, **52**(12), pp. 4510–4519.

51. Yu, R., Peh, E., Mavliutova, L., Weber, N., and Tauer, K. (2016). Borderline particles: Approaching the limit between colloidal stability

and instability during heterophase polymerization. *Macromolecular Chemistry and Physics*, **217**(7), pp. 901–911.

52. Ramsden, W. (1904). Separation of solids in the surface-layers of solutions and 'suspensions' (observations on surface-membranes, bubbles, emulsions, and mechanical coagulation). Preliminary account. *Proceedings of the Royal Society of London*, **72**(477–486), pp. 156–164.

53. Pickering, S. U. (1907). CXCVI—Emulsions. *Journal of the Chemical Society, Transactions*, **91**, pp. 2001–2021.

54. Schrade, A., Landfester, K., and Ziener, U. (2013). Pickering-type stabilized nanoparticles by heterophase polymerization. *Chemical Society Reviews*, **42**(16), pp. 6823–6839.

55. Lotierzo, A. and Bon, S. A. (2017). A mechanistic investigation of Pickering emulsion polymerization. *Polymer Chemistry*, **8**(34), pp. 5100–5111.

56. Brunier, B., Sheibat-Othman, N., Chevalier, Y., and Bourgeat-Lami, É. (2018). Effect of Pickering stabilization on radical entry in emulsion polymerization. *AIChE Journal*, **64**(7), pp. 2612–2624.

57. Li, L., Yang, J., and Zhou, J. (2013). Linear-, cyclic-, and multiblock amphiphilic polyelectrolytes as surfactants in emulsion polymerization: Role of topological structure. *Macromolecules*, **46**(7), pp. 2808–2817.

58. Urbani, C. N. and Monteiro, M. J. (2008). RAFT-mediated polymerization in heterogeneous systems. In: Barner-Kowollik, C., *Handbook of RAFT Polymerization* (Wiley-VCH, Germany), pp. 285–314.

59. Ferguson, C. J., Hughes, R. J., Pham, B. T., Hawkett, B. S., Gilbert, R. G., Serelis, A. K., and Such, C. H. (2002). Effective *ab initio* emulsion polymerization under RAFT control. *Macromolecules*, **35**(25), pp. 9243–9245.

60. Tauer, K., Antonietti, M., Rosengarten, L., and Müller, H. (1998). Initiators based on poly (ethylene glycol) for starting heterophase polymerizations: Generation of block copolymers and new particle morphologies. *Macromolecular Chemistry and Physics*, **199**(5), pp. 897–908.

61. Tauer, K., Müller, H., Rosengarten, L., and Riedelsberger, K. (1999). The use of polymers in heterophase polymerizations. *Colloids and Surfaces A: Physicochemical and Engineering Aspects*, **153**(1–3), pp. 75–88.

62. Topp, M. D. C., Leunen, I. H., Dijkstra, P. J., Tauer, K., Schellenberg, C., and Feijen, J. (2000). Quasi-living polymerization of N-isopropylacrylamide onto poly (ethylene glycol). *Macromolecules*, **33**(14), pp. 4986–4988.

63. Tauer, K., Khrenov, V., Shirshova, N., and Nassif, N. (2005). Preparation and application of double hydrophilic block copolymer particles. *Macromolecular Symposia*, **226**(1), pp. 187–202.

64. Tauer, K., Weber, N., Nozari, S., Padtberg, K., Sigel, R., Stark, A., and Völkel, A. (2009). Heterophase polymerization as synthetic tool in polymer chemistry for making nanocomposites. *Macromolecular Symposia*, **281**(1), pp. 1–13.

65. Tauer, K. (2010). Polymer nanoparticles with surface active initiators and polymeric stabilizers. In: Mittal, V., *Advanced Polymer Nanoparticles: Synthesis and Surface Modifications* (CRC Press, USA), pp. 329–359.

66. Wan, W. M. and Pan, C. Y. (2010). One-pot synthesis of polymeric nanomaterials via RAFT dispersion polymerization induced self-assembly and re-organization. *Polymer Chemistry*, **1**(9), pp. 1475–1484.

67. Warren, N. J. and Armes, S. P. (2014). Polymerization-induced self-assembly of block copolymer nano-objects via RAFT aqueous dispersion polymerization. *Journal of the American Chemical Society*, **136**(29), pp. 10174–10185.

68. Canning, S. L., Smith, G. N., and Armes, S. P. (2016). A critical appraisal of RAFT-mediated polymerization-induced self-assembly. *Macromolecules*, **49**(6), pp. 1985–2001.

69. Derry, M. J., Fielding, L. A., and Armes, S. P. (2016). Polymerization-induced self-assembly of block copolymer nanoparticles via RAFT non-aqueous dispersion polymerization. *Progress in Polymer Science*, **52**, pp. 1–18.

70. Brotherton, E. E., Hatton, F. L., Cockram, A. A., Derry, M. J., Czajka, A., Cornel, E. J., Topham, P. D., Mykhaylyk, O. O., and Armes, S. P. (2019). In Situ SAXS studies during RAFT aqueous emulsion polymerization. *Journal of the American Chemical Society*, **141**, pp. 13664–13675.

71. Chernikova, E. V., Lysenko, E. A., Serkhacheva, N. S., and Prokopov, N. I. (2018). Self-assembly of amphiphilic block copolymers during reversible addition-fragmentation chain transfer heterophase polymerization: Problems, achievements, and outlook. *Polymer Science, Series C*, **60**(1), pp. 192–218.

72. Wang, X. and An, Z. (2019). New insights into RAFT dispersion polymerization-induced self-assembly: From monomer library, morphological control, and stability to driving forces. *Macromolecular Rapid Communications*, **40**(2), 1800325.

73. Chernikova, E. V., Plutalova, A. V., Mineeva, K. O., Nasimova, I. R., Kozhunova, E. Y., Bol'shakova, A. V., Tolkachev, A. V., Serkhacheva, N. S., Zaitsev, S. D., Prokopov, N. I., and Zezin, A. B. (2015). Homophase and heterophase polymerizations of butyl acrylate mediated by poly (acrylic acid) as a reversible addition–fragmentation chain-transfer agent. *Polymer Science Series B*, **57**(6), pp. 547–559.

74. Li, Y. and Armes, S. P. (2010). RAFT synthesis of sterically stabilized methacrylic nanolatexes and vesicles by aqueous dispersion polymerization. *Angewandte Chemie International Edition*, **49**(24), pp. 4042–4046.

75. Fuchs, A. V. and Thurecht, K. J. (2015). Interfacial RAFT miniemulsion polymerization: Architectures from an interface. *Macromolecular Chemistry and Physics*, **216**(12), pp. 1271–1281.

76. Li, Y., Christian-Tabak, L., Fuan, V. L. F., Zou, J., and Cheng, C. (2014). Crosslinking-induced morphology change of latex nanoparticles: A study of RAFT-mediated polymerization in aqueous dispersed media using amphiphilic double-brush copolymers as reactive surfactants. *Journal of Polymer Science Part A: Polymer Chemistry*, **52**(22), pp. 3250–3259.

77. Palmiero, U. C., Singh, J., and Moscatelli, D. (2018). The RAFT polymerization and its application to aqueous dispersed systems. *Current Organic Chemistry*, **22**(13), pp. 1285–1296.

78. Ballard, N., Aguirre, M., Simula, A., Leiza, J. R., van Es, S., and Asua, J. M. (2017). High solids content nitroxide mediated miniemulsion polymerization of n-butyl methacrylate. *Polymer Chemistry*, **8**(10), pp. 1628–1635.

79. Ballard, N., Aguirre, M., Simula, A., Leiza, J. R., van Es, S., and Asua, J. M. (2017). Nitroxide mediated suspension polymerization of methacrylic monomers. *Chemical Engineering Journal*, **316**, pp. 655–662.

80. Charleux, B. and Nicolas, J. (2007). Water-soluble SG1-based alkoxyamines: A breakthrough in controlled/living free-radical polymerization in aqueous dispersed media. *Polymer*, **48**(20), pp. 5813–5833.

81. Dellaitre, G. (2016). Nitroxide-mediated polymerization in dispersed media. In: Gigmes, D. *Nitroxide Mediated Polymerization: From Fundamentals to Applications in Materials Science* (Royal Society of Chemistry, UK), pp. 199–263.

82. Yildiz, U., Hazer, B., and Tauer, K. (2012). Tailoring polymer architectures with macromonomer azoinitiators. *Polymer Chemistry*, **3**(5), pp. 1107–1118.

83. Laurino, P., Hernandez, H. F., Bräuer, J., Krüger, K., Grützmacher, H., Tauer, K., and Seeberger, P. H. (2012). Snowballing radical generation leads to ultrahigh molecular weight polymers. *Macromolecular Rapid Communications*, **33**(20), pp. 1770–1774.

84. Carmean, R. N., Becker, T. E., Sims, M. B., and Sumerlin, B. S. (2017). Ultra-high molecular weights via aqueous reversible-deactivation radical polymerization. *Chem*, **2**(1), pp. 93–101.

85. Liu, D., He, J., Zhang, L., and Tan, J. (2019). 100th anniversary of macromolecular science viewpoint: Heterogenous reversible deactivation radical polymerization at room temperature. Recent advances and future opportunities. *ACS Macro Letters*, **8**, pp. 1660–1669.

86. Pérez-García, M. G., Torres, E. V., Ceja, I., Pérez-Carrillo, L. A., López, R. G., López-Serrano, F., Mendizábal, E., and Puig, J. E. (2012). Semicontinuous heterophase copolymerization of styrene and acrylonitrile. *Journal of Polymer Science Part A: Polymer Chemistry*, **50**(16), pp. 3332–3339.

87. Pérez-García, M. G., Pérez-Carrillo, L. A., Mendizábal, E., Puig, J. E., and López-Serrano, F. (2013). Modeling the semicontinuous heterophase polymerization for synthesizing poly (n-butyl methacrylate) nanoparticles. *Macromolecular Reaction Engineering*, **7**(9), pp. 440–452.

88. Ovando-Medina, V. M., Díaz-Flores, P. E., Peralta, R. D., Mendizábal, E., and Cortez-Mazatan, G. Y. (2013). Semicontinuous heterophase copolymerization of vinyl acetate and butyl acrylate. *Journal of Applied Polymer Science*, **127**(4), pp. 2458–2464.

89. Ovando-Medina, V. M., Corona-Rivera, M. A., Márquez-Herrera, A., Lara-Ceniceros, T. E., Manríquez-González, R., and Peralta, R. D. (2014). Heterophase polymerization of different methacrylates: Effect of alkyl ester group on kinetics and colloidal behavior. *Journal of Applied Polymer Science*, **131**(8), 40191.

90. Pérez García, M. G., Alvarado, A. G., Pérez-Carrillo, L. A., Puig, J. E., López-Serrano, F., Garcia Sandoval, J. P., and Mendizábal, E. (2015). On the modeling of the semicontinuous heterophase polymerization of alkyl methacrylates with different water solubilities. *Macromolecular Reaction Engineering*, **9**(2), pp. 114–124.

91. Pérez-García, M. G., Rabelero, M., Nuño-Donlucas, S. M., Mendizábal, E., Martínez-Richa, A., López, R. G., Arellano, M., and Puig, J. E. (2012). Semi-continuous heterophase polymerization of n-butyl methacrylate: Effect of monomer feeding rate. *Journal of Macromolecular Science, Part A*, **49**(7), pp. 539–546.

92. López-Serrano, F., Silva, J. M., Sánchez-Díaz, J. C., Vargas, R. O., and Puig, J. E. (2018). Understanding the semicontinuous-heterophase, surfactant-supported, acrylamide inverse polymerization. *Polymer Engineering & Science*, **58**(12), pp. 2219–2226.
93. Saade, H., Torres, S., Barrera, C., Sánchez, J., Garza, Y., and López, R. G. (2015). Effect of pH and monomer dosing rate in the anionic polymerization of ethyl cyanoacrylate in semicontinuous operation. *International Journal of Polymer Science*, **2015**, 827059.
94. Fernández, A., Ruiz-Bermejo, M., and José, L. (2018). Modelling the kinetics and structural property evolution of a versatile reaction: Aqueous HCN polymerization. *Physical Chemistry Chemical Physics*, **20**(25), pp. 17353–17366.
95. Ferapontov, N. B., Tokmachev, M. G., Gagarin, A. N., Strusovskaya, N. L., and Khudyakova, S. N. (2013). Influence of the environment on swelling of hydrophilic polymers. *Reactive and Functional Polymers*, **73**(8), pp. 1137–1143.
96. Tauer, K., Wei, C., Tripathi, A., and Kiryutina, O. (2017). Experiments and thoughts on mass transfer during emulsification. *Advances in Polymer Science*, **281**, pp. 23–43.
97. Wei, C. and Tauer, K. (2016). Features of emulsion polymerization: How does the monomer move from the droplets into the latex particles? *Macromolecular Symposia*, **370**(1), pp. 99–109.
98. Tripathi, A., Wei, C., and Tauer, K. (2017). Swelling of latex particles: Towards a solution of the riddle. *Colloid and Polymer Science*, **295**(1), pp. 189–196.
99. Höhne, P., Krüger, K., and Tauer, K. (2013). Vapor phase composition and radical polymerization: How the gas phase influences the kinetics of heterophase polymerization. *Colloid and Polymer Science*, **291**(3), pp. 483–500.
100. Kühn, I. and Tauer, K. (1995). Nucleation in emulsion polymerization: A new experimental study. 1. Surfactant-free emulsion polymerization of styrene. *Macromolecules*, **28**(24), pp. 8122–8128.
101. Kozempel, S., Tauer, K., and Rother, G. (2005). Aqueous heterophase polymerization of styrene: A study by means of multi-angle laser light scattering. *Polymer*, **46**(4), pp. 1169–1179.
102. Wei, C., Esposito, D., and Tauer, K. (2016). Thermal properties of thermoplastic polymers: Influence of polymer structure and procedure of radical polymerization. *Polymer Degradation and Stability*, **131**, pp. 157–168.
103. Pauer, W. (2018). Reactor concepts for continuous emulsion polymerization. *Advances in Polymer Science*, 280, pp. 1–18.

104. Frauendorfer, E., Wolf, A., and Hergeth, W. D. (2010). Polymerization online monitoring. *Chemical Engineering & Technology*, **33**(11), pp. 1767–1778.

105. Reed, W. F. and Alb, A. M. (2014). *Monitoring Polymerization Reactions: From Fundamentals to Applications* (John Wiley & Sons, USA).

106. Haven, J. J. and Junkers, T. (2017). Online monitoring of polymerizations: Current status. *European Journal of Organic Chemistry*, **2017**(44), pp. 6474–6482.

107. Bloch, D., Bröge, P., and Pauer, W. (2017). Inline turbidity measurements of batch emulsion polymerization. *Macromolecular Reaction Engineering*, **11**(4), 1600063.

108. Feng, H., Dan, Y., and Zhao, Y. (2010). A sensitive fluorescence method for monitoring the kinetics of microemulsion polymerization. *Canadian Journal of Chemistry*, **88**(3), pp. 185–191.

109. Ghasemi, S., Darestani, M. T., Abdollahi, Z., and Gomes, V. G. (2015). Online monitoring of emulsion polymerization using electrical impedance spectroscopy. *Polymer International*, **64**(1), pp. 66–75.

110. Abdollahi, Z., Darestani, M. T., Ghasemi, S., and Gomes, V. G. (2015). Monitoring inverse-phase emulsion polymerization using electrical impedance spectroscopy. *Polymer International*, **64**(6), pp. 787–794.

111. Houben, C., Nurumbetov, G., Haddleton, D., and Lapkin, A. A. (2015). Feasibility of the simultaneous determination of monomer concentrations and particle size in emulsion polymerization using in situ Raman spectroscopy. *Industrial & Engineering Chemistry Research*, **54**(51), pp. 12867–12876.

112. Yamamoto, H., Ishigami, H., and Uchimura, T. (2017). Online monitoring of a styrene monomer and a dimer in an emulsion via laser ionization time-of-flight mass spectrometry. *Analytical Sciences*, **33**(6), pp. 731–733.

113. Tauer, K. and Deckwer, R. (1998). Polymer end groups in persulfate-initiated styrene emulsion polymerization. *Acta Polymerica*, **49**(8), pp. 411–416.

114. Veloso, A., Leal, G. P., Agirre, A., and Leiza, J. R. (2018). Combining SEC & MALDI for characterization of the continuous phase in dispersion polymerization. *European Polymer Journal*, **105**, pp. 265–273.

115. Kajero, O. T., Chen, T., Yao, Y., Chuang, Y. C., and Wong, D. S. H. (2017). Meta-modelling in chemical process system engineering. *Journal of the Taiwan Institute of Chemical Engineers*, **73**, pp. 135–145.

116. Madhuranthakam, C. M. R. and Penlidis, A. (2016). Surrogate models for online monitoring and process troubleshooting of NBR emulsion copolymerization. *Processes*, **4**(1), 6.

117. Kang, J., Shao, Z., Chen, X., Gu, X., and Feng, L. (2017). Fast and reliable computational strategy for developing a rigorous model-driven soft sensor of dynamic molecular weight distribution. *Journal of Process Control*, **56**, pp. 79–99.

118. Kang, J., Shao, Z., and Chen, X. (2018). Dynamic reduced order models for polymerization process based on molecular weight distribution. *Computer Aided Chemical Engineering*, **44**, pp. 559–564.

119. Kang, J., Shao, Z., Chen, X., and Biegler, L. T. (2019). Reduced order models for dynamic molecular weight distribution in polymerization processes. *Computers & Chemical Engineering*, **126**, pp. 280–291.

120. Irzhak, V. I., Solov'ev, M. E., and Irzhak, T. F. (2018). Architecture of polymers: Topological structure–properties relationship. *Review Journal of Chemistry*, **8**(1), pp. 76–182.

121. Chaloupka, T., Zubov, A., and Kosek, J. (2017). Real-time hybrid Monte Carlo method for modelling of 4 monomer semi-batch emulsion copolymerization. *Computer Aided Chemical Engineering*, **40**, pp. 259–264.

122. Marien, Y. W., Van Steenberge, P. H., R. D'hooge, D., and Marin, G. B. (2019). Particle by particle kinetic Monte Carlo tracking of reaction and mass transfer events in miniemulsion free radical polymerization. *Macromolecules*, **52**(4), pp. 1408–1423.

123. Urrea-Quintero, J. H., Ochoa, S., and Hernández, H. (2019). A reduced-order multiscale model of a free-radical semibatch emulsion polymerization process. *Computers & Chemical Engineering*, **127**, pp. 11–24.

124. Zubov, A., Naeem, O., Hauger, S. O., Bouaswaig, A., Gjertsen, F., Singstad, P., Hungenberg, K.-D., and Kosek, J. (2017). Bringing the on-line control and optimization of semibatch emulsion copolymerization to the pilot plant. *Macromolecular Reaction Engineering*, **11**(4), 1700014.

125. Colegrove, B., Deshpande, K., Harner, R., Mikolajczyk, L., Stephenson, S. K., Tate, J. D., and Weston, J. (2017). Use of online spectroscopy to control polymerization in industrial processes. *Macromolecular Reaction Engineering*, **11**(4), 1600056.

126. Gerlinger, W., Asua, J. M., Chaloupka, T., Faust, J. M., Gjertsen, F., Hamzehlou, S., Hauger, S. O., Jahns, E., Joy, P. J., Kosek, J., Lapkin, A., Leiza, J. R., Mhamdi, A., Mitsos, A., Naeem, O., Rajabalinia, N., Singstad, P., and Suberu, J. (2019). Dynamic optimization and non-linear model predictive control to achieve targeted particle morphologies. *Chemie Ingenieur Technik*, **91**(3), pp. 323–335.

127. Cho, Y. S., Shin, C. H., and Han, S. (2016). Dispersion polymerization of polystyrene particles using alcohol as reaction medium. *Nanoscale Research Letters*, **11**(1), 46.

128. Nauman, N., Zaquen, N., Junkers, T., Boyer, C., and Zetterlund, P. B. (2019). Particle size control in miniemulsion polymerization via membrane emulsification. *Macromolecules*, **52**, pp. 4492–4499.

129. Antonietti, M. and Tauer, K. (2003). 90 years of polymer latexes and heterophase polymerization: More vital than ever. *Macromolecular Chemistry and Physics*, **204**(2), pp. 207–219.

130. Anastas, P. T. and Warner, J. C. (1998). *Green Chemistry: Theory and Practice* (Oxford University Press, USA).

131. Anastas, P. T. and Zimmerman, J. B. (2003). Peer reviewed: Design through the 12 principles of green engineering. *Environmental Science & Technology*, **37**(5), pp. 94A–101A.

132. Zhang, Y. and Dubé, M. A. (2018). Green emulsion polymerization technology. In polymer reaction engineering of dispersed systems. *Advances in Polymer Science*, **280**, pp. 65–100.

133. Schmidt, B. V., Molinari, V., Esposito, D., Tauer, K., and Antonietti, M. (2017). Lignin-based polymeric surfactants for emulsion polymerization. *Polymer*, **112**, pp. 418–426.

134. Tan, H. W., Aziz, A. A., and Aroua, M. K. (2013). Glycerol production and its applications as a raw material: A review. *Renewable and Sustainable Energy Reviews*, **27**, pp. 118–127.

135. Il'in, V. M. and Rezova, A. K. (2015). Styrene butadiene rubber: Production worldwide. *International Polymer Science and Technology*, **42**(10), pp. 35–44.

136. Camia, A., Robert, N., Jonsson, R., Pilli, R., Garcia-Condado, S., Lopez-Lozano, R., Marijn, V. D. V., Tevecia, R., Patricia, G. A., Robert, M., Saulius, T., Gianluca, F., Rita, D. S. F. D. A., Nicolas, H., Luisa, M., and Giuntoli, J. (2018). *Biomass Production, Supply, Uses and Flows in the European Union. First Results from an Integrated Assessment.* EUR 28993 EN. (Publications Office of the European Union, Luxembourg).

137. Chalmin, P. (2019). The history of plastics: From the Capitol to the Tarpeian Rock. Field Actions Science Reports. *The Journal of Field Actions*, **19**, pp. 6–11.

138. Vobecka, Z., Wei, C., Tauer, K., and Esposito, D. (2015). Poly(α-methylene-γ-valerolactone) 1. Sustainable monomer synthesis and radical polymerization studies. *Polymer*, **74**, pp. 262–271.

139. Hatton, F. L. (2020). Recent advances in RAFT polymerization of monomers derived from renewable resources. *Polymer Chemistry*, **11**(2), pp. 220–229.

140. Wang, X., Shen, L., and An, Z. (2018). Dispersion polymerization in environmentally benign solvents via reversible deactivation radical polymerization. *Progress in Polymer Science*, **83**, pp. 1–27.

141. Leibovitz, B. E. and Siegel, B. V. (1980). Aspects of free radical reactions in biological systems: Aging. *Journal of Gerontology*, **35**(1), pp. 45–56.

142. De San Luis, A., Santini, C. C., Chalamet, Y., and Dufaud, V. (2019). Removal of volatile organic compounds from bulk and emulsion polymers: A comprehensive survey of the existing techniques. *Industrial & Engineering Chemistry Research*, **58**(27), pp. 11601–11623.

143. Silva, J. M., Ortega-Gudiño, P., Rabelero, M., Sánchez-Díaz, J. C., Guerrero-Ramírez, L. G., and Puig, J. E. (2013). Increasing solid content in latexes of polyacrylamide nanoparticles made by semi-continuous inverse heterophase polymerization. *Journal of Macromolecular Science, Part A*, **50**(6), pp. 596–601.

144. Pérez-García, M. G., Alvarado, A. G., Rabelero, M., Arellano, M., Pérez-Carrillo, L. A., López-Serrano, F., López, R. G., Mendizábal, E., and Puig, J. E. (2014). Semicontinuous heterophase polymerization of methyl and hexyl methacrylates to produce latexes with high nanoparticles content. *Journal of Macromolecular Science, Part A*, **51**(2), pp. 144–155.

145. Texter, J. (2012). Anion responsive imidazolium-based polymers. *Macromolecular Rapid Communications*, **33**(23), pp. 1996–2014.

146. Rozik, N., Antonietti, M., Yuan, J., and Tauer, K. (2013). Polymerized ionic liquid as stabilizer in aqueous emulsion polymerization enables a hydrophilic–hydrophobic transition during film formation. *Macromolecular Rapid Communications*, **34**(8), pp. 665–671.

147. Yu, R. and Tauer, K. (2014). From particles to stabilizing blocks–polymerized ionic liquids in aqueous heterophase polymerization. *Polymer Chemistry*, **5**(19), pp. 5644–5655.

148. Alves, R. C., Agner, T., Rodrigues, T. S., Machado, F., Neto, B. A., da Costa, C., de Araújo, P. H. H., and Sayer, C. (2018). Cationic miniemulsion polymerization of styrene mediated by imidazolium based ionic liquid. *European Polymer Journal*, **104**, pp. 51–56.

149. Zhao, D., Liao, Y., and Zhang, Z. (2007). Toxicity of ionic liquids. *Clean–soil, air, water*, **35**(1), pp. 42–48.

150. Hahladakis, J. N., Velis, C. A., Weber, R., Iacovidou, E., and Purnell, P. (2018). An overview of chemical additives present in plastics: Migration, release, fate and environmental impact during their use, disposal and recycling. *Journal of Hazardous Materials*, **344**, pp. 179–199.

151. Fricke, M. and Lahl, U. (2005). Risikobewertung von Perfluortensiden als Beitrag zur aktuellen Diskussion zum REACH-Dossier der EU-Kommission. *Umweltwissenschaften und Schadstoff-Forschung*, **17**(1), pp. 36–49.

152. Stahl, T., Ackmann, R., Georgii, S., Wohlfarth, R., and Brunn, H. (2007). Perfluorierte Tenside. *Ernährung-Wissenschaft und Praxis*, **1**(1), pp. 27–35.

153. Martin, J. W., Smithwick, M. M., Braune, B. M., Hoekstra, P. F., Muir, D. C., and Mabury, S. A. (2004). Identification of long-chain perfluorinated acids in biota from the Canadian Arctic. *Environmental Science & Technology*, **38**(2), pp. 373–380.

154. Giesy, J. P. and Kannan, K. (2001). Global distribution of perfluorooctane sulfonate in wildlife. *Environmental Science & Technology*, **35**(7), pp. 1339–1342.

155. Istratov, V. V., Milushkova, E. V., Levachev, S. M., Gusev, S. A., Gritskova, I. A., and Vasnev, V. A. (2015). Heterophase polymerization of styrene in the presence of biodegradable surfactants. *Polymer Science Series B*, **57**(6), pp. 567–575.

156. Yang, W. and Hutchinson, R. A. (2016). Investigating the effectiveness of reactive dispersants in non-aqueous dispersion polymerization. *Macromolecular Reaction Engineering*, **10**(1), pp. 71–81.

157. Prokopov, N. I., Gritskova, I. A., Kiryutina, O. P., Khaddazh, M., Tauer, K., and Kozempel, S. (2010). The mechanism of surfactant-free emulsion polymerization of styrene. *Polymer Science Series B*, **52**(5–6), pp. 339–345.

158. Chernikova, E. V., Serkhacheva, N. S., Smirnov, O. I., Prokopov, N. I., Plutalova, A. V., Lysenko, E. A., and Kozhunova, E. Y. (2016). Emulsifier-free polymerization of n-butyl acrylate involving trithiocarbonates based on oligomer acrylic acid. *Polymer Science Series B*, **58**(6), pp. 629–639.

159. Bilgin, S., Tomovska, R., and Asua, J. M. (2017). Surfactant-free high solids content polymer dispersions. *Polymer*, **117**, pp. 64–75.

160. Errezma, M., Mabrouk, A. B., Magnin, A., Dufresne, A., and Boufi, S. (2018). Surfactant-free emulsion Pickering polymerization stabilized

by aldehyde-functionalized cellulose nanocrystals. *Carbohydrate Polymers*, **202**, pp. 621–630.

161. da Costa, J. P., Santos, P. S., Duarte, A. C., and Rocha-Santos, T. (2016). (Nano)plastics in the environment–sources, fates and effects. *Science of the Total Environment*, **566**, pp. 15–26.
162. Laycock, B., Nikolić, M., Colwell, J. M., Gauthier, E., Halley, P., Bottle, S., and George, G. (2017). Lifetime prediction of biodegradable polymers. *Progress in Polymer Science*, **71**, pp. 144–189.
163. Lendlein, A. and Sisson, A. (2011). *Handbook of Biodegradable Polymers: Synthesis, Characterization and Applications* (Wiley-VCH, Germany).
164. Thakur, V. K. and Thakur, M. K. (2016). *Handbook of Sustainable Polymers: Processing and Applications* (Jenny Stanford Publishing, Singapore).
165. Takashima, Y. and Harada, A. (2013). Self-healing polymers. In: Kobayashi, S. and Müllen, K., *Encyclopedia of Polymeric Nanomaterials* (Springer, Germany).
166. Francis, R. (2017). *Recycling of Polymers: Methods, Characterization and Applications* (Wiley-VCH, Germany).
167. Yang, L., Tan, X., Wang, Z., and Zhang, X. (2015). Supramolecular polymers: Historical development, preparation, characterization, and functions. *Chemical Reviews*, **115**(15), pp. 7196–7239.
168. Liao, H., Liao, S., Tao, X., Liu, C., and Wang, Y. (2018). Intrinsically recyclable and self-healable conductive supramolecular polymers for customizable electronic sensors. *Journal of Materials Chemistry C*, **6**(47), pp. 12992–12999.
169. Wei, M., Gao, Y., Li, X., and Serpe, M. J. (2017). Stimuli-responsive polymers and their applications. *Polymer Chemistry*, **8**(1), pp. 127–143.
170. Aguilar, M. R. and San Román, J. (2019). *Smart Polymers and Their Applications* (Woodhead Publishing, UK).
171. Ramos, J., Imaz, A., and Forcada, J. (2012). Temperature-sensitive nanogels: Poly(N-vinylcaprolactam) versus poly (N-isopropylacrylamide). *Polymer Chemistry*, **3**(4), pp. 852–856.
172. Yu, R., Hartmann, J., and Tauer, K. (2013). Stabile, thermoresponsive colloidal clusters: An unusual morphology of polymer dispersions. *Macromolecular Rapid Communications*, **34**(20), pp. 1629–1634.
173. Aguilar, J., Moscoso, F., Rios, O., Ceja, I., Sánchez, J. C., Bautista, F., Puig, J. E., and Fernández, V. V. A. (2014). Swelling behavior of poly

(N-isopropylacrylamide) nanogels with narrow size distribution made by semi-continuous inverse heterophase polymerization. *Journal of Macromolecular Science, Part A*, **51**(5), pp. 412–419.

174. Chen, Y. and Sajjadi, S. (2014). Temperature-triggered disintegrable poly(N-isopropylacrylamide) nanoparticles via heterophase polymerization in the presence of tetramethylethylenediamine and sodium dodecyl sulfate. *Journal of Applied Polymer Science*, **131**(18), 40781.

175. Meléndez-Ortiz, H. I., Peralta, R. D., Bucio, E., and Zerrweck-Maldonado, L. (2014). Preparation of stimuli-responsive nanogels of poly[2-(dimethylamino) ethyl methacrylate] by heterophase and microemulsion polymerization using gamma radiation. *Polymer Engineering & Science*, **54**(7), pp. 1625–1631.

176. Alvarado, A. G., Ortega, A., Pérez-Carrillo, L. A., Ceja, I., Arellano, M., López, R. G., and Puig, J. E. (2017). Synthesis, characterization, and drug delivery from pH-and thermoresponsive poly (N-isopropylacrylamide)/chitosan core/shell nanocomposites made by semicontinuous heterophase polymerization. *Journal of Nanomaterials*, **2017**, 6796412.

177. Iyisan, B. and Landfester, K. (2019). Modular approach for the design of smart polymeric nanocapsules. *Macromolecular Rapid Communications*, **40**(1), 1800577.

178. Filippov, A., Tarabukina, E., Simonova, M., Kirila, T., Fundueanu, G., Harabagiu, V., Constantin, M., and Popescu, I. (2015). Synthesis and investigation of double stimuli-responsive behavior of N-isopropylacrylamide and maleic acid copolymer in solutions. *Journal of Macromolecular Science, Part B*, **54**(9), pp. 1105–1121.

179. Soll, S., Antonietti, M., and Yuan, J. (2011). Double stimuli-responsive copolymer stabilizers for multiwalled carbon nanotubes. *ACS Macro Letters*, **1**(1), pp. 84–87.

180. Men, Y., Drechsler, M., and Yuan, J. (2013). Double-stimuli-responsive spherical polymer brushes with a poly (ionic liquid) core and a thermoresponsive shell. *Macromolecular Rapid Communications*, **34**(21), pp. 1721–1727.

181. Guragain, S., Bastakoti, B. P., Malgras, V., Nakashima, K., and Yamauchi, Y. (2015). Multi-stimuli-responsive polymeric materials. *Chemistry–A European Journal*, **21**(38), pp. 13164–13174.

182. Cao, Z. Q. and Wang, G. J. (2016). Multi-stimuli-responsive polymer materials: Particles, films, and bulk gels. *The Chemical Record*, **16**(3), pp. 1398–1435.

183. Fu, X., Hosta-Rigau, L., Chandrawati, R., and Cui, J. (2018). Multi-stimuli-responsive polymer particles, films, and hydrogels for drug delivery. *Chem*, **4**(9), pp. 2084–2107.

184. González, E., Bonnefond, A., Barrado, M., Barrasa, A. M. C., Asua, J. M., and Leiza, J. R. (2015). Photoactive self-cleaning polymer coatings by TiO_2 nanoparticle Pickering miniemulsion polymerization. *Chemical Engineering Journal*, **281**, pp. 209–217.

185. He, J. (2017). *Self-cleaning Coatings: Structure, Fabrication and Application* (Royal Society of Chemistry, UK).

186. Dararatana, N., Seidi, F., and Crespy, D. (2018). pH-sensitive polymer conjugates for anticorrosion and corrosion sensing. *ACS Applied Materials & Interfaces*, **10**(24), pp. 20876–20883.

187. Mishra, R. K. and Majeed, A. B. A. (2015). Controlled drug delivery: Smart and natural polymer systems. In: Mishra, M., *Encyclopedia of Biomedical Polymers and Polymeric Biomaterials* (CRC Press, USA), pp. 2147–2154

188. Moghanjoughi, A. A., Khoshnevis, D., and Zarrabi, A. (2016). A concise review on smart polymers for controlled drug release. *Drug Delivery and Translational Research*, **6**(3), pp. 333–340.

189. Chakraborty, D. D., Nath, L. K., and Chakraborty, P. (2018). Recent progress in smart polymers: Behavior, mechanistic understanding and application. *Polymer-Plastics Technology and Engineering*, **57**(10), pp. 945–957.

190. Jose, A. J., Rajan, M. S., Anjalipriya, S., and John, S. (2018). Smart polymers: A smart approach to life. In: Haghi, A. K., Balköse, D., and Thomas, S., *Applied Physical Chemistry with Multidisciplinary Approaches* (Apple Academic Press, USA), pp. 23–42.

191. Gonçalves, V. S. S., Rodríguez-Rojo, S., Matias, A. A., Nunes, A. V. M., Nogueira, I. D., Nunes, D., Fortunato, E., de Matos, A. P. A., Cocero, M. J., and Duarte, C. M. M. (2015). Development of multicore hybrid particles for drug delivery through the precipitation of CO_2 saturated emulsions. *International Journal of Pharmaceutics*, **478**(1), pp. 9–18.

192. Stubbs, J. M. and Sundberg, D. C. (2008). The dynamics of morphology development in multiphase latex particles. *Progress in Organic Coatings*, **61**(2–4), pp. 156–165.

193. Vonka, M. and Kosek, J. (2012). Modelling the morphology evolution of polymer materials undergoing phase separation. *Chemical Engineering Journal*, **207**, pp. 895–905.

194. Hamzehlou, S., Leiza, J. R., and Asua, J. M. (2016). A new approach for mathematical modeling of the dynamic development of particle morphology. *Chemical Engineering Journal*, **304**, pp. 655–666.

195. Pei, X., Zhai, K., Tan, Y., Xu, K., Lu, C., Wang, P., Wang, T., Chen, C., Tao, Y., Dai, L., and Li, H. (2017). Synthesis of monodisperse starch-polystyrene core-shell nanoparticles via seeded emulsion polymerization without stabilizer. *Polymer*, **108**, pp. 78–86.

196. Sahagún Aguilar, L. O., Ceja Andrade, I., Alvarado Mendoza, A. G., Puig Arevalo, J. E., and Rabelero Velasco, M. (2019). Tunable mechanical properties of core-shell polymers of poly(hexyl methacrylate) and poly(methyl methacrylate) by semicontinuous heterophase polymerization. *Polymer Engineering & Science*, **59**(2), pp. 365–371.

197. Shevchenko, N., Pankova, G., Shabsels, B., Baigildin, V., Khoshkin, A., Ukleev, T., and Sel'kin, A. (2019). Fluorescent core-shell polymer particles containing luminophore dyes: Synthesis and optical response to acetone. *Journal of Dispersion Science and Technology*, **40**(6), pp. 802–810.

198. Chen, C., Zhang, L., Sheng, M., Guan, Y., Dong, H., and Fu, S. (2019). Robust raspberry-like all-polymer particles for the construction of superhydrophobic surface with high water adhesive force. *Journal of Materials Science*, **54**(2), pp. 1898–1912.

199. Gao, S., Song, S., Wang, J., Mei, S., Yuan, J., Liu, G., and Pan, M. (2019). Self-assembled heteromorphous raspberry-like colloidal particles from Pickering-like emulsion polymerization. *Colloids and Surfaces A: Physicochemical and Engineering Aspects*, **577**, pp. 360–369.

200. Li, J., Sihler, S., and Ziener, U. (2019). Double in situ preparation of raspberry-like polymer particles. *Langmuir*, **35**(18), pp. 6161–6168.

201. Ghiasi, F. F., Taromi, F. A., and Eslami, H. (2018). Synthesis of raspberry- and confetti-like hybrid poly(styrene-co-2-vinyl pyrrolidone)/silica nanocomposite particles via alcoholic dispersion polymerization. *Polymer Composites*, **39**(9), pp. 3363–3376.

202. Kikuchi, S., Yoshida, S., Kanehashi, S., Ma, G. H., and Ogino, K. (2019). Fabrication of core-shell, Janus, dumbbell, snowman-like and confetti-like structured microspheres of blends of poly (4-butyl triphenylamine) and poly(methyl methacrylate) by solvent evaporation method. *Journal of Fiber Science and Technology*, **75**(3), pp. 22–28.

203. Zhai, W., Wang, B., Wang, Y., He, Y. F., Song, P., and Wang, R. M. (2016). An efficient strategy for preparation of polymeric Janus particles with controllable morphologies and emulsifiabilities. *Colloids and Surfaces A: Physicochemical and Engineering Aspects*, **503**, pp. 94–100.

204. Zhang, J., Grzybowski, B. A., and Granick, S. (2017). Janus particle synthesis, assembly, and application. *Langmuir*, **33**(28), pp. 6964–6977.

205. Lotierzo, A., Longbottom, B. W., Lee, W. H., and Bon, S. A. (2018). Synthesis of Janus and patchy particles using nanogels as stabilizers in emulsion polymerization. *ACS Nano*, **13**(1), pp. 399–407.

206. Wang, Z., Qiu, T., Guo, L., Ye, J., He, L., and Li, X. (2019). The building of molecularly imprinted single hole hollow particles: A miniemulsion polymerization approach. *Chemical Engineering Journal*, **357**, pp. 348–357.

207. Sosa, D., Guillén, L., Saade, H., Mendizábal, E., Puig, J., and López, R. (2015). Effect of monomer dosing rate in the preparation of mesoporous polystyrene nanoparticles by semicontinuous heterophase polymerization. *Molecules*, **20**(1), pp. 52–69.

208. Esquivel, O., Treviño, M. E., Saade, H., Puig, J. E., Mendizábal, E., and López, R. G. (2011). Mesoporous polystyrene nanoparticles synthesized by semicontinuous heterophase polymerization. *Polymer Bulletin*, **67**(2), pp. 217–226.

209. Taherzadeh, H., Sotowa, S., and Ogino, K. (2015). Morphology control of polymer microspheres containing block copolymers with seed polymerization. *Open Journal of Organic Polymer Materials*, **5**, pp. 43–50.

210. Yu, B., Cong, H., Peng, Q., Gu, C., Tang, Q., Xu, X., Tian, C., and Zhai, F. (2018). Current status and future developments in preparation and application of nonspherical polymer particles. *Advances in Colloid and Interface Science*, **256**, pp. 126–151.

211. Pei, X., Zhai, K., Liang, X., Deng, Y., Xu, K., Tan, Y., Yao, X., and Wang, P. (2018). Fabrication of shape-tunable macroparticles by seeded polymerization of styrene using non-cross-linked starch-based seed. *Journal of Colloid and Interface Science*, **512**, pp. 600–608.

212. Zhai, W., Song, Y., Gao, Z., Fan, J. B., and Wang, S. (2019). Precise synthesis of polymer particles spanning from anisotropic Janus particles to heterogeneous nanoporous particles. *Macromolecules*, **52**(9), pp. 3237–3243.

213. Rangelov, S. and Petrov, P. (2016). Template-assisted approaches for preparation of nano-sized polymer structures. In: Fakirov, S., *Nano-size Polymers* (Springer, Germany), pp. 367–396.

214. Lan, Y., Liu, J., Eiser, E., and Scherman, O. A. (2019). Polymeric raspberry-like particles via template-assisted polymerisation. *Polymer Chemistry*, **10**, pp. 3772–3777.

215. Weber, N., Tiersch, B., Unterlass, M. M., Heilig, A., and Tauer, K. (2011). "Schizomorphic" emulsion copolymerization particles. *Macromolecular Rapid Communications*, **32**(23), pp. 1925–1929.

216. Sheu, H. R., El-Aasser, M. S., and Vanderhoff, J. W. (1987). Phase domain formation in latex interpenetrating polystyrene networks. *Polymeric Materials Science and Engineering*, **57**, pp. 911–915.

217. Sheu, H. R., El-Aasser, M. S., and Vanderhoff, J. W. (1990). Phase separation in polystyrene latex interpenetrating polymer networks. *Journal of Polymer Science Part A: Polymer Chemistry*, **28**(3), pp. 629–651.

218. Kim, J. W., Larsen, R. J., and Weitz, D. A. (2006). Synthesis of nonspherical colloidal particles with anisotropic properties. *Journal of the American Chemical Society*, **128**(44), pp. 14374–14377.

219. Wei, C., Plucinski, A., Nuasaen, S., Tripathi, A., Tangboriboonrat, P., and Tauer, K. (2017). Swelling-induced deformation of spherical latex particles. *Macromolecules*, **50**(1), pp. 349–363.

220. Dendukuri, D., Tsoi, K., Hatton, T. A., and Doyle, P. S. (2005). Controlled synthesis of nonspherical microparticles using microfluidics. *Langmuir*, **21**(6), pp. 2113–2116.

221. Xu, S., Nie, Z., Seo, M., Lewis, P., Kumacheva, E., Stone, H. A., Garstecki, P., Weibel, D. B., Gitlin, I., and Whitesides, G. M. (2005). Generation of monodisperse particles by using microfluidics: Control over size, shape, and composition. *Angewandte Chemie International Edition*, **44**(5), pp. 724–728.

222. Keville, K. M., Caruthers, J. M., and Franses, E. I. (1986). Characterization of dimensions of ellipsoidal microparticles via electron microscopy. *Journal of Microscopy*, **142**(3), pp. 327–340.

223. Nagy, M. and Keller, A. (1989). Ellipsoidal polymer particles with predesigned axial ratio. *Polymer Communications*, **30**(5), pp. 130–132.

224. Ho, C. C., Keller, A., Odell, J. A., and Ottewill, R. H. (1993). Preparation of monodisperse ellipsoidal polystyrene particles. *Colloid and Polymer Science*, **271**(5), pp. 469–479.

225. Ho, C. C., Ottewill, R. H., Keller, A., and Odell, J. A. (1993). Monodisperse ellipsoidal polystyrene latex particles: Preparation and characterisation. *Polymer International*, **30**(2), pp. 207–211.

226. Keville, K. M., El Franses, I., and Caruthers, J. M. (1991). Preparation and characterization of monodisperse polymer microspheroids. *Journal of Colloid and Interface Science*, **144**(1), pp. 103–126.

227. Gabrielson, L. and Folkes, M. J. (2001). Manufacture of colloidal polymer ellipsoids for anisotropic conducting nano-composites. *Journal of Materials Science*, **36**(1), pp. 1–6.

228. Okubo, M., Yamashita, T., and Ise, E. (1999). Thermodynamics for preparation of micron-sized, monodispersed, monomer-"adsorbed" polymer particles having "snowman" shape by utilizing the dynamic swelling method and the seeded polymerization. *Proceedings of the Japan Academy, Series B*, **75**(7), pp. 195–200.

229. Okubo, M. and Minami, H. (2000). Production of micron-sized monodispersed anomalous polymer particles having red blood corpuscle shape. *Macromolecular Symposia*, **150**, pp. 201–210.

230. Okubo, M., Konishi, Y., Takebe, M., and Minami, H. (2000). Preparation of micron-sized, monodispersed, anomalous polymer particles by utilizing the solvent-absorbing/releasing method. *Colloid and Polymer Science*, **278**(10), pp. 919–926.

231. Okubo, M., Miya, T., Minami, H., and Takekoh, R. (2002). Morphology of micron-sized, monodisperse, nonspherical polystyrene/poly(n-butyl methacrylate) composite particles produced by seeded dispersion polymerization. *Journal of Applied Polymer Science*, **83**(9), pp. 2013–2021.

232. Champion, J. A., Katare, Y. K., and Mitragotri, S. (2007). Particle shape: A new design parameter for micro-and nanoscale drug delivery carriers. *Journal of Controlled Release*, **121**(1–2), pp. 3–9.

233. Faucheu, J., Gauthier, C., Chazeau, L., Cavaillé, J. Y., Mellon, V., and Lami, E. B. (2010). Miniemulsion polymerization for synthesis of structured clay/polymer nanocomposites: Short review and recent advances. *Polymer*, **51**(1), pp. 6–17.

234. Mabrouk, A. B., Vilar, M. R., Magnin, A., Belgacem, M. N., and Boufi, S. (2011). Synthesis and characterization of cellulose whiskers/polymer nanocomposite dispersion by mini-emulsion polymerization. *Journal of Colloid and Interface Science*, **363**(1), pp. 129–136.

235. Silva-Jara, J. M., Manríquez-González, R., López-Dellamary, F. A., Puig, J. E., and Nuño-Donlucas, S. M. (2015). Semi-continuous heterophase polymerization to synthesize nanocomposites of poly(acrylic acid)-functionalized carbon nanotubes. *Journal of Macromolecular Science, Part A*, **52**(9), pp. 732–744.

236. Kakhramanov, N. T., Azizov, A. G., Osipchik, V. S., Mamedli, U. M., and Arzumanova, N. B. (2017). Nanostructured composites and polymer materials science. *International Polymer Science and Technology*, **44**(2), pp. 37–48.

237. Ghiasi, F. F., Taromi, F. A., and Eslami, H. (2018). Synthesis of raspberry- and confetti-like hybrid poly(styrene-co-2-vinyl pyrrolidone)/silica nanocomposite particles via alcoholic dispersion polymerization. *Polymer Composites*, **39**(9), pp. 3363–3376.
238. Delibaş, A., Yıldız, U., and Tauer, K. (2019). Composite latex production with high solid content. *Journal of Applied Polymer Science*, **136**(18), 47423.
239. Morgen, T. O., Luttikhedde, H., and Mecking, S. (2019). Aqueous dispersions of ethylene copolymers and their laponite clay nanocomposites from free-radical dispersion polymerization. *Macromolecules*, **52**(11), pp. 4270–4277.
240. Nguyen, D., Duguet, E., Bourgeat-Lami, E., and Ravaine, S. (2010). An easy way to control the morphology of colloidal polymer-oxide supraparticles through seeded dispersion polymerization. *Langmuir*, **26**(9), pp. 6086–6090.
241. Serkhacheva, N. S., Prokopov, N. I., Gerval'd, A. Y., Lobanov, N. A., and Gritskova, I. A. (2017). The heterophase polymerisation of styrene in the presence of zinc oxide nanoparticles. *International Polymer Science and Technology*, **44**(10), pp. 13–18.
242. Schrade, A., Mikhalevich, V., Landfester, K., and Ziener, U. (2011). Synthesis and characterization of positively charged, alumina-coated silica/polystyrene hybrid nanoparticles via pickering miniemulsion polymerization. *Journal of Polymer Science Part A: Polymer Chemistry*, **49**(22), pp. 4735–4746.
243. Schoth, A., Adurahim, E. S., Bahattab, M. A., Landfester, K., and Muñoz-Espí, R. (2016). Waterborne polymer/silica hybrid nanoparticles and their structure in coatings. *Macromolecular Reaction Engineering*, **10**(1), pp. 47–54.
244. Yao, J., Cao, Z., Shang, Y., Chen, Q., Yang, L., Zhang, Y., and Qi, D. (2017). Preparation of polymeric/inorganic nanocomposite particles in miniemulsions: II. Narrowly size-distributed polymer/SiO_2 nanocomposite particles. *Colloids and Surfaces A: Physicochemical and Engineering Aspects*, **530**, pp. 104–116.
245. Chanra, J., Budianto, E., and Soegijono, B. (2018). Synthesis of polymer hybrid latex poly(methyl methacrylate-co-butyl acrylate) with organo montmorillonite via miniemulsion polymerization method for barrier paper. *Journal of Physics: Conference Series*, **985**(1), 012029.
246. Grüneberger, F., Huch, A., Geiger, T., Zimmermann, T., and Tingaut, P. (2016). Fibrillated cellulose in heterophase polymerization of nanoscale poly(methyl methacrylate) spheres. *Colloid and Polymer Science*, **294**(9), pp. 1393–1403.

247. Saboori, R. and Sabbaghi, S. (2017). Synthesis and characterization of carboxymethyl cellulose/polystyrene core-shell nanoparticles by miniemulsion polymerization. *International Journal of Nanoscience and Nanotechnology*, **13**(3), pp. 275–281.

248. Zhenqian, Z., Bo, X., Pei, W., and Ninyyi, Y. (2015). Hybrid latex particles preparation with seeded semibatch emulsion polymerization. *Colloids and Surfaces A: Physicochemical and Engineering Aspects*, **482**, pp. 422–430.

249. Hood, M., Mari, M., and Muñoz-Espí, R. (2014). Synthetic strategies in the preparation of polymer/inorganic hybrid nanoparticles. *Materials*, **7**(5), pp. 4057–4087.

250. Shragin, D. I., Gritskova, I. A., Kopylov, V. V., Milushkova, E. V., Zlydneva, L. A., and Levachev, S. M. (2015). Novel approach to synthesis of monodisperse polymeric microspheres: Heterophase polymerization of styrene and methyl methacrylate in presence of water-insoluble functional PDMSs. *Silicon*, **7**(2), pp. 217–227.

251. Gritskova, I. A., Lakhtin, V. G., Shragin, D. I., Ezhova, A. A., Sokolskaya, I. B., Krizhanovsky, I. N., Storozhenko, P.A., and Muzafarov, A. M. (2018). Synthesis of oligosiloxanes with 3-aminopropyl groups and their testing as surfactants in the preparation of polymer microspheres. *Russian Chemical Bulletin*, **67**(10), pp. 1908–1914.

252. Jensen, A. T., Neto, W. S., Ferreira, G. R., Glenn, A. F., Gambetta, R., Gonçalves, S. B., Valadares, L.F., and Machado, F. (2017). Synthesis of polymer/inorganic hybrids through heterophase polymerizations. In: Visakh, P. M., Markovic, G., and Pasquini, D., *Recent Developments in Polymer Macro, Micro and Nano Blends: Preparation and Characterisation* (Woodhead Publishing, UK), pp. 207–235.

253. Sierra-Martin, B. and Fernandez-Barbero, A. (2016). Inorganic/polymer hybrid nanoparticles for sensing applications. *Advances in Colloid and Interface Science*, **233**, pp. 25–37.

254. Cao, Z., Schrade, A., Landfester, K., and Ziener, U. (2011). Synthesis of raspberry-like organic–inorganic hybrid nanocapsules via Pickering miniemulsion polymerization: Colloidal stability and morphology. *Journal of Polymer Science Part A: Polymer Chemistry*, **49**(11), pp. 2382–2394.

255. Kalia, S. and Haldorai, Y. (2015). *Organic–Inorganic Hybrid Nanomaterials* (Springer, Switzerland).

256. Mir, S. H., Nagahara, L. A., Thundat, T., Mokarian-Tabari, P., Furukawa, H., and Khosla, A. (2018). Organic–inorganic hybrid functional materials: An integrated platform for applied technologies. *Journal of The Electrochemical Society*, **165**(8), pp. B3137–B3156.

257. Mobin, R., Rangreez, T. A., Chisti, H. T. N., and Rezakazemi, M. (2019). Organic–inorganic hybrid materials and their applications. In: Mazumder, M. A. J., Sheardown, H., and Al-Ahmed, A. *Functional Polymers* (Springer Nature, Switzerland), pp. 1135–1156.

258. Das, T. K. and Prusty, S. (2012). Review on conducting polymers and their applications. *Polymer-Plastics Technology and Engineering*, **51**(14), pp. 1487–1500.

259. Pecher, J. and Mecking, S. (2010). Nanoparticles of conjugated polymers. *Chemical Reviews*, **110**(10), pp. 6260–6279.

260. Liu, Y., Chu, Y., and Yang, L. (2006). Adjusting the inner-structure of polypyrrole nanoparticles through microemulsion polymerization. *Materials Chemistry and Physics*, **98**(2–3), pp. 304–308.

261. Ovando-Medina, V. M., Peralta, R. D., Mendizábal, E., Martínez-Gutiérrez, H., Lara-Ceniceros, T. E., and Ledezma-Rodríguez, R. (2011). Synthesis of polypyrrole nanoparticles by oil-in-water microemulsion polymerization with narrow size distribution. *Colloid and Polymer Science*, **289**(7), pp. 759–765.

262. Ruppert, M., Ziener, U., and Landfester, K. (2011). Oxidative polymerization of ethylenedioxythiophene with Fenton's reagent by the miniemulsion technique. *Colloid and Polymer Science*, **289**(12), pp. 1321–1328.

263. Huber, J., Jung, C., and Mecking, S. (2012). Nanoparticles of low optical band gap conjugated polymers. Macromolecules, **45**(19), pp. 7799–7805.

264. Kuehne, A. J., Gather, M. C., and Sprakel, J. (2012). Monodisperse conjugated polymer particles by Suzuki–Miyaura dispersion polymerization. *Nature Communications*, **3**, 1088.

265. Corona-Rivera, M. A., Ovando-Medina, V. M., Martínez-Gutiérrez, H., Silva-Aguilar, F. E., Pérez, E., and Antonio-Carmona, I. D. (2015). Morphology and conductivity tuning of polyaniline using short-chain alcohols by heterophase polymerization. *Colloid and Polymer Science*, **293**(2), pp. 605–615.

266. González-Iñiguez, J. C., Ovando-Medina, V. M., Jasso-Gastinel, C. F., González, D. A., Puig, J. E., and Mendizábal, E. (2014). Synthesis of polypyrrole nanoparticles by batch and semicontinuous heterophase polymerizations. *Colloid and Polymer Science*, **292**(6), pp. 1269–1275.

267. Ülkü, G., Güvel, E. A., Köken, N., and Kızılcan, N. (2015). Heterophase polymerization of pyrrole and thienyl end capped ethoxylated nonyl phenol by Iron (III) Chloride. *International Scholarly and Scientific Research & Innovation*, **9**(2), pp. 248–252.

268. Castillo-Reyes, B. E., Ovando-Medina, V. M., González-Ortega, O., Alonso-Dávila, P. A., Juárez-Ramírez, I., Martínez-Gutiérrez, H., and Márquez-Herrera, A. (2015). TiO_2/polypyrrole nanocomposites photoactive under visible light synthesized by heterophase polymerization in the presence of different surfactants. *Research on Chemical Intermediates*, **41**(11), pp. 8211–8231.

269. Schellenberg, C., Auschra, C., Peter, W., Heinrich, F. O., and Tanabe, J. (2016). Water based concentrated product forms of photoinitiators made by a heterophase polymerization technique. U.S. Patent Application No. 15/006,329.

270. Schellenberg, C., Auschra, C., Peter, W., Kimpel, D., and Hayoz, P. (2010). Water based concentrated product forms of light stabilizers made by a heterophase polymerization technique. U.S. Patent No. 7,687,554.

271. Ghasimi, S., Prescher, S., Wang, Z. J., Landfester, K., Yuan, J., and Zhang, K. A. (2015). Heterophase photocatalysts from water-soluble conjugated polyelectrolytes: An example of self-initiation under visible light. *Angewandte Chemie International Edition*, **54**(48), pp. 14549–14553.

272. Hu, F., Guo, L., Qiu, T., and Li, X. (2016). Synthesis of polypyrrole–polystyrene composite microspheres via pseudo-multicomponent heterophase polymerization and the potential application on Cr (VI) removal. *RSC Advances*, **6**(52), pp. 46900–46907.

273. Dou, J., Zhang, Q., Jian, L., and Gu, J. (2010). Magnetic nanoparticles encapsulated latexes prepared with photo-initiated miniemulsion polymerization. *Colloid and Polymer Science*, **288**(18), pp. 1751–1756.

274. Bonnefond, A., Ibarra, M., Mecerreyes, D., and Leiza, J. R. (2016). Adding magnetic ionic liquid monomers to the emulsion polymerization tool-box: Towards polymer latexes and coatings with new properties. *Journal of Polymer Science Part A: Polymer Chemistry*, **54**(8), pp. 1145–1152.

275. Li, J., Fan, J., Cao, R., Zhang, Z., Du, J., and Peng, X. (2018). Encapsulated dye/polymer nanoparticles prepared via miniemulsion polymerization for inkjet printing. *ACS Omega*, **3**(7), pp. 7380–7387.

276. Bao, Z. and Smith, K. W. (2018). Miniemulsion polymerization to prepare drag reducers. U.S. Patent No. 9,951,151.

277. Generalova, A. N. and Zubov, V. P. (2018). Design of polymer particle dispersions (latexes) in the course of radical heterophase polymerization for biomedical applications. *Colloids and Surfaces B: Biointerfaces*, **166**, pp. 303–322.

278. Thomas, J., Thomas, S., Kalarikkal, N., and Jose, J. (2019). *Nanoparticles in Polymer Systems for Biomedical Applications* (Apple Academic Press, Canada).

279. Deb, P. K., Kokaz, S. F., Abed, S. N., Paradkar, A., and Tekade, R. K. (2019). Pharmaceutical and biomedical applications of polymers. In: Tekade, R. K., *Basic Fundamentals of Drug Delivery* (Academic Press, UK), pp. 203–267.

280. Ahmed, T., Saleem, A., Ramyakrishna, P., Rajender, B., Gulzar, T., Khan, A., and Asiri, A. M. (2019). Nanostructured polymer composites for bio-applications. In: Swain, S. K. and Jawaid, M. *Nanostructured Polymer Composites for Biomedical Applications* (Elsevier, The Netherlands), pp. 167–188.

281. Generalova, A. N., Asharchuk, I. M., and Zubov, V. P. (2018). Multifunctional polymer dispersions for biomedical assays obtained by heterophase radical polymerization. *Russian Chemical Bulletin*, **67**(10), pp. 1759–1780.

282. Yang, J., Li, Y., Wang, J., Sun, X., Cao, R., Sun, H., Huang, C., and Chen, J. (2015). Molecularly imprinted polymer microspheres prepared by Pickering emulsion polymerization for selective solid-phase extraction of eight bisphenols from human urine samples. *Analytica Chimica Acta*, **872**, pp. 35–45.

283. Wang, Z., Qiu, T., Guo, L., Ye, J., He, L., and Li, X. (2018). The synthesis of molecular recognition polymer particles via miniemulsion polymerization. *Reactive and Functional Polymers*, **126**, pp. 1–8.

284. Chen, G., Qiu, H., Prasad, P. N., and Chen, X. (2014). Upconversion nanoparticles: Design, nanochemistry, and applications in theranostics. *Chemical Reviews*, **114**(10), pp. 5161–5214.

285. Generalova, A., Mironova, K., Sholina, N., Rocheva, V., Nechaev, A., Grebenik, E., Guller, A., Zvyagin, A., Deyev, S., Zubov, V., and Khaydukov, E. (2018). Upconversion nanoparticles: On the way from diagnostics to theranostics. *EPJ Web of Conferences*, **190**, 03001.

286. Saade, H., Barrera, C., Espinoza, A., López-Quintanilla, M. L., Fernández, S., and López, R. G. (2013). Ultrafine nanoparticles of ibuprofen-poly (methyl methacrylate) by a polymerization-loading method. *Drug Delivery Letters*, **3**(1), pp. 54–60.

287. Saade, H., Garza, J. C., Lopez-Quintanilla, M. L., Enriquez-Medrano, F., Fernandez, S., Lopez, R. G., and Rodriguez, H. (2014). Semicontinuous heterophase polymerization as a tool for preparing ultrafine nanoparticles of ibuprofen-poly(methyl methacrylate). *Nano*, **9**(6), 1450060.

288. Saade, H., Barrera, C., Guerrero, R., Mendizábal, E., Puig, J. E., and López, R. G. (2016). Preparation and loading with Rifampicin of sub-50 nm polyethyl cyanoacrylate nanoparticles by semicontinuous heterophase polymerization. *Journal of Nanomaterials*, **2016**, 25.

289. Bohrey, S., Chourasiya, V., and Pandey, A. (2016). Preparation, optimization by 2^3 factorial design, characterization and in vitro release kinetics of lorazepam loaded PLGA nanoparticles. *Polymer Science Series A*, **58**(6), pp. 975–986.

290. Saltel, D. A. and Vachon, M. (2016). Sustained release formulations of lorazepam. U.S. Patent Application No. 14/757,465.

291. Pokharkar, V., Patil, V., and Mandpe, L. (2015). Engineering of polymer–surfactant nanoparticles of doxycycline hydrochloride for ocular drug delivery. *Drug Delivery*, **22**(7), pp. 955–968.

292. Neto, W. S., Pena, L. I., Ferreira, G. R., Souza Junior, F. G., and Machado, F. (2017). Target delivery from modified polymers to cancer treatment. *Current Organic Chemistry*, **21**(1), pp. 4–20.

293. Kalepu, S. and Nekkanti, V. (2015). Insoluble drug delivery strategies: Review of recent advances and business prospects. *Acta Pharmaceutica Sinica B*, **5**(5), pp. 442–453.

294. Ghaffar, A., Yameen, B., Latif, M., and Malik, M. I. (2019). pH-sensitive drug delivery systems. In: Shah, M. R., Imran, M., and Ullah, S., *Metal Nanoparticles for Drug Delivery and Diagnostic Applications* (Elsevier), pp. 259–278.

295. Pundir, S., Badola, A., and Sharma, D. (2013). Sustained release matrix technology and recent advance in matrix drug delivery system: A review. *International Journal of Drug Research and Technology*, **3**(1), pp. 12–20.

296. Tibbitt, M. W., Dahlman, J. E., and Langer, R. (2016). Emerging frontiers in drug delivery. *Journal of the American Chemical Society*, **138**(3), pp. 704–717.

297. Asghari, F., Samiei, M., Adibkia, K., Akbarzadeh, A., and Davaran, S. (2017). Biodegradable and biocompatible polymers for tissue engineering application: A review. *Artificial cells, Nanomedicine, and Biotechnology*, **45**(2), pp. 185–192.

298. Saade, H., Guillén, M. D. L., Romero, J. C., Cepeda, J., Ilyna, A., Fernández, S., Enríquez-Medrano, F. J., and López, R. G. (2016). Biocompatible and biodegradable ultrafine nanoparticles of poly(methyl methacrylate-co-methacrylic acid) prepared via semicontinuous heterophase polymerization: Kinetics and product characterization. *International Journal of Polymer Science*, **2016**, 7674620.

299. Mansour, H. M., Sohn, M., Al-Ghananeem, A., and DeLuca, P. P. (2010). Materials for pharmaceutical dosage forms: Molecular pharmaceutics and controlled release drug delivery aspects. *International Journal of Molecular Sciences*, **11**(9), pp. 3298–3322.

300. Nezakati, T., Seifalian, A., Tan, A., and Seifalian, A. M. (2018). Conductive polymers: Opportunities and challenges in biomedical applications. *Chemical Reviews*, **118**(14), pp. 6766–6843.

301. Kaur, G., Adhikari, R., Cass, P., Bown, M., and Gunatillake, P. (2015). Electrically conductive polymers and composites for biomedical applications. *RSC Advances*, **5**(47), pp. 37553–37567.

302. Zhang, X., Yu, J., Wu, C., Jin, Y., Rong, Y., Ye, F., and Chiu, D. T. (2012). Importance of having low-density functional groups for generating high-performance semiconducting polymer dots. *ACS Nano*, **6**(6), pp. 5429–5439.

303. Yu, J., Rong, Y., Kuo, C. T., Zhou, X. H., and Chiu, D. T. (2016). Recent advances in the development of highly luminescent semiconducting polymer dots and nanoparticles for biological imaging and medicine. *Analytical Chemistry*, **89**(1), pp. 42–56.

304. Guo, L., Ge, J., and Wang, P. (2018). Polymer dots as effective phototheranostic agents. *Photochemistry and Photobiology*, **94**(5), pp. 916–934.

305. Ruiz Perez, J. D. and Mecking, S. (2017). Anisotropic polymer nanoparticles with tunable emission wavelengths by intersegmental chain packing. *Angewandte Chemie International Edition*, **56**(22), pp. 6147–6151.

306. Meng, X., Wen, T., Qiang, L., Ren, J., and Tang, F. (2013). Luminescent electrophoretic particles via miniemulsion polymerization for night-vision electrophoretic displays. *ACS Applied Materials & Interfaces*, **5**(9), pp. 3638–3642.

Index